杨青平 ◎ 著

Android Telephony
原理解析与开发指南

Android Telephony
Principle Analysis and Development Guide

人民邮电出版社
北　京

图书在版编目（CIP）数据

Android Telephony 原理解析与开发指南 / 杨青平著. -- 北京：人民邮电出版社，2018.9（2020.6重印）
ISBN 978-7-115-48915-9

Ⅰ. ①A… Ⅱ. ①杨… Ⅲ. ①移动终端－应用程序－程序设计 Ⅳ. ①TN929.53

中国版本图书馆CIP数据核字(2018)第182123号

内 容 提 要

随着 Android 平台的应用越来越广泛，越来越多的人加入到 Android 系统的定制研发中来。Android 的基本通信功能是 Android 系统定制的核心模块，本书主要围绕 Android Telephony 关键业务流程展开，从接打电话、网络服务、数据上网三方面解析 Telephony。

全书共 10 章，主要内容包括初识 Android、搭建 Android 源代码编译调试环境、深入解析通话流程、详解 Telecom、详解 TeleService、Voice Call 语音通话模型、ServiceState 网络服务、Data Call 移动数据业务、SMS & MMS 业务、Radio Interface Layer。

本书适合计算机科学技术、信息技术、通信工程、软件工程等专业的研究生、本科学生、高职高专学生使用。

◆ 著　　杨青平
　　责任编辑　祝智敏
　　责任印制　马振武

◆ 人民邮电出版社出版发行　北京市丰台区成寿寺路 11 号
　　邮编　100164　电子邮件　315@ptpress.com.cn
　　网址　http://www.ptpress.com.cn
　　涿州市京南印刷厂印刷

◆ 开本：787×1092　1/16
　　印张：19　　　　　　　　　2018 年 9 月第 1 版
　　字数：519 千字　　　　　　2020 年 6 月河北第 2 次印刷

定价：59.80 元

读者服务热线：(010)81055256　印装质量热线：(010)81055316
反盗版热线：(010)81055315
广告经营许可证：京东市监广登字 20170147 号

前　　言

随着智能手机的日益普及，移动互联网正深刻改变着我们的日常生活，iOS 和 Android 两大移动操作系统功不可没。由于 Android 的开源，各大厂商纷纷通过定制 Android 系统降低研发成本和周期，从而快速推出智能产品。

Telephony 作为 Android 系统的核心业务，包括接打电话、手机上网、短信、彩信等应用场景，也是终端用户使用最多、最频繁的业务。因此，在定制 Android 系统的开发过程中，对 Telephony 业务的定制和修改也非常多，如优化通话界面、过滤骚扰短信和电话，分析短信内容和优化展示关键信息等，既能方便用户操作和使用，提升用户体验，也能增加广告收入。

Android Telephony 业务跨度大，涉及多个层之间的交互：应用层、系统框架层、HAL 硬件抽象层和 BP Modem。理解和掌握 Android Telephony 业务和技术，方能更好地完成 Android 系统定制、业务扩展和数据挖掘。

本书主要特点如下。

1．解析源代码并总结 Android Telephony 业务模型、设计原理和系统架构

Android 的源代码汇集了大量 Google 工程师的设计思想和理念。本书对 Telephony 源代码中的关键设计模式、设计原理和实现方式做了详细的分析和总结，可帮助读者拓展思路和加强训练，以提升软件设计水平和编码能力。

本书由浅入深地详细讲解了搭建 Android 编译环境、分析源代码并推导出 Telephony 业务模型，可帮助读者提升 Ubuntu 系统动手能力、Java 语言编程能力、Android 开发能力、UML 阅读能力，理解和掌握常用的设计模式，以及 Android 系统平台定制的核心工作。

2．合理、有效的组织

本书以 Telephony 关键业务流程为主线，分析和总结主要业务、核心实现和业务模型。首先分析和总结 Telephony 业务模型中 Voice Call 语音通话、ServiceState 网络服务、Data Call 移动数据和 SMS&MMS 的关键业务流程和核心实现机制，然后扩展 Telephony Phone、Voice Call、Data Call 等关键业务模型，同时输出时序图、类图、状态图等 UML 图和架构图，以帮助读者理解和掌握 Telephony 业务。

3．内容充实、实用并结合实践

本书紧紧围绕 Android Telephony 关键业务展开，主要包括 Voice Call 语音通话、ServiceState 网络服务、Data Call 移动数据和 SMS&MMS 中需要优化和定制的关键业务，以及 Radio Interface Layer（无线通信接口抽象层）的消息处理机制和系统架构。

本书总结的 Telephony 业务模型可在华为 Nexus 6P 手机上进行运行和验证，使读者对 Telephony 业务和实现有更加直观的认识和理解，并掌握调试 Android 代码的方法和技巧。

尽管作者在写作过程中力求准确、完善，但书中不妥或错误之处仍在所难免，殷切希望广大读者批评指正！恳请读者一旦发现错误，于百忙之中及时与作者联系，以便尽快更正，作者将不胜感激。E-mail：android_tele@163.com。

杨青平

2018 年 4 月

目 录

第1章 初识 Android ··· 1
1.1 智能手机的系统结构 ··· 1
1.2 Android 系统架构 ··· 2
1.2.1 应用层 ··· 3
1.2.2 应用框架层 ··· 3
1.2.3 系统运行库层 ··· 3
1.2.4 核心层 ··· 4
1.3 Android Telephony 框架结构 ··· 5
1.3.1 系统运行库层的 HAL ··· 6
1.3.2 简析 HAL 结构 ··· 6
1.3.3 Android 为什么引入 HAL ··· 7
1.3.4 Android 中 HAL 的运行结构 ··· 7
本章小结 ··· 8

第2章 搭建 Android 源代码编译调试环境 ··· 9
2.1 Ubuntu Linux 操作系统及工具安装 ··· 10
2.1.1 PC 配置建议 ··· 10
2.1.2 Ubuntu 安装光盘制作 ··· 10
2.1.3 Ubuntu 安装过程 ··· 10
2.1.4 安装 OpenJDK ··· 12
2.1.5 Ubuntu 系统工具包更新升级 ··· 13
2.2 Android 源代码下载及编译过程 ··· 13
2.2.1 工作目录设置 ··· 13
2.2.2 源代码下载 ··· 13
2.2.3 开始编译 Android 源代码 ··· 14
2.2.4 编译单个模块 ··· 16
2.3 Android Studio 及 SDK ··· 17
2.3.1 下载和配置 Android Studio ··· 17
2.3.2 Android SDK 下载及配置和使用 ··· 17
2.3.3 使用 Android SDK 启动 Android 虚拟设备 ··· 19
2.3.4 Android 调试工具 adb 的使用方法 ··· 20
2.3.5 相关技巧汇总 ··· 20
2.4 在 Google 手机上调试 Android 源码 ··· 21
2.4.1 Google 手机对应编译选项 ··· 21

- 2.4.2 Google 手机刷入工厂镜像 ················· 21
- 2.4.3 编译本地镜像并刷入 Google 手机 ················· 22
- 2.4.4 Google 手机上调试 Android 源码 ················· 25
- 2.4.5 关键问题总结 ················· 26
- 本章小结 ················· 27

第 3 章 深入解析通话流程 ················· 29

- 3.1 拨号流程分析 ················· 29
  - 3.1.1 打开 Nexus 6P 手机的拨号盘 ················· 30
  - 3.1.2 进入拨号界面 DialtactsActivity ················· 30
  - 3.1.3 DialpadFragment 拨号盘 ················· 32
  - 3.1.4 ITelecomService 接收拨号请求服务 ················· 33
  - 3.1.5 CallsManager 拨号流程处理 ················· 35
  - 3.1.6 IInCallService 服务的响应过程 ················· 40
  - 3.1.7 继续分析 CallsManager.placeOutgoingCall ················· 46
  - 3.1.8 Telecom 应用拨号流程回顾与总结 ················· 50
  - 3.1.9 IConnectionService 服务的响应过程 ················· 51
  - 3.1.10 TelecomAdapter 接收消息回调 ················· 55
  - 3.1.11 拨号流程总结 ················· 56
- 3.2 来电流程分析 ················· 57
  - 3.2.1 分析 radio 来电日志 ················· 58
  - 3.2.2 UNSOL_RESPONSE_CALL_STATE_CHANGED 消息处理 ················· 58
  - 3.2.3 扩展 RegistrantList 消息处理机制 ················· 59
  - 3.2.4 GsmCdmaCallTracker 消息处理 ················· 61
  - 3.2.5 ITelecomService 处理来电消息 ················· 63
  - 3.2.6 来电流程总结 ················· 66
- 3.3 通话总结 ················· 66
  - 3.3.1 通话关键代码汇总 ················· 66
  - 3.3.2 通话状态更新消息上报流程 ················· 68
  - 3.3.3 控制通话消息下发流程 ················· 69
- 3.4 建立 Android 通话模型 ················· 70
- 本章小结 ················· 71

第 4 章 详解 Telecom ················· 73

- 4.1 Telecom 应用加载入口 ················· 73
  - 4.1.1 TelecomManager 类核心逻辑分析 ················· 74
  - 4.1.2 Telecom 应用代码汇总 ················· 76
  - 4.1.3 ITelecomService 的 onBind 过程 ················· 77
  - 4.1.4 第二个拨号入口 ················· 79
- 4.2 Telecom 交互模型 ················· 79

- 4.2.1 汇总 frameworks/base/telecomm 代码 ··· 80
- 4.2.2 绑定 IInCallService 机制 ··· 81
- 4.2.3 绑定 IConnectionService 机制 ··· 82
- 4.2.4 演进 Telecom 交互模型 ··· 85
- 4.3 核心 Listener 回调消息处理 ··· 86
  - 4.3.1 CallsManagerListener ··· 86
  - 4.3.2 Call.Listener ··· 88
  - 4.3.3 CreateConnectionResponse ··· 90
  - 4.3.4 总结 Listener 消息 ··· 90
- 4.4 扩展 CallsManager ··· 92
  - 4.4.1 记录通话日志 ··· 92
  - 4.4.2 耳机 Hook 事件 ··· 93
  - 4.4.3 通知栏信息同步 ··· 93
- 本章小结 ··· 94

第 5 章 详解 TeleService ··· 95

- 5.1 加载过程分析 ··· 95
  - 5.1.1 应用基本信息 ··· 96
  - 5.1.2 PhoneGlobals.onCreate ··· 97
  - 5.1.3 TelephonyGlobals.onCreate ··· 98
- 5.2 Telephony Phone ··· 98
  - 5.2.1 GsmCdmaPhone ··· 99
  - 5.2.2 Composition（组合）关系 ··· 101
  - 5.2.3 Facade Pattern ··· 102
  - 5.2.4 Handler 消息处理机制 ··· 103
- 5.3 扩展 PhoneAccount ··· 105
  - 5.3.1 PhoneAccount 初始化过程 ··· 105
  - 5.3.2 PhoneAccount 注册响应 ··· 108
  - 5.3.3 PhoneAccount 在拨号流程中的作用分析 ··· 109
  - 5.3.4 小结 ··· 112
- 5.4 TeleService 服务 ··· 113
  - 5.4.1 phone 系统服务 ··· 113
  - 5.4.2 isub 系统服务 ··· 115
  - 5.4.3 IConnectionService 应用服务 ··· 118
- 本章小结 ··· 123

第 6 章 Voice Call 语音通话模型 ··· 125

- 6.1 详解 GsmCdmaCallTracker ··· 125
  - 6.1.1 代码结构解析 ··· 126
  - 6.1.2 Handler 消息处理方式 ··· 127

- 6.1.3 与 RILJ 对象的交互机制 ·· 130
- 6.2 handlePollCalls 方法 ·· 134
  - 6.2.1 准备阶段 ·· 134
  - 6.2.2 更新通话相关信息 ·· 135
  - 6.2.3 发出通知 ·· 140
  - 6.2.4 更新 mState ·· 141
- 6.3 通话管理模型分析 ··· 142
  - 6.3.1 GsmCdmaCall ··· 143
  - 6.3.2 GsmCdmaConnection ·· 143
  - 6.3.3 DriverCall、Call、Connection ··· 146
- 6.4 补充通话连接断开处理机制 ··· 149
  - 6.4.1 本地主动挂断通话 ·· 149
  - 6.4.2 远端断开通话连接 ·· 152
- 6.5 区分 Connection ·· 154
- 6.6 扩展 InCallUi ·· 155
  - 6.6.1 初始化过程 ·· 155
  - 6.6.2 addCall ··· 158
  - 6.6.3 InCallUi 通话界面 ·· 160
  - 6.6.4 updateCall ·· 165
- 6.7 验证 Call 运行模型 ··· 166
  - 6.7.1 Telephony Voice Call ·· 167
  - 6.7.2 Telecom Call ·· 170
  - 6.7.3 InCallUi Call ·· 171
- 本章小结 ··· 173

第 7 章 ServiceState 网络服务 ·· 175

- 7.1 ServiceState ··· 176
  - 7.1.1 ServiceState 类的本质 ··· 176
  - 7.1.2 关键常量信息 ··· 177
  - 7.1.3 关键属性 ·· 177
  - 7.1.4 关键方法 ·· 178
- 7.2 ServiceStateTracker 运行机制详解 ··· 179
  - 7.2.1 核心类图 ·· 179
  - 7.2.2 代码结构 ·· 180
  - 7.2.3 Handler 消息处理机制 ·· 181
  - 7.2.4 与 RILJ 对象的交互机制 ·· 184
- 7.3 handlePollStateResult 方法 ·· 186
  - 7.3.1 异常处理 ·· 186
  - 7.3.2 handlePollStateResultMessage ·· 187
  - 7.3.3 继续更新 mNewSS ··· 190

- 7.3.4 完成收尾工作 ···191
- 7.4 *#*#4636#*#*测试工具 ··193
  - 7.4.1 网络服务信息 ··194
  - 7.4.2 扩展 ITelephonyRegistry ·····································196
  - 7.4.3 展示小区信息 ··197
  - 7.4.4 小区信息更新源头 ··198
  - 7.4.5 信号强度实时变化 ··199
- 7.5 飞行模式 ··201
  - 7.5.1 飞行模式开启关闭入口逻辑 ····································201
  - 7.5.2 Radio 模块开启关闭 ···202
  - 7.5.3 WiFi 模块开启关闭 ··202
  - 7.5.4 蓝牙模块开启关闭 ··202
- 7.6 扩展 SIM 卡业务 ··203
  - 7.6.1 SIM 卡业务分析 ···203
  - 7.6.2 驻网过程分析 ···204
  - 7.6.3 SoftSim 业务实现分析 ···205
- 本章小结 ··206

第 8 章 Data Call 移动数据业务 ·······································207

- 8.1 DcTracker 初始化过程 ···207
  - 8.1.1 Handler 消息注册 ···208
  - 8.1.2 初始化 ApnContext ··208
  - 8.1.3 认识 APN ··210
  - 8.1.4 创建 DcController ··212
  - 8.1.5 注册 Observer ··213
  - 8.1.6 广播接收器 ··213
  - 8.1.7 加载 ApnSetting ··213
- 8.2 解析 StateMachine ··215
  - 8.2.1 State 设计模式 ···215
  - 8.2.2 StateMachine 核心类 ··215
  - 8.2.3 初始化流程 ···216
  - 8.2.4 运行流程 ···217
  - 8.2.5 小结 ···218
- 8.3 DataConnection ···219
  - 8.3.1 关键属性 ···220
  - 8.3.2 关键方法 ···220
  - 8.3.3 StateMachine 初始化流程 ······································221
- 8.4 开启移动数据业务 ···222
  - 8.4.1 流程分析 ···222
  - 8.4.2 前置条件分析 ···227

8.4.3 DcActiveState 收尾工作	231
8.4.4 Suspend 挂起状态	232
8.4.5 查看手机上网基本信息	232
8.5 关闭移动数据业务	233
8.6 DataConnection 状态转换	233
8.7 获取 Android 手机上网数据包	234
8.7.1 使用 tcpdump 工具抓取 TCP/IP 数据包	234
8.7.2 使用 Wireshark 软件分析 TCP/IP 数据包	235
本章小结	235

第 9 章 SMS&MMS 业务 236

- 9.1 短信发送流程 236
  - 9.1.1 进入短信应用 236
  - 9.1.2 短信编辑界面 237
  - 9.1.3 Action 处理机制 239
  - 9.1.4 继续跟进短信发送流程 241
  - 9.1.5 phone 进程中的短信发送流程 243
- 9.2 扩展短信发送业务 245
  - 9.2.1 确认短信发送结果 245
  - 9.2.2 重发机制 246
  - 9.2.3 状态报告 247
- 9.3 短信接收流程 247
  - 9.3.1 RIL 接收短信消息 247
  - 9.3.2 GsmInboundSmsHandler 248
  - 9.3.3 Messaging 应用接收新短信 250
  - 9.3.4 PDU 251
  - 9.3.5 短信业务小结 252
- 9.4 彩信关键业务逻辑 253
  - 9.4.1 彩信发送入口 253
  - 9.4.2 imms 系统服务 254
  - 9.4.3 彩信发送流程 255
  - 9.4.4 Data Call 256
  - 9.4.5 doHttp 259
  - 9.4.6 接收彩信 259
  - 9.4.7 MmsService 小结 260
- 本章小结 261

第 10 章 Radio Interface Layer 262

- 10.1 解析 RILJ 263
  - 10.1.1 认识 RIL 类 263

- 6 -

10.1.2	RILRequest	265
10.1.3	IRadio 关联的服务	266
10.1.4	RIL 消息分类	270
10.1.5	Solicited Request	270
10.1.6	Solicited Response	271
10.1.7	UnSolicited	274

10.2 详解 rild ················· 274
    10.2.1 RIL_startEventLoop ················· 275
    10.2.2 获取 RIL_RadioFunctions ················· 275
    10.2.3 注册 RIL_RadioFunctions ················· 277

10.3 libril 初始化流程 ················· 278
    10.3.1 RIL_startEventLoop ················· 278
    10.3.2 RIL_register ················· 280

10.4 扩展 hal 接口 ················· 281
    10.4.1 增加接口定义 ················· 282
    10.4.2 验证生成的代码 ················· 282
    10.4.3 实现新增接口 ················· 285
    10.4.4 运行结果验证 ················· 286

10.5 RILC 运行机制 ················· 287
    10.5.1 Solicited 消息 ················· 287
    10.5.2 UnSolicited 消息 ················· 291

本章小结 ················· 293

第 1 章
初识 Android

学习目标

- 学习智能手机基本硬件体系结构。
- 掌握基于 Linux Kernel 的 Android 系统分层架构。
- 掌握 Telephony 在 Android 系统中的结构。

Android 中文意思为"机器人",中文译名为"安卓",是谷歌公司于 2007 年 11 月 5 日发布的基于 Linux 平台的开源手机操作系统,其由操作系统、中间件、用户界面和应用软件组成,号称首个为移动终端打造的真正开放和完整的移动软件。谷歌公司通过与电信运营商、手机设备制造商、芯片开发商及其他有关方面结成深层次的合作伙伴关系,希望借助建立标准化、开放式的移动电话软件平台,在移动产业内形成一个开放式的生态系统。

从 2007 年至今,经过长时间的考验,Android 已经成为全球最热门的手机操作系统之一。本章主要从智能手机的基本硬件结构、Android 手机操作系统整体架构和 Android 的 Telephony 模块的体系结构三个方面逐步认识 Android,特别将 Android 手机操作系统平台下 Telephony 模块作为本章讲解的重点内容。

1.1 智能手机的系统结构

Android 手机的基本硬件结构符合智能手机的基本硬件结构,我们要学习 Android 移动开发,首先需要了解智能手机的硬件系统基本结构。

随着通信领域的快速发展,移动终端的发展和变化也非常巨大,已经由原来单一的通话功能、短信功能,向彩信、数据上网、图像处理、音乐和多媒体方向演变。到目前为止,市面上的移动手机基本上可以分成两大类:一类是功能手机(Feature Phone);另一类是智能手机(Smart Phone)。

这两类手机如何区分呢?智能手机具有传统手机的基本功能,如打电话、发短信、照相等。智能手机的特点:具有开放的操作系统、硬件和软件的可扩充性和支持第三方的二次开发。相对于功能手机,智能手机就像计算机一样,可通过安装第三方软件来扩展其功能和应用,因此,智能手机越来越受到人们的青睐,已成为手机终端市场的一种潮流。

那么先来看看智能手机较多采用的硬件基本结构,如图1-1所示。

智能手机的基本硬件结构大多采用双处理器架构:主处理器和从处理器。主处理器运行开放式操作系统以及操作系统之上的各种应用,负责整个系统的控制;从处理器负责无线通信基本能力,主要包括DBB(Digital Baseband,数字基带)和ABB(Analog Baseband,模拟基带),完成语音信号和数字信号调制解调、信道编码解码和无线Modem控制。

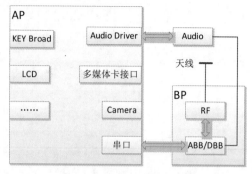

图1-1　智能手机硬件结构图

主处理器也叫AP(Application Processor,应用处理器),从处理器也叫BP(Baseband Processor,基带处理器),它们之间通过串口、总线或USB等方式进行通信。不同手机芯片生产厂家采用的集成方式都不一样,目前市面上仍以串口通信为主。

不难发现,在智能手机的基本硬件结构中,BP部分只要再加一定的外围电路,如音频芯片、LCD控制、摄像机控制器、扬声器、天线等,就是一个完整的普通手机的硬件结构。

现在我们能区分功能手机与智能手机吗?回顾手机终端的发展历程,不难发现这样一条规律:随着手机芯片处理能力的提升、上网能力的扩展和发展(蓝牙、WiFi、3G网络),手机应用得到非常广泛的扩大和发展。在智能手机的硬件设计上,采用处理能力比较强大的处理器作为AP,来支持开放手机操作系统及操作系统之上的扩展应用,由此可见智能手机发展的趋势和方向。

1.2　Android系统架构

前面学习了智能手机的基本硬件结构,可通过功能手机与智能手机的特点和区别从本质上去认识它们。Android作为一款运行在AP上的开源智能手机操作系统,其系统架构是什么样的呢?我们先来看看图1-2。

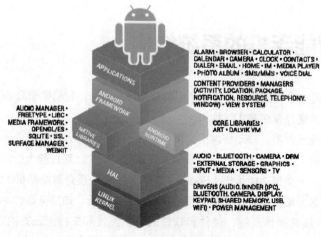

图1-2　Android系统架构

通过图 1-2 不难发现，Android 是一个分层的基于 Linux Kernel 的智能手机操作系统，共分为四层，从上到下依次是应用层（Applications）、应用框架层（Framework）、系统运行库层（Libraries）和核心层（Linux Kernel），下面将对这四层进行简要的分析和介绍。

1.2.1 应用层

Android 近几年的发展可谓是非常迅猛，有一个很重要的原因，那就是它的应用非常多。安卓市场已发布的软件个数和软件下载量目前仅次于苹果的应用商店，并且保持着快速增长态势。只有这些第三方开发的应用（如游戏、导航、播放器、桌面主题等）日益丰富，手机终端用户才能不断地发展和壮大，而这些应用均在应用层运行。

应用层包括了各种 Android 应用程序，这些应用程序是使用 Java 语言开发，并运行在 Dalvik 虚拟机上，处于 Android 系统架构中的第一层。在 Android 源码和 SDK 中，谷歌公司已经捆绑和发布了一些核心应用及源代码，如 Dialer、MMS、日历、谷歌地图、浏览器、联系人等。

1.2.2 应用框架层

如图 1-2 所示，Android 系统架构中的第二层是应用框架层，是用 Java 语言实现和开发的。有了应用框架层，开发者使用该层提供的 API 便可非常方便地完成访问设备硬件、获取位置信息、向状态栏添加通知消息、设置闹铃等操作，而不必关心具体的底层实现机制和硬件实现方式。这样，简化了 Android 应用开发者开发程序时的架构设计，从而能够快速开发新的应用程序。

应用框架层是谷歌公司发布核心应用时所使用的 API 框架，开发人员可以使用这些框架提供的 API 来快速开发自己的应用程序。下面是对 Android 中一些主要的组件的简要总结及相关说明。

- 视图（View）

在 Android SDK（Software Development Kit，软件开发工具包）中介绍了其丰富的视图的使用方法及相关属性，所有的 Android 应用程序都由这些视图构成，主要包括列表（List）、网格（Grid）、文本框（Text）、按钮（Buttons）等基础 Android 应用的界面控件。

- 资源管理器（Resource Manager）

提供非代码资源转换和访问，如本地字符串（xml 文件配置）、图片和布局文件（Layout File，使用 xml 文件配置）。

- 通知管理器（Notification Manager）

应用可以在状态栏中显示自定义的提示信息，如新短信通知、未接来电通知、手机信号量通知等。

- Activity 管理器（Activity Manager）

用来管理 Android 应用程序界面的生命周期（onCreate 创建、onResume 显示、onPause 暂停、onStop 停止等），一个手机屏幕界面可对应一个 Activity。

1.2.3 系统运行库层

如图 1-2 所示，Android 系统架构中的第三层为系统运行库层，这一层主要包含了手机操作系统平台必备的 C/C++核心库、Dalvik 虚拟机运行环境和 HAL 子层。我们跳过 HAL，先简单地介绍和分析 C/C++核心库和 Dalvik 虚拟机运行环境。

1. C/C++核心库

系统运行库层包含一个C/C++库的集合，当使用Android应用框架的一些接口时，系统运行库层通过C/C++核心库来支持对应的组件使用，使其能更好地为Android应用开发者服务。下面是一些主要的核心C/C++库及其简要说明。

- libc（系统C库）

C语言标准库，处于系统最底层的系统库，由Linux系统来调用。

- Media Framework（多媒体库）

Android系统多媒体库，支持当前手机平台上主流的音频和视频格式播放和录制，以及静态图像。如MPEG-4、MP3、AAC、JPG、PNG等多媒体格式。

- SGL

2D图形引擎库。

- OpenGL

3D效果的支持。

- SQLite

轻量级关系数据库引擎，可用来增、删、改、查通话记录、联系人等信息。

- WebKit

新式的Web浏览器引擎，支持当前非常流行的HTML 5。

- SSL

基于TCP/IP网络协议，为数据安全通信提供支持。

2. Dalvik虚拟机运行环境

系统运行库层包含了Android Runtime，其核心为Dalvik虚拟机。每一个Android应用程序都运行在Dalvik虚拟机之上，且每一个应用程序都有自己独立运行的进程空间；Dalvik虚拟机只执行DEX可执行文件。

DEX格式是专为Dalvik设计的一种压缩格式，适合内存和处理器速度有限的系统。要生成DEX格式文件，首先通过Java程序编译生成class文件，然后通过Android提供的dx工具将class文件格式转换成DEX格式。

Dalvik虚拟机的特性总结如下。

- 每一个Android应用运行在一个Dalvik虚拟机实例中，而每一个虚拟机实例都是一个独立的进程空间。
- 虚拟机的线程机制、内存分配和管理、Mutex（进程同步）等的实现都依赖底层Linux操作系统。
- 所有Android应用的线程都对应一个Linux线程，因而虚拟机可以更多地使用Linux操作系统的线程调度和管理机制。

注意

因为Android的编程语言是Java的缘故，我们很容易将Dalvik虚拟机与Java虚拟机误认为是同一个东西。但Dalvik虚拟机并不是按照Java虚拟机的规范来实现的，两者并不兼容；它们之间最大的不同在于Java虚拟机运行的是Java字节码，而Dalvik虚拟机运行的是其专有的文件格式——DEX（Dalvik Executable）文件。

1.2.4 核心层

Android 4.0基于Linux Kernel 3.0.8提供核心系统服务，如文件管理、内存管理、进程管理、网络堆栈、驱动模型等操作系统的基本服务能力。Linux Kernel 同时也作为硬件和软件之间的抽象层，需要一些与移动设备相关的驱动程序来支持，主要的驱动如下。

- 显示驱动（Display Driver）

基于 Linux 的帧缓冲驱动。

- 键盘驱动（Keyboard Driver）

输入设备的键盘驱动，如 Home（待机）、Menu（菜单）、Return（返回）、Power（电源）等设备按键。

- 音频驱动（Audio Driver）

常用的基于 ALSA（Advanced Linux Sound Architecture）的高级 Linux 声音体系驱动。

- 电源管理（Power Management）

如电池电量、充电、屏幕开启关闭管理等。

- Binder IPC 驱动

Android 平台上一个特殊的驱动程序，具有单独设备访问节点，用来提供 IPC 进程间的通信功能。

- 蓝牙驱动（Bluetooth Driver）

基于 IEEE 802.15.1 标准的蓝牙无线传输技术。

- WiFi 驱动（WIFI Driver）

基于 IEEE 802.11 标准的 WiFi 驱动。

- 照相机驱动（Camera Driver）

常用的基于 Linux 的照相机驱动。

1.3 Android Telephony 框架结构

通过前面对 Android 手机操作系统整体框架结构及每一层的简单分析和说明，相信大家对 Android 智能手机操作系统已经有了一些基本的了解和认识。结合 Android 手机操作系统的整体框架，我们接着学习 Android Telephony 涉及的框架结构，首先看图 1-3。

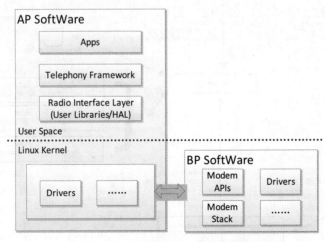

图 1-3 Android Telephony 框架结构

通过图 1-3 不难发现 Android Telephony 框架结构的一些规律，总结如下：

- Android Telephony 的业务应用跨越 AP 和 BP，AP 与 BP 相互通信，符合前面介绍的智能手机的硬件基本结构。
- Android 系统在 AP 上运行，而 Telephony 运行在 Linux Kernel 之上的用户空间。
- Android Telephony 也采用了分层结构的设计，共跨越了三层：应用层、应用框架层和系统运行库层，与 Android 操作系统整体分层结构保持一致；
- Android Telephony 从上到下共分三层：Telephony 应用、Telephony 框架、RIL（Radio Interface Layer，无线通信接口层，主要位于系统运行库层的 HAL 中，什么是 HAL，接下来会详细介绍）。
- BP SoftWare 在 BP 上运行，主要负责实际的无线通信能力处理，不在本书讨论的范围。

1.3.1 系统运行库层的 HAL

HAL（Hardware Abstraction Layer，硬件抽象层）在 Linux 和 Windows 操作系统平台下有不同的实现方式。

Windows 下的 HAL 位于操作系统的最底层，它直接操作物理硬件设备，用来隔离与不同硬件相关的信息，为上层的操作系统和设备驱动程序提供一个统一接口，起到对硬件的抽象作用。这样更换硬件后编写硬件的驱动时，只要实现符合 HAL 定义的标准接口即可，而上层应用并不会受到影响，也不必关心具体实现的是什么硬件。

Linux 下的 HAL 与 Windows 下的 HAL 不太一样，HAL 并不是位于操作系统的最底层，直接操作硬件；相反，它位于操作系统核心层和驱动程序之上，是一个运行在用户空间中的服务程序。

1.3.2 简析 HAL 结构

通过前面的学习，我们知道 Android 是基于 Linux Kernel 的开源智能手机操作系统，所以这里重点介绍基于 Linux 下的 HAL 结构，就不再单独介绍 Windows 下的 HAL 结构。

要想知道 HAL 结构，先看看来源于 HAL 0.4.0 Specification 的框图，如图 1-4 所示。

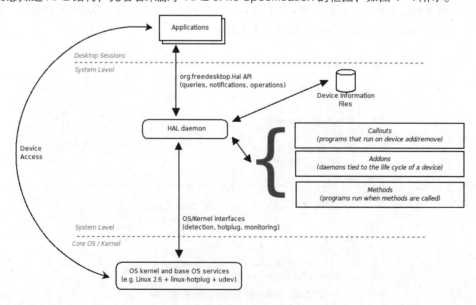

图 1-4　HAL 0.4.0 Specification

HAL 是一个位于操作系统和驱动程序之上,运行在用户空间中的服务程序。其目的是对上层应用提供一个统一的查询硬件设备的接口。我们都知道,抽象就是为了隔离变化,那么这里的 HAL 可以带给我们什么?首先,有了 HAL 接口,可以提前开始应用的开发,而不必关心具体实现的是什么硬件;其次,硬件厂家需要更改硬件设备,只要按照 HAL 接口规范和标准提供对应的硬件驱动,而不需要改变应用;最后,HAL 简化了应用程序查询硬件的逻辑,把这一部分的复杂性转移到由 HAL 统一处理,这样当一些应用程序使用 HAL 的时候,可以把对不同硬件的实际操作的复杂性也交给不同硬件厂家提供的库函数来处理。

总之,HAL 所谓的抽象并不提供对硬件的实际操作,对硬件的操作仍然由具体的驱动程序来完成。

1.3.3 Android 为什么引入 HAL

HAL 的一些优势在前面已经提到,这里回顾一下。Android 引入 HAL 不仅看重其自身的优势,而且还有一个非常重要的因素,为了保障在 Android 平台基于 Linux 开发的硬件驱动和应用程序,不必遵循 GPL(General Public License)许可而保持封闭,以保障更多厂家的利益。我们都知道,Linux Kernel 开源而且遵循 GPL 许可证,根据 GPL 许可证规定,对代码的任何修改都必须向社会开源。

那么 Android 是如何做到的呢?Linux Kernel 和 Android 的许可证不一样,Linux Kernel 是 GPL 许可证,Android 是 ASL(Apache Software License)许可证。ASL 许可证规定,可以随意使用源码,但不必开源,所以构筑在 Android 之上的硬件驱动和应用程序都可以保持封闭。也就是说,只要把关键的与驱动处理相关的主要逻辑转移到 Android 平台内,在 Linux Kernel 中仅保留基础的通信功能,即使开源一部分代码,对厂家来讲也不会有什么损失。

谷歌选择了这样做,并且特意修改了 Kernel,原本应该包括在 Linux Kernel 中的某些驱动关键处理逻辑,被转移到了 HAL 层中,从而达到了不必开源的目的。

本书不再对 GPL、ASL 或其他的开源许可证做深入探讨,有兴趣的读者可以上网搜索详细资料。

1.3.4 Android 中 HAL 的运行结构

由图 1-2 中可以知道 Android 源码中已经实现了一部分 HAL,包括 Wi-Fi、GPS、RIL、Sensor 等,这些代码主要存储于以下目录:

- Android_src/hardware/libhardware_legacy

老式 HAL 结构,采用直接调用 so 动态链接库方式。

- Android_src /hardware/libhardware

新式 HAL 结构,采用 Stub 代理方式调用。

- Android_src /hardware/ril

RIL(Radio Interface Layer,无线通信接口层)作为本书重点关注和学习的内容,后面会采用单独章节详细讲解。

在 Android 中,HAL 的运行机制是什么样的呢?它有两种运行机制,老式 HAL 和新式 HAL,如图 1-5 所示。

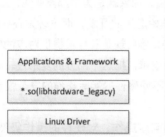
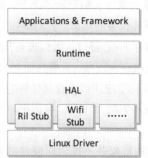

图 1-5　Android 中 HAL 两种运行结构

从图中不难看出，左边是老式 HAL 结构。如图中所示，应用或框架通过 so 动态链接库调用从而达到对硬件驱动的访问。在 so 动态链接库里，实现了对驱动的访问逻辑处理。我们重点学习和理解 HAL Stub 方式，RIL 也采用了此方式的设计思想。

HAL Stub 是一种代理概念，虽然 Stub 仍是以 *.so 的形式存在，但 HAL 已经将 *.so 的具体实现隐藏了起来。Stub 向 HAL 提供 operations 方法，Runtime 通过 Stub 提供的 so 获取它的 operations 方法，并告知 Runtime 的 callback 方法。这样 Runtime 和 Stub 都有对方调用的方法，一个应用的请求通过 Runtime 调用 Stub 的 operations 方法，而 Stub 响应 operations 方法并完成后，再调用 Runtime 的 callback 方法返回。根据前面的描述再结合图 1-6 会更容易理解。

上层调用底层，通过底层 HAL 提供的函数，而底层在处理完上层请求后或硬件状态发生变化时回调上层，则通过 Runtime 提供的 callback 接口完成。

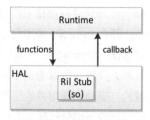

图 1-6　HAL Stub 结构

HAL Stub 有一种包含关系，即 HAL 里包含了很多的 Stub。Runtime 只要说明请求类型，就可以取得并操作 Stub 对应的 operations 方法。其实现主要在 hardware.c 和 hardware.h 文件中。实质也是通过 dlopen 方法加载 .so 动态链接库，从而调用*.so 里的符号（symbol）实现。

本 章 小 结

还记得智能手机基本硬件中的 AP 和 BP 主从处理器结构吗？基于 Android 平台的手机也符合智能手机的体系结构。Android 是基于 Linux Kernel 运行在 AP 上的智能手机操作系统，同时也是一个分层的操作系统平台，从上到下主要分为四层；在 AP 上运行的 Telephony 相关应用与 Android 整体分层结构保持一致，每层中的 Telephony 相关内容都是本书要重点讲述的内容。

Android 手机中的基本通信能力跨越了 AP 和 BP，又由 AP 和 BP 的相互协作完成基本的通信功能。AP 上的各种 Telephony 应用，通过丰富的界面展示了通信相关的各种形式和状态，也可通过界面向 BP 发起通信能力相关的控制，且由 BP 负责具体的通信能力实施。

这里提出一个问题，请读者结合本章内容思考：Android Telephony 处于整个 Android 智能手机平台的什么位置，HAL 是什么？通过本章的学习，你清楚了吗？

第 2 章
搭建 Android 源代码编译调试环境

学习目标

- 学习 Ubuntu 系统安装及相关依赖工具安装。
- 下载和编译 Android 源码。
- 掌握 Android Studio 和 SDK 的使用。
- 掌握在 Google 手机上运行调试 Android 源码。

第 1 章介绍了 Android 及其 Telephony 功能模块的基本结构，在本章中将搭建 Android 编译环境。如果你已经能够在自己的计算机上成功编译 Android 源码，还是希望你能阅读本章内容，因为其中不乏一些技巧的总结和 Android 相关实用工具的使用介绍。

Android 的编译环境作为深入学习 Android 的基础，不可缺失，能够完整下载 Android 源代码并编译成功，对深入学习 Android 是非常关键和重要的一步。Android 环境搭建比较繁琐和枯燥，花费时间长，从 Ubuntu 17.10 操作系统安装、编译所需工具包的下载、更新和安装，到 JDK、Android SDK 下载、配置以及关键工具的使用，再到 Android 8.1.0 源代码下载、编译和"刷机"等。其中有一步未完成或稍有偏差，都可能导致 Android 源码编译不能通过或是无法在 Google 手机上调试 Android 源码。

注意

Android 源代码编译调试环境在深入学习 Android 的过程中非常关键和重要，为什么这么说呢？因为有了这样的环境，我们在面对大量的代码不知道如何下手时，可以尝试着修改源码并在 Google Nexus 6P 或是其他 Google 手机上调试、运行修改源码后的效果，特别是在没有开发板或工程手机的条件下（本书将告诉你如何使用 Google 手机替代），对我们学习 Android 系统至关重要。

因此本章内容尽量做到直观和简洁，关键步骤均有截图或说明，并且每步操作均做了真实、严格的验证，只要读者按照本书步骤一步一步对照操作，都能够成功搭建起自己本地的 Android 编译、调试和开发环境。

2.1 Ubuntu Linux 操作系统及工具安装

要搭建 Android 的编译环境，首先选择操作系统。本书选择 Ubuntu 17.10 桌面版 64 位 Linux 操作系统，作为编译 Android 源码的操作系统。

2.1.1 PC 配置建议

CPU 类型：英特尔酷睿 i5 处理器或 i7 处理器
内存：8GB 或更大容量
硬盘容量：500GB 或 1TB
显卡：集成显卡或其他独立显卡

2.1.2 Ubuntu 安装光盘制作

首先下载 Ubuntu 安装镜像文件，进入 Ubuntu 官方网站，发现有如下 Ubuntu 版本：
- Ubuntu 17.10 (Artful Aardvark)
- Ubuntu 17.04 (Zesty Zapus)
- Ubuntu 16.04.3 LTS (Xenial Xerus)
- Ubuntu 14.04.5 LTS (Trusty Tahr)
- Ubuntu 12.04.5 LTS (Precise Pangolin)

本书选用 64 位 Ubuntu 17.10 桌面版操作系统，选择下载 ubuntu-17.10-desktop-amd64.iso 镜像文件，然后使用刻录软件工具将 ISO 镜像文件刻录到光盘上，制作成系统安装光盘。

注意

如果没有刻录光盘的条件，可以采用制作 USB 系统启动盘的方式。目前大多数计算机支持 USB 引导系统，加上目前 USB 闪存容量大且价格相对较低，可选择 2GB 或以上容量的 U 盘，使用 Universal USB Installer 在 Windows 下制作 Ubuntu 17.10 系统安装启动盘，具体的方法可上网搜索制作 U 盘安装盘详情。

2.1.3 Ubuntu 安装过程

Ubuntu 安装盘制作完成后，就可以开始安装 Ubuntu Linux 操作系统了。没有使用过 Linux 的读者也不必担心，Ubuntu Linux 操作系统的安装和使用绝大部分采用图形化界面，对中文的支持也已经做得非常好了。下面开始 Ubuntu 系统的安装。

将前期制作的光盘或 USB 系统安装盘放入计算机光驱或插入 USB 接口，启动电脑后按 F12 键选择引导方式，选择从光驱或 U 盘引导系统，然后进入 Ubuntu 17.10 的系统安装欢迎界面，如图 2-1 所示。

图 2-1 的界面非常友好，选择语言为英语或中文简体，其安装过程与 Windows 相似，均采用图形化安装向导形式。安装 Ubuntu 17.10，进入下一步，提示安装 Ubuntu 需要两个必备的条件，一是至少有 4.6GB 的磁盘空间，二是笔记本电脑需要插入电源方式供电。满足这两个条件，不必接入网络，

继续进入下一步。

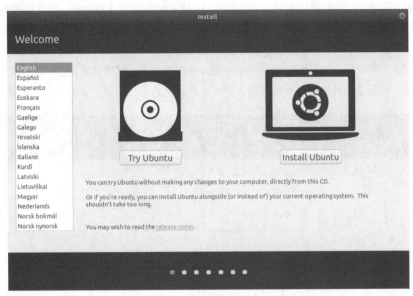

图 2-1　Ubuntu 安装欢迎界面

这一步非常关键，是对磁盘的分区。如何分配磁盘空间？这里建议手工划分磁盘空间大小，划分两个 Linux 操作系统必备的磁盘空间即可：/根目录挂载分区和 swap 交换分区，如图 2-2 所示。

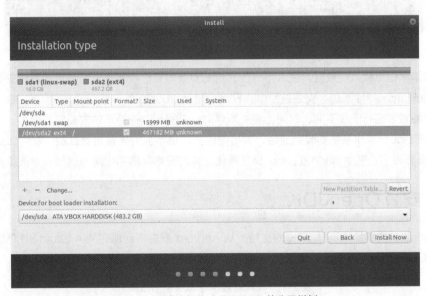

图 2-2　磁盘空间大小为 500GB 的分区样例

根据图 2-2，它们分别是：
- swap 交换分区

根据经验来讲，交换分区以内存大小×2 为最佳。本例中分配 16GB 磁盘空间大小。
- / 根目录挂载分区

不论磁盘大小，除去 swap 交换分区占用的磁盘空间，剩余的都分配在/根目录挂载分区。本例中分配 480GB 磁盘空间大小。

目前市场上的硬盘都较大，2TB 已成为基本配置。本例中的/ 根目录挂载分区包括了/home 用户数据分区（读者也可以分配独立的/home 分区挂载点）。在 Android 编译环境中，源代码的保存和编译都在此进行，所以需要较大空间，因此将除去 swap 交换分区占用的磁盘空间剩余的磁盘空间都分配到此挂载分区。

分区完成后，进入下一步，时区选择 Shanghai，进入下一步，然后选择键盘布局，这里使用默认的 USA 键盘布局；最后进入计算机基本信息输入界面，输入用户名、密码、计算机名，如图 2-3 所示。

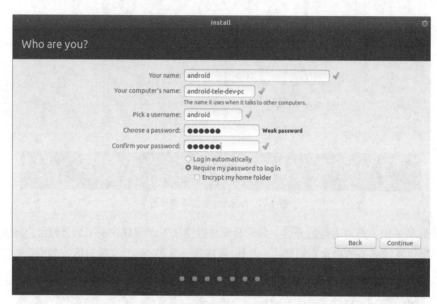

图 2-3　输入计算机基本信息

计算机基本信息输入完成后，进入下一步，到此安装向导已收集完安装信息，系统安装程序开始安装 Ubuntu Linux 操作系统。安装过程中，界面会显示 Ubuntu 的一些特性，时间在 20 分钟左右。根据计算机配置和运行速度不同，时间上会有所浮动，只需耐心等待系统安装完成即可。

Ubuntu Linux 操作系统安装完成后，弹出安装完成提示，需要重启计算机，就可以进入图形化 Ubuntu 登录界面了。整个 Ubuntu Linux 操作系统安装过程都是图形化向导过程，非常简单。

2.1.4　安装 OpenJDK

Android 源码的编译离不开 JDK 的支持。从 Android 刚发布到现在的 Android 8.1.0 版本，不同版本对 JDK 的需求是不一样的。表 2-1 中列举了 Android 版本与 JDK 的对应关系。

表 2-1　Android 版本与 JDK 的对应关系

Android 版本	JDK	备注
Android 2.3.x (Gingerbread) ~ Android 4.4.x (KitKat)	Java JDK 6	
Android 5.x (Lollipop) ~ Android 6.0 (Marshmallow)	OpenJDK 7	从 Android 5.0 开始，使用 OpenJDK 编译源码
Android 7.0 (Nougat) ~ Android 8.0 (O release)	OpenJDK 8	

因此，我们要编译 Android 8.1.0 的源码，对 JDK 的需求是 OpenJDK 8。在 Ubuntu 17.10 系统中，安装和验证 OpenJDK 的命令如下：

```
$ sudo apt-get update
$ sudo apt-get install openjdk-8-jdk

$ java -version
openjdk version "1.8.0_151"
OpenJDK Runtime Environment (build 1.8.0_151-8u151-b12-0ubuntu0.17.10.2-b12)
OpenJDK 64-Bit Server VM (build 25.151-b12, mixed mode)
```

2.1.5　Ubuntu 系统工具包更新升级

前面的步骤已经完成了 Ubuntu 操作系统的安装，现在需要更新和安装 Android 编译环境需要的系统工具包，其 Linux 命令为：

```
$ sudo apt-get install git-core gnupg flex bison gperf build-essential zip curl \
zlib1g-dev gcc-multilib g++-multilib libc6-dev-i386 lib32ncurses5-dev \
x11proto-core-dev libx11-dev lib32z-dev ccache libgl1-mesa-dev libxml2-utils xsltproc unzip
```

因网络异常或其他异常，这些工具包可能不能完全下载和更新，那就需要在执行完此命令以后，再次执行此命令来验证工具包是否完整安装和更新。

```
xsltproc is already the newest version (1.1.28-2.1ubuntu0.1).
zlib1g-dev is already the newest version (1:1.2.8.dfsg-2ubuntu4.1).
0 upgraded, 0 newly installed, 0 to remove and 247 not upgraded.
```

说明所有的包都没有遗漏，已经安装完成。如每个工具包提示均已完成更新和安装，便可进入下一步操作，否则继续执行此命令更新和安装剩余未完成的 Ubuntu 系统工具包。

2.2　Android 源代码下载及编译过程

前面已经完成了 Ubuntu Linux、OpenJDK 和编译 Android 源码所需的系统工具包的安装和配置，接下来开始下载和编译 Android 8.1.0 源代码，这个过程简单但花费的时间比较长。

2.2.1　工作目录设置

Android 8.0 的代号为 Oreo，简称 O，中文名称奥利奥。采用名称 Oreo 作为本书选用 Android 8.1.0 版本源代码的根目录，在用户根目录下新建代码根目录 Oreo 文件夹，并设置此目录为工作目录，在.bashrc 中增加 export $oreo=~/code/Oreo。

在 Android 开发过程中，配置工作目录的环境变量有利于提升工作效率，主要体现在一些 Android 工具命令、代码路径、编译结果路径等配合工作目录的环境变量使用会简化操作，读者在使用过程中可逐步体会；后续涉及的$oreo 即是 Android 8.1.0 源代码的根目录路径。

2.2.2　源代码下载

Android 8.1.0 源代码的本地目录已经建立，接下来就要开始下载代码了，相关操作及说明如下：

repo 脚本是 Android 项目编写的 Python 脚本,用来统一管理 Android 项目的代码仓库

```
$ sudo apt-get install python
$ curl https://storage.googleapis.com/git-repo-downloads/repo > repo
$ chmod a+x repo
```

在~/用户主目录下新建一个 bin 目录,并将此目录设置在 PATH 目录中;我们将在此目录下保存常用的一些脚本或二进制可执行程序,以后不必更新系统环境变量就能在任意目录执行这些脚本或可执行程序。

```
$ mkdir ~/bin
$ vi ~/.bashrc
//在文件最后一行增加 PATH=~/bin:$PATH,保存退出
$ source .bashrc//立即生效配置的 PATH 目录
$ mv repo ~/bin/
$ cd $oreo
//配置 git 个人信息
$ git config --global user.name android_tele
$ git config --global user.email android_tele @163.com
//查看配置的 git 信息
$ cat ~/.gitconfig
[user]
        name = android_tele
        email = android_tele @163.com
[color]
        ui = auto
//获取 Android 源码分支信息
$ repo init -u https://android.googlesource.com/platform/manifest
 ......
  * [new tag]           android-8.0.0_r32 -> android-8.0.0_r32
  * [new tag]           android-8.0.0_r33 -> android-8.0.0_r33
  * [new tag]           android-8.0.0_r34 -> android-8.0.0_r34
  * [new tag]           android-8.0.0_r35 -> android-8.0.0_r35
  * [new tag]           android-8.0.0_r36 -> android-8.0.0_r36
  * [new tag]           android-8.0.0_r4 -> android-8.0.0_r4
  * [new tag]           android-8.0.0_r7 -> android-8.0.0_r7
  * [new tag]           android-8.0.0_r9 -> android-8.0.0_r9
  * [new tag]           android-8.1.0_r1 -> android-8.1.0_r1
 ......
//读者可根据实际情况,选择最新的 Android 源码分支下载,本书选择 android-8.1.0_r1 分支下载
$ repo init -u https://android.googlesource.com/platform/manifest -b android-8.1.0_r1
 repo has been initialized in /home/android/Oreo
$ repo sync -j8 //开始下载 Android O 源码,工作进程数量,本例中使用 8 个,读者可以根据网络带宽进行调整

//这个过程花费的时间很长,视网络情况而定;建议读者在晚上下载,如果中途代码下载中断了,也不必担心,
//repo sync 支持续传
```

2.2.3 开始编译 Android 源代码

Android 8.1.0 源代码下载完成后,可以开始编译源码,详情见如下操作及相关说明。

```
$ cd $oreo
$ source build/envsetup.sh //或者. build/envsetup.sh //加载编译脚本
//使用第二种方法需要注意,build 前有一个空格
including device/asus/fugu/vendorsetup.sh
including device/generic/car/vendorsetup.sh
including device/generic/mini-emulator-arm64/vendorsetup.sh
......
```

```
including device/huawei/angler/vendorsetup.sh
including device/lge/bullhead/vendorsetup.sh
including sdk/bash_completion/adb.bash
$ lunch   //选择编译的产品信息

You're building on Linux

Lunch menu...... pick a combo:
     1. aosp_arm-eng
     2. aosp_arm64-eng
     3. aosp_mips-eng
     4. aosp_mips64-eng
     5. aosp_x86-eng
     6. aosp_x86_64-eng
     ......
     28. aosp_angler-userdebug
     29. aosp_bullhead-userdebug
     30. aosp_bullhead_svelte-userdebug
     31. hikey-userdebug
     32. hikey960-userdebug

Which would you like? [aosp_arm-eng] aosp_arm64-eng
============================================
PLATFORM_VERSION_CODENAME=REL
PLATFORM_VERSION=8.1.0  //Android O 版本
TARGET_PRODUCT=aosp_arm64  //lunch 选择 aosp_arm64-eng
TARGET_BUILD_VARIANT=eng
......
BUILD_ID=OPM1.171019.011  //编译号
OUT_DIR=out
AUX_OS_VARIANT_LIST=
============================================
$ make -j8
//编译时间长,编译过程中输出的日志很大,这里对编译日志进行了省略
//以下是编译成功日志信息,可以看出成功编译出 system.img 镜像文件
encoding RS(255, 253) to '/tmp/tmpGZPyrs_verity_images/verity_fec.img' for input files:
        1: 'out/target/product/angler/obj/PACKAGING/systemimage_intermediates/system.img'
        2: '/tmp/tmpGZPyrs_verity_images/verity.img'
appending /tmp/tmpGZPyrs_verity_images/verity_fec.img to
 /tmp/tmpGZPyrs_verity_images/verity.img
Running: append2simg
out/target/product/angler/obj/PACKAGING/systemimage_intermediates/system.img/tmp/
tmpGZPyrs_verity_images/verity.img

 [100% 89721/89721] Install system fs image: out/target/product/generic_arm64/system.img
 out/target/product/angler/system.img+out/target/product/generic_arm64/obj/PACKAGING/
recovery_patch_intermediates/recovery_from_boot.p maxsize=3288637440 blocksize=135168
total=1048456853 reserve=33251328

#### build completed successfully (02:26:35 (hh:mm:ss)) ####
```

第一次编译时间较长,不同的计算机花费的时间不同。作者使用较老的笔记本电脑(酷睿 2 代 i7+8GB)编译 Android O 源码,共使用了 6.5 小时。如果计算机处理能力较强,可使用更多的工作进程,比如:make –j16,增加编译工作进程数从而减少编译时间。编译完成后,进入 $oreo/out/target/product/generic_arm64 目录,关注此目录下的 system.img、ramdisk.img、userdata.img 三个 IMG 镜像文件,以及 data、obj、root、system 等目录。请读者自己查看,编译完成后究竟生成了一些什么文件,这里重点关注 system 目录,其主要目录结构如下。

- app/priv-app(应用 apk 文件,如 TeleService.apk、Mms.apk 等)

- bin（可执行文件，app_process32/64、toybox、netd 等）
- etc（系统配置信息）
- fonts（字体文件）
- framework（主要保存一些 jar 包，framework.jar、telephony-common.jar 等）
- lib/lib64（主要保存一些 so 动态链接库文件，libbrillo.solibsurfaceflinger.so 等）
- usr（用户配置信息）
- xbin（系统的一些可执行文件）

2.2.4 编译单个模块

整个 Android 编译环境搭建已经完成了 60%。在前面曾经谈到为什么要搭建这样的编译环境，那就是能够调试、运行修改的内容。如果在 Telephony 应用里修改了 Android 源码增加日志打印功能，是不是也要通过 make 来编译呢？这样的话，在编译方面就需要花很多时间。在 Android 中能够按照模块进行模块的单独编译，可减少不必要的编译时间开销。

在进行分模块编译之前，必须先完成整体编译后才能进行，否则不能成功编译需要单个编译的模块。

分模块编译主要有三种方式。第一种，在$oreo 代码根目录下执行 mmm module path 命令；第二种，进入对应的应用模块代码所在目录执行 mm 命令；第三种，在$oreo 代码根目录下执行 make module name 命令，详情见如下操作及相关说明。

```
$ cd $oreo
$ source build/envsetup.sh           //或者. build/envsetup.sh
//使用第二种方法需要注意 build 前有一个空格
$ mmm packages/service/Telephony/    //编译 TeleService 应用
$ mmm frameworks/base/               //编译 framework.jar

$ cd packages/service/Telephony      //TeleService 应用代码目录
$ mm                                 //编译 TeleService 模块
$ cd $oreo
$ cd frameworks/base                 //进入 framework 代码目录
$ mm                                 //编译 framework
$ cd $oreo
$ make TeleService                   //编译 TeleService 应用
$ make framework                     //编译 framework.jar 应用
```

不论采用什么方式编译单个模块，编译成功后，均有类似如下的日志。

```
[100% 10/10] Install:
out/target/product/generic_arm64/system/priv-app/TeleService/TeleService.apk
#### build completed successfully (01:16 (mm:ss)) ####
[100% 131/31] Install:
out/target/product/generic_arm64/system/framework/arm64/boot.art
#### build completed successfully (05:28 (mm:ss)) ####
```

建议使用 make module name 和 mmm 方式分模块编译，编译过程不涉及目录的切换，可以减少工作量。而这两种方式中优先选择 mmm 的编译方式，因为比起 make module name 方式，它更加省时。

2.3 Android Studio 及 SDK

2.3.1 下载和配置 Android Studio

Google 提供了 Windows 32/64、Mac 和 Linux 四个不同平台的 Android Studio 版本供我们下载和使用。选择 Linux 版本，下载 android-studio-ide-171.4443003-linux.zip 文件到本地，约 740MB 大小。

```
$ unzip android-studio-ide-171.4443003-linux.zip
$ tree -L 1 android-studio
android-studio
├── bin
├── build.txt
├── gradle
├── Install-Linux-tar.txt
├── jre
├── lib
├── license
├── LICENSE.txt
├── NOTICE.txt
└── plugins

6 directories, 4 files
$ cd android-studio/bin
$ ./studio.sh  //启动 Android Studio
```

（1）第一次启动 Android Studio 将启动安装向导，可在线更新 Android SDK、Gradle 等工具包。要关闭启动安装向导，可修改 bin 目录下的 idea.properties 配置文件，增加一行配置信息如下：

```
disable.android.first.run=true
```

（2）每次启动 Android Studio 都需要进入 android-studio/bin 目录，再运行 studio.sh 脚本，操作较多。比较省事的办法是增加一个 desktop 图标，在 Ubuntu Activity 菜单中可方便启动，具体的操作如下：

```
$ sudo vi /usr/share/applications/androidstudio.desktop  //需要 root 权限
//增加以下配置信息
[Desktop Entry]
Name=androidstudio
Comment=androidstudio
Type=Application
Terminal=false
Icon=/home/android/tools/android-studio/bin/studio.png
Exec=/home/android/tools/android-studio/bin/studio.sh
```

单击 Ubuntu Activity，输入 androidstudio 即可匹配出 Android Studio 应用，再单击 Android Studio 图标即可启动它了。

2.3.2 Android SDK 下载及配置和使用

Android SDK 的下载页面与 Android Studio 在同一个页面，谷歌也提供了 Windows、Mac 和 Linux

三个平台的版本供我们下载和使用。

同样选择 Linux 版本，下载 sdk-tools-linux-3859397.zip 到本地，约 130MB 大小。

```
$ unzip sdk-tools-linux-3859397.zip
$ tree -L 1 tools/
tools/
├── android
├── bin
├── emulator
├── emulator-check
├── lib
├── mksdcard
├── monitor
├── NOTICE.txt
├── proguard
├── source.properties
└── support

4 directories, 7 files
```

注意

（1）此版本的 Android SDK 是基本的 Android 命令行工具，可以使用此工具中包含的 sdkmanager 工具下载和更新其他的 SDK 工具包。

（2）使用 Android Studio 通过界面的方式可以更加方便地管理 Android SDK。

启动 Android Studio，在 Configure 菜单中选择 SDK Manager 或者通过 File 菜单打开的 Settings 界面，进入 Android SDK 管理界面，选择 Edit 编辑 Android SDK 需要保存的路径，如图 2-4 所示。

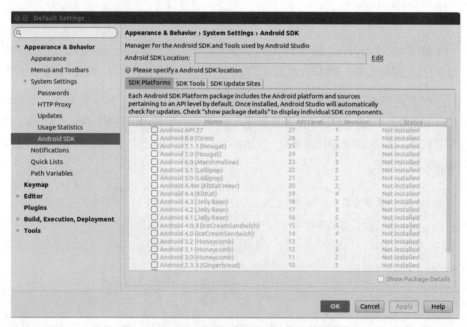

图 2-4　Android Studio 中 Android SDK 管理界面

后续都是界面向导式的操作，我们只需关注 Android SDK 的保存路径即可。本例中的路径为：~/Android/Sdk。接受 License，开始下载，完成后，我们再查看两个 SDK 的异同。

```
$ cd ~/Android
$ tree -L 1 Sdk/
```

```
Sdk/
├── build-tools
├── emulator
├── extras
├── patcher
├── platforms
├── platform-tools
├── sources
└── tools

8 directories, 0 files
```

根据图 2-4 所示，其实我们仅安装了 SDK 框架，并未包含具体的 SDK 内容。接着通过 Android Studio 中的 Android SDK Manager 增加 Android SDK 的内容；在 SDK Platforms 和 SDK Tools 两个选项卡中下载和安装自己想要的内容。本书选择了 Android 8.0(Oreo)和 SDK Tools 中的基础工具。

（1）SDK Manager 界面列出了 Android 已经发布的所有 SDK 版本，选择安装的 Android SDK 版本越多，下载和安装的时间就越长，需要的磁盘空间就越大。读者可以根据自己的兴趣和需要选择下载。

（2）Android SDK 安装完成后，还请记录 SDK 中对应工具的 PATH 路径，具体操作如下：

```
$ vi ~/.bashrc
//增加以下配置信息
export ANDROID_HOME=~/Android/Sdk
export PATH=$PATH:$ANDROID_HOME/tools:$ANDROID_HOME/platform-tools
$ source ~/.bashrc//重新加载配置信息
$ adb devices //验证 Android SDK 的 adb 工具是否可以使用
List of devices attached
```

2.3.3 使用 Android SDK 启动 Android 虚拟设备

启动 Android Studio，单击 Tools 菜单→Android 选项→AVD Manager，打开 AVD Manager,通过界面向导式的操作创建虚拟设备。

本例中选择硬件为 Phone，平台为 Nexus 6P，API 为 26 即 Android O，详情如图 2-5 所示。

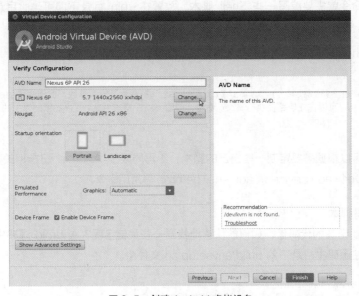

图 2-5 创建 Android 虚拟设备

2.3.4 Android 调试工具 adb 的使用方法

Android 模拟设备启动完成后，会默认开启 5554 端口连接到计算机，这时可运行 adb devices 命令查看连接到本地计算机上的 Android 设备列表，详情如下。

```
$ adb devices
List of devices attached
emulator-5554 device     //设备状态为正常情况，可查看模拟器运行日志
```

查看虚拟机运行日志：

- adb logcat

查看应用日志，可结合"|grep"过滤匹配自己想查看的日志内容。

- adb logcat –b radio

查看 RIL 日志。

adb 无法连接 Android 设备时，设备连接状态为 offline 或 no permissions，可尝试使用 root 用户执行 adb kill-server 命令重启设备连接。一般来说，adb kill-server 重启设备连接的有效期与 Ubuntu 系统运行结束的周期一致，开机或重启一次 Ubuntu 便需要重启一次设备连接。

2.3.5 相关技巧汇总

1．adb logcat 日志输出脚本

我们在开发、调试 Android 代码的过程中使用最多的就是 adb 命令，通过它我们可以获取到想要的日志信息，来帮助我们分析、定位程序逻辑。

查看 main 日志的命令：adb logcat –vthreadtime

查看 radio 日志的命令：adb logcat –vtime –b radio

查看 event 日志的命令：adb logcat –vtime –b events

要简化上述命令就需要编写一些 Shell 脚本，放置在~/bin 目录下即可，此目录之前已经加入了 PATH 环境变量，因此新加入的 Shell 脚本不用再配置环境变量，随时可以使用。

```
$ vi ~/bin/mlog
#!/bin/sh

adb logcat -vthreadtime $1 $2 $3  //加入$1 $2 $3 是为了传入参数，方便命令的扩展
$ mlog -s *:W//可以在任何路径执行 mlog 命令，查看 main 日志中 warning 级别以上的日志
- waiting for device -
```

读者也可以根据需要定制一些自己的脚本，不再局限于 Android 日志相关命令。比如，adb kill-server、adb devices、adb shell getprop 等常用命令。

2．查找代码脚本

面对 Android 浩瀚的代码量，如何能快速地找到一些关键字呢？Android 源码中已经准备了对应的脚本工具，就在源码主目录下的 build/envsetup.sh 文件中。

```
$ gedit build/envsetup.sh
function jgrep()  //查找并匹配 java 文件中对应的关键字
```

```
{
    find . -name .repo -prune -o -name .git -prune -o -name out -prune -o -type f
-name "*\.java" \ -exec grep --color -n "$@" {} +
}
function cgrep() //查找并匹配c相关文件中对应的关键字
{
    find . -name .repo -prune -o -name .git -prune -o -name out -prune -o -type f \
 (-name '*.c' -o -name '*.cc' -o -name '*.cpp' -o -name '*.h' -o -name '*.hpp' \) \
        -exec grep --color -n "$@" {} +
}
......
```

当然还有 ggrep、resgrep、mangrep、sepgrep 等脚本信息可以供我们使用，但要使用这些脚本首先需要执行 source build/envsetup.sh。一劳永逸的方法就是将这些需要的脚本写入 ~/.bashrc 用户配置文件中，这样无论当前用户在什么路径都可以执行这些命令了。

2.4 在 Google 手机上调试 Android 源码

2.4.1 Google 手机对应编译选项

Google 手机至今已经发布了很多款，目前最新发布的是 Google Pixel 2。本书选择 Nexus 6P 作为实例，读者也可以选择其他 Google 手机调试 Android 源码。

Google 手机对应的代码名称、编译选项详情见表 2-2。

表 2-2 Google 手机编译选项表

终端	代码名称	编译选项
Pixel XL	marlin	aosp_marlin-userdebug
Pixel	sailfish	aosp_sailfish-userdebug
HiKey（开发板）	hikey	hikey-userdebug
Nexus 6P	angler	aosp_angler-userdebug
Nexus 5X	bullhead	aosp_bullhead-userdebug
Nexus 6	shamu	aosp_shamu-userdebug

2.4.2 Google 手机刷入工厂镜像

首先下载 Google Nexus 和 Pixel 系列手机的工厂镜像文件。Nexus 6P 手机对应的工厂镜像是 angler 8.1.0 (OPM1.171019.011, Dec 2017)，即 Nexus 6P 手机 Android 8.1.0 工厂刷机镜像。

```
$ unzip angler-opm1.171019.011-factory-39448337.zip
$ tree angler-opm1.171019.011
angler-opm1.171019.011
├── bootloader-angler-angler-03.78.img
├── flash-all.bat
├── flash-all.sh
├── flash-base.sh
├── image-angler-opm1.171019.011.zip
└── radio-angler-angler-03.85.img
```

```
0 directories, 6 files
```

image-angler-opm1.171019.011.zip 压缩包中包含了 android-info.txt、boot.img、recovery.img、system.img、vendor.img 等镜像文件。

先将 Nexus 6P 手机关机,然后同时按下电源键和音量键两个按键,持续几秒不要松开,就可以进入 fastboot 刷机模式。

```
$ sudo -s //一定要使用 root 账号刷机,否则没有权限,fastboot 刷机将失败
# fastboot flashing unlock//或者 fastboot flashing unlock_critical 对 Pixel 2 XL 有效
# ./flash-all.sh
......
sending sparse 'system' 5/5 (51472 KB)...
OKAY [  1.704s]
writing 'system' 5/5...
OKAY [  0.743s]
sending 'vendor' (192545 KB)...
OKAY [  6.143s]
finished. total time: 104.914s
```

等待 Nexus 6P 手机重启完成,我们便可以体验和使用原汁原味的 Android 8.1.0 系统了,详情如图 2-6 所示。

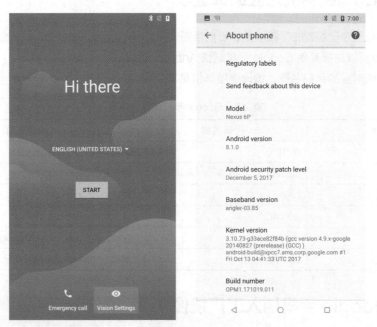

图 2-6 Nexus 6P 开机向导和 Android 8.1.0 版本信息

2.4.3 编译本地镜像并刷入 Google 手机

Google 手机刷入工厂镜像文件是如此得简单方便,但我们需要调试 Android 源码,仅刷入工厂镜像是无法办到的,因为它是用户版本的,无法调试系统级源码;因此还需要刷入本地编译出来的 userdebug 版本镜像文件,主要的步骤如下:

1. 下载 Google 手机对应的驱动文件(Driver Binaries)

到官网下载 Google 手机对应的驱动文件,选择 Nexus 6P ("angler") binaries for Android 8.1.0

第 2 章　搭建 Android 源代码编译调试环境

(OPM1.171019.011)的两个 Driver 文件：Vendor image 和 Qualcomm，对应的文件名分别是 huawei-angler-opm1.171019.011-41db8ed5.tgz 和 qcom-angler-opm1.171019.011-f7e511bb.tgz，解压后是两个 Shell 脚本：extract-huawei-angler.sh 和 extract-qcom-angler.sh，将这两个文件复制到 Android O 源码的主目录下。

2．将驱动文件导入到 Android 8.1.0 源码工程中

```
$ cd $oreo
$ ./extract-huawei-angler.sh  //运行自解压脚本，并接受 License
$ ./extract-qcom-angler.sh    //运行自解压脚本，并接受 License
//将在当前目录下生成 vendor 目录，其中包括了华为和高通的二进制文件和对应的编译脚本
$ tree vendor -L 3
vendor
├── huawei
│   └── angler
│       ├── android-info.txt
│       ├── BoardConfigPartial.mk
│       ├── BoardConfigVendor.mk
│       ├── device-partial.mk
│       ├── device-vendor.mk
│       └── proprietary
└── qcom
    └── angler
        ├── BoardConfigPartial.mk
        ├── device-partial.mk
        └── proprietary
```

3．使用 angler 编译选项重新编译

前面编译 Android 源码时，lunch 选项选择的是 aosp_arm64-eng。而现在导入 Nexus 6P 的驱动文件后，编译 Nexus 6P 手机对应的镜像文件时，lunch 需要选择 aosp_angler-userdebug，并以全新的方式编译整个代码，最简单的方式就是删除保存编译结果的 out 目录。

```
$ rm -rf out
$ source build/envsetup.sh  //或者. build/envsetup.sh,
//使用第二种方法需要注意 build 前有一个空格
including device/asus/fugu/vendorsetup.sh
including device/generic/car/vendorsetup.sh
including device/generic/mini-emulator-arm64/vendorsetup.sh
......
including device/huawei/angler/vendorsetup.sh
including device/lge/bullhead/vendorsetup.sh
including sdk/bash_completion/adb.bash
$ lunch

You're building on Linux

Lunch menu... pick a combo:
      ......
      28. aosp_angler-userdebug
      29. aosp_bullhead-userdebug
      30. aosp_bullhead_svelte-userdebug
      31. hikey-userdebug
      32. hikey960-userdebug

Which would you like? [aosp_arm-eng] aosp_angler-userdebug//选择 Nexus 6P 对应的编译选项
============================================
PLATFORM_VERSION_CODENAME=REL
PLATFORM_VERSION=8.1.0  //Android O 版本
TARGET_PRODUCT=aosp_angler  //lunch 选择 aosp_angler-userdebug
```

```
TARGET_BUILD_VARIANT=eng
......
BUILD_ID=OPM1.171019.011  //编译号
OUT_DIR=out
AUX_OS_VARIANT_LIST=
============================================
$ make -j8
```

4．fastboot 刷入本地编译出的镜像文件

首先进入 fastboot 刷机模式，然后使用 Android SDK 中的 fastboot 工具刷入镜像文件，详情操作如下：

```
$ sudo -s //一定要使用 root 账号刷机，否则没有权限，fastboot 刷机将失败
# fastboot flash boot boot.img
# fastboot flash system system.img
# fastboot flash vendor vendor.img
# fastboot flash userdata userdata.img
# fastboot reboot
```

等待手机完成重启后，验证本地编译的 Android O 系统：

```
Android Version:8.1.0
Builder Number:aosp_angler-userdebug 8.1.0 OPM1.171019.011 eng.androi20171210.
102134test-keys
```

- aosp_angler_userdebug——即 Nexus 6P 手机的 userdebug 版本。
- eng.androi engineer——工程模式，编译环境的用户名 android（因长度限制少了字母 d）。
- 20171210.102134——2017 年 12 月 10 日 10 点 21 分 34 秒开始编译。
- test-keys——系统镜像的签名使用 test-keys 密钥。

详情如图 2-7 所示。

图 2-7　Nexus 6P 刷入本地编译的 Android 8.1.0 userdebug 系统

2.4.4　Google 手机上调试 Android 源码

1．Android Studio 导入 Android O 源码

首先，编译出 idegen.sh 脚本依赖的 jar 包：idegen.jar，操作详情如下。

```
$ mmm development/tools/idegen/ //编译 idegen.jar
[100% 3/3] Install: out/host/linux-x86/framework/idegen.jar

$ development/tools/idegen/idegen.sh //当前代码主目录下将生成 android.iml 和 android.ipr 两
                                     //个 Android Studio 的工程配置文件
Read excludes: 15ms
Traversed tree: 70465ms
```

接着，打开 Android Studio，进入 Open an existing Android Studio project，选择 android.ipr 文件，开始导入 Android O 源码。

不同的计算机处理能力不同，导入的时间也不同，需要耐心等待一段时间，由 Android Studio 准备 Android 等相关插件工具，以及建立工程的代码文件索引，以提升后续的操作性能。

2．修改代码模块编译

本例选择 com.android.phone 进程加载的代码入口文件 PhoneApp.java 作为修改实例，其相对路径为：packages/services/Telephony/src/com/android/phone/PhoneApp.java。

在 Android Studio 连续快速地两次按下右 Shift 键，输入 PhoneApp.java 将快速匹配出该文件。在代码的 onCreate 方法中增加一行打印日志的代码，来验证代码修改后是否能成功运行在 Google 手机上，代码修改和编译详情如下。

```
@Override
public void onCreate() {
    android.util.Log.d("Android", "My Code run on the Nexus 6P");
    ......
}
$ cd $oreo
$ mmm packages/services/Telephony/
[100% 10/10] Install: out/target/product/angler/system/priv-app/TeleService/TeleService.apk
```

3．挂载手机

前面成功编译了 TeleService.apk 文件，需要将此文件 push 到 Nexus 6P 手机上运行，在此之前还需要挂载手机，只有挂载成功以后才能 push apk 系统应用、系统 jar 包、so 动态链接库等文件到手机/system 挂载点。具体操作如下：

```
$ adb root
restarting adbd as root
$ adb remount
dm_verity is enabled on the system partition.
Use "adb disable-verity" to disable verity.
If you do not, remount may succeed, however, you will still not be able to write to
these volumes.
remount succeeded
$ adb disable-verity
Verity disabled on /system
Now reboot your device for settings to take effect//需要重启手机
$ adb reboot
```

```
$ adb root
restarting adbd as root
$ adb remount
$ remount succeeded
```

挂载成功的手机系统重启以后，挂载的状态会失效，需要再做一次挂载操作。

4. push 模块并重启应用

手机挂载成功以后，接着就要开始 push 应用到手机上；TeleService 模块编译成功后的日志，只提示我们安装 out 目录下的 TeleService.apk 文件，其实还要安装 dex 相关的文件，修改才能在手机上生效。

进入 out/target/product/angler/system/priv-app/TeleService/目录查证编译后文件更新的情况，除 TeleService.apk 文件更新了，oat 目录也同时更新了，该目录下的 TeleService.odex 和 TeleService.vdex 这两个文件同样需要安装到 Nexus 6P 手机上对应的目录，否则我们的修改不会生效。具体操作如下：

```
$ tree out/target/product/angler/system/priv-app/TeleService/oat
out/target/product/angler/system/priv-app/TeleService/oat
└── arm64
    ├── TeleService.odex
    └── TeleService.vdex
$ adb push out/target/product/angler/system/priv-app/TeleService/TeleService.apk
 /system/priv-app/TeleService/
out/target/product/angler/system/priv-app/TeleService/TeleService.apk: 1 file pushed.
21.0 MB/s (7691558 bytes in 0.350s)
$ adb push out/target/product/angler/system/priv-app/TeleService/oat
 /system/priv-app/TeleService/  //重点关注 push 的目录没有 oat
out/target/product/angler/system/priv-app/TeleService/oat/: 2 files pushed. 10.9 MB/
s (2032058 bytes in 0.178s)
$ adb reboot //重启手机或是"杀死" com.android.phone 进程重启应用
```

请读者使用 adb shell 命令进入手机中对应的目录，通过修改时间和文件大小来查看 push 的文件是否已经成功更新，同时关注 push 的 oat 目录是否在/system/priv-app/TeleService/oat 目录下再次生成了 oat 目录。

5. 日志验证代码修改内容

```
$ mlog |grep -i "nexus 6p"//mlog -s Android
02-26 00:01:28.077  4671  4671 D Android : My Code run on the Nexus 6P
```

"02-26 00:01:28.077"：以手机上时间为准的时间戳。

"Android"：打印日志的 TAG。

"My Code run on the Nexus 6P"：日志打印内容。

至此，我们已经成功搭建了 Android 源代码调试环境，从 Android 官网下载 Android 8.1.0 源码、Android Studio、Android SDK 以及 Nexus 6P 手机对应的工厂镜像和驱动文件，然后成功编译本地镜像文件，刷入 Android 8.1.0 userdebug 版本的 angler 手机镜像文件，最后修改 TeleService 模块代码，编译后 push 到 Nexus 6P 手机上验证我们的代码修改是否生效。

2.4.5 关键问题总结

搭建 Android 源代码调试环境的环节众多，时间开销大，任何小细节出现问题都将无法满足要求，

>>>>>>>>>> 第 2 章　搭建 Android 源代码编译调试环境

因此，作者总结出编译 Android 源码过程中自己碰到的两个非常关键的问题：Jack 内存溢出和模块编译失败。

1. Jack 编译器内存溢出

```
[  0% 38/54562] Building with Jack:
 out/target/common/obj/JAVA_LIBRARIES/framework_intermediates/with-local/classes.dex
FAILED: out/target/common/obj/JAVA_LIBRARIES/framework_intermediates/with-local/classes.dex
/bin/bash
out/target/common/obj/JAVA_LIBRARIES/framework_intermediates/with-local/classes.dex.rsp
Out of memory error (version 1.3-rc6 'Douarn'
(441800 22a11d4b264ae70e366aed3025ef47362d1522bb by android-jack-team@google.com)).
```

Jack 编译器内存不足，需要修改配置文件以增加编译器内存大小。修改源码目录下的 prebuilts/sdk/tools/jack-admin 文件，其中有两个参数：JACK_SERVER_VM_ARGUMENT 和 JACK_SERVER_COMMAND，本例中添加-Xmx4096M（配置的内存大小可根据自己计算机实际内存大小进行调整），最后执行 make clean、make -j8 命令重新编译。

```
JACK_SERVER_VM_ARGUMENTS="${JACK_SERVER_VM_ARGUMENTS:-Dfile.encoding=UTF-8 -Xmx4096M}"
JACK_SERVER_COMMAND="java -XX:MaxJavaStackTraceDepth=-1 -Xmx4096M -Djava.io.tmpdir=
$TMPDIR $JACK_SERVER_VM_ARGUMENTS -cp $LAUNCHER_JAR $LAUNCHER_NAME"

$ ps -ef|grep jack
android   25499  2315 99 17:11 pts/0    00:01:59 java -XX:MaxJavaStackTraceDepth=
-1 -Djava.io.tmpdir=/tmp -Dfile.encoding=UTF-8 -XX:+TieredCompilation -Xmx4096M -cp
/home/android/.jack-server/launcher.jar com.android.jack.launcher.ServerLauncher
//修改了配置仍然没有生效，可以通过ps命令查看jack进程信息，手工"杀死"对应的进程，再重新执行编译命令
```

2. 模块编译失败

Android 8.1.0 源码在使用 mmm 方式编译 TeleService 单个模块时，packages/services/Telephony/目录下包含了与测试相关的应用，但因为 android-support-test Libraries 依赖关系的缺失，将导致编译失败，错误信息如下：

```
ninja: error:
'out/target/common/obj/JAVA_LIBRARIES/android-support-test_intermediates/classes.dex.
toc', needed by 'out/target/common/obj/APPS/TeleServiceTests_intermediates/with-local/
classes.dex', missing and no known rule to make it
```

因此，我们只需要做简单的修改，比如在 packages/services/Telephony/目录下将 tests/和 testapps/两个目录移除或是将这两个目录下的 Android.mk 文件改名为 Android.mk.bak，总之与测试相关的应用或 jar 包不进行编译。

或者不做任何改动，直接使用 make TeleService 的方式编译单个模块。

本 章 小 结

本章主要介绍了如何搭建 Android 源码下载、编译和调试环境，以及相关工具的使用技巧，包括 Ubuntu Linux 系统、源代码编译、Android SDK、Android Studio 等的使用方法，最后是对 Google Nexus 6P 手机的刷机和调试环境的构建。

这些内容是我们深入学习 Android 源码的基础，希望读者能够根据本章内容搭建起自己的编译调试环境。如果没有 Linux 的基础，还要多多学习 Linux 常用命令，可在 Ubuntu Linux 系统中实际动手

操作一下。

　　还记得本章最重要的内容是哪些吗？那就是根据下载到本地的Android 8.1.0源代码编译出system.img、boot.img、userdata.img、vendor.img等IMG镜像文件，并将这些镜像文件通过fastboot刷入Nexus 6P手机，从而能够在手机上运行我们本地编译出来的userdebug系统，这样就可以修改、调试Android 8.1.0源码并在手机上运行和验证，方便我们更好地深入研究和学习Android最新源码。

第 3 章

深入解析通话流程

学习目标

- 跟踪和学习 Android 通话关键流程。
- 掌握本地主动发起通话请求核心流程。
- 掌握本地被动接受通话请求核心流程。
- 创建 Android Telephony 通话模型。

通过前面两章的学习和 Android 编译调试环境的准备，相信读者已经对 Android 有了初步、直观的认识和理解。从本章开始，将深入解析 Android Telephony 的核心机制、关键流程以及相关的设计方法和思想。

Android Telephony 主要包括四方面的手机通信能力。

- ServiceState/SIM（网络服务，包含 SIM）
- Call（通话能力）
- DataCall（移动数据上网能力）
- SMS/MMS（短信/彩信能力）

通话能力是 Telephony 中最核心的业务，我们先从它入手分析。

Android 手机打电话、接电话功能是如何实现的？其业务流程是怎么流转的？

本章选择常见的拨号和接听来电——用户使用最多也是最频繁的两个业务场景，通过阅读和分析源代码，再结合 Nexus 6P 手机的实际运行流程，解析 Android Telephony 通话的核心业务流程；最后总结出 Android 通话模型，站在更高的层次去认识和理解 Android 通话。

▎3.1 拨号流程分析

在没有任何参考资料的情况下，我们面对超过 50GB 的 Android 源码，该如何入手，寻找拨号的代码入口呢？

在第 2 章中，我们已经将下载到本地的 Android 8.1.0 源码，编译出 Nexus 6P 的镜像文件，并刷入手机成功运行系统，插入 SIM 卡后能接打电话、收发短信和使用移动数据上网。"实践是检验真理的唯一标准。"作为实践的开始，使用 Nexus 6P 手机打开拨号盘，同时查看手机的运行日志，这样便能找到拨号键盘的入口程序。

3.1.1 打开 Nexus 6P 手机的拨号盘

通过数据线将 Nexus 6P 手机连接到计算机，打开 Ubuntu 命令行。首先使用 adb devices 命令查看和确认手机与计算机连接是否成功，然后使用 adb logcat 命令查看 Nexus 6P 手机的运行日志，最后操作手机，点击 Home 界面最下面一排最左边带有电话图标的应用按钮，打开拨号界面。计算机上的操作详情如下。

```
//使用 root 用户，如果没有权限可执行 adb kill-server 命令以 root 用户身份重启 adb 计算机上的链接服务
# adb devices
List of devices attached
* daemon not running. starting it now at tcp:5037 *
* daemon started successfully *
84B7N15A2000XXXX     device   //84B7N15A2000 开头的内容是手机的序列号
//查看 Android 系统的 events 日志,可方便我们找出对应当前操作,Android 系统启动了什么 Activity
$ adb logcat -vtime -b events
I/am_new_intent(797):[0,69578539,51,com.android.dialer/.app.DialtactsActivity,android.intent.action.MAIN,NULL,NULL,270532608]
I/am_proc_start(797):[0,1726,10013,com.android.dialer,activity,com.android.dialer/.app.DialtactsActivity]
I/am_proc_bound(797): [0,1726,com.android.dialer]
I/am_restart_activity(797): [0,69578539,51,com.android.dialer/.app.DialtactsActivity]
I/am_on_resume_called(1726): [0,com.android.dialer.app.DialtactsActivity,LAUNCH_ACTIVITY]
I/am_activity_launch_time(797): [0,69578539,com.android.dialer/.app.DialtactsActivity,488,488]
```

上面是 Android 系统的 events 日志，其中关键的信息已经加粗。点击电话按钮进入拨号界面，ActivityManagerService 将启动 com.android.dialer 包下的 DialtactsActivity。

797 进程是 system_server，即 ActivityManagerService 所在的系统进程；DialtactsActivity 运行的进程编号是 1726，通过 ps 查看进程信息命令可以确认相关的进程信息，详情如下：

```
$ adb shell ps -ef
USER            PID    PPID    VSZ        NAME
system          797    583     4741680    system_server
u0_a13          1726   583     4357400    com.android.dialer
```

PID 为 797 和 1726 的两个进程编号是由系统生成的，不同手机、不同时段，其进程编号也不相同，因此，读者重点关注进程名即可。

3.1.2 进入拨号界面 DialtactsActivity

要在 Android Studio 中快速打开 DialtactsActivity.java 文件，可连续两次按下右边的 Shift 按键，打开 Search Everywhere 对话框，输入 DialtactsActivity，在输入过程中有逐个英文字母匹配的过程，输入完成后便可完整匹配 DialtactsActivity.java 文件，按回车键或鼠标单击操作即可。

DialtactsActivity.java 文件的详细路径是 packages/apps/Dialer/java/com/android/dialer/app/DialtactsActivity.java。packages/apps/Dialer 是我们涉及的第一个代码库，可查看对应的 Android.mk

文件，此代码库将编译出 Dialer.apk Android 应用程序，以后统一称其为 Dialer 应用，其运行的进程名是 com.android.dialer。

接着使用 Android Studio 快捷键 Ctrl+F12 快速打开当前类属性和方法列表浮动菜单，输入 oncreate，快速匹配到 onCreate 方法，按回车键即可快速定位到 DialtactsActivity 类的 onCreate 方法。

在 Android Studio 快速匹配过程中，可使用*（匹配多个字符）进行模糊匹配，并且输入的字符不区分大小写。

查看 DialtactsActivity 类的 onCreate 方法，找到拨号界面对应的 layout 界面布局文件 dialtacts_activity.xml，其中就包含了打开拨号盘的浮动按钮，其定义如下。

```xml
<android.support.design.widget.FloatingActionButton
    android:id="@+id/floating_action_button"
    android:layout_width="@dimen/floating_action_button_width"
    android:layout_height="@dimen/floating_action_button_height"
    android:layout_gravity="center_horizontal|bottom"
    android:layout_marginBottom="@dimen/floating_action_button_margin_bottom"
    android:contentDescription="@string/action_menu_dialpad_button"
    android:src="@drawable/quantum_ic_dialpad_white_24"
    android:scaleType="center"
    app:elevation="@dimen/floating_action_button_translation_z"
    app:backgroundTint="@color/dialer_secondary_color"/>
```

代码中关键的信息已用粗体字表示，其中 quantum_ic_dialpad_white_24 是对应拨号盘浮动按钮中带有键盘样式的图片，读者可以用图片管理器打开查看；而 dialer_secondary_color 作为打开拨号盘浮动按钮的背景样式，定义为红色，和我们在手机上看到的红色背景的拨号盘按钮是一致的。

拨号盘浮动按钮作为用户拨号盘的打开入口，查找此按钮的点击响应事件即可找到拨号盘对应的处理逻辑。首先确定 id 为 floating_action_button 的拨号盘浮动按钮在 DialtactsActivity.java 代码中的处理信息，找到对应的关键信息，详情如下：

```java
//获取 layout 中的 View 对象
FloatingActionButton floatingActionButton =
    (FloatingActionButton) findViewById(R.id.floating_action_button);
//将浮动按钮的 click 事件交给 this 处理，即 DialtactsActivity
floatingActionButton.setOnClickListener(this);
mFloatingActionButtonController =
    new FloatingActionButtonController(this, floatingActionButton);
```

接着，继续查看 DialtactsActivity 类的 onClick 方法，发现拨号盘浮动按钮的响应事件为调用当前类的 showDialpadFragment 方法，此方法中的关键信息如下：

```java
private void showDialpadFragment(boolean animate) {
    ......
    final FragmentTransaction ft = getFragmentManager().beginTransaction();
    if (mDialpadFragment == null) {
        //通过 FragmentTransaction 增加 DialpadFragment
        mDialpadFragment = new DialpadFragment();
        ft.add(R.id.dialtacts_container, mDialpadFragment, TAG_DIALPAD_FRAGMENT);
    } else {
        ft.show(mDialpadFragment);
    }
    ......//动画操作
    setTitle(R.string.launcherDialpadActivityLabel);
}
```

原来我们点击拨号盘浮动按钮，弹出拨号盘对应的代码是 DialpadFragment.java。

3.1.3　DialpadFragment 拨号盘

在拨号盘发现界面元素的一些规律，可以将其分为三个区域。

- 已输入电话号码显示和控制区域

包括拨号控制菜单、已输入电话号码输入框和删除按钮。

- 12 个键盘区域

包括 0~9 共 10 个数字按键、*按键和#按键。

- 拨号按钮

打电话的过程就是在此界面输入电话号码后，单击拨号按键，因此，需要找到拨号按键的 onClick 事件中的响应逻辑。查找 DialpadFragment.java 源文件的 onClick 方法，发现响应拨号按钮点击事件时将调用当前类的 handleDialButtonPressed 方法，此方法非常简单，用来判断输入的呼叫号码是否合法，非法将弹出异常信息提示用户，合法则执行正常的拨号流程。

（1）创建拨号请求 intent 对象

CallIntentBuilder.build 方法构造 Action 为 Intent.ACTION_CALL 的 intent 对象，并添加 Extra 附加信息，如 isVideoCall、phoneAccountHandle、createTime 等。

（2）发送 intent 拨号请求消息

继续跟踪 DialerUtils.startActivityWithErrorToast 方法，根据 intent 的 action 将有两个分支处理逻辑，分别执行 placeCallOrMakeToast(context, intent)和 context.startActivity(intent)。

placeCallOrMakeToast 方法将继续执行拨号请求，最终将调用 TelecomManager 类的 placeCall 方法。TelecomManager 类在 frameworks/base 代码库下，其完整路径为：frameworks/base/telecomm/java/android/telecom/TelecomManager.java。

注意

① TelecomManager.java 程序在 framework/base 路径下，将编译出 framework.jar 包，是程序的静态关系；本例中它将运行在 PID 为 1726 的进程中，即 com.android.dialer 进程。

② 如果其他运行进程空间的应用获取了 TelecomManager 对象并调用了其方法，同理，TelecomManager 将运行在对应的应用进程中。

③ 假如 TelecomManager 是以单例的方式获取其对象，那么在 Android 系统中是不是只有一个该对象存在呢？请读者思考并给出自己的答案。

作者的答案是：一个进程只有一个 TelecomManager 对象。对于整个 Android 系统来说，支持多个进程，也就是可以创建多个 TelecomManager 对象。

```
public void placeCall(Uri address, Bundle extras) {
    ITelecomService service = getTelecomService();
    if (service != null) {
        try {
            //拨号流程中的第一次跨进程的服务调用
            service.placeCall(address, extras == null ? new Bundle() :
            extras,
                    mContext.getOpPackageName());
        } catch (RemoteException e) {...... }
    }
}
//获取服务名为 Context.TELECOM_SERVICE 的系统服务
private ITelecomService getTelecomService() {
    return ITelecomService.Stub.asInterface(
            ServiceManager.getService(Context.TELECOM_SERVICE));
}
```

ITelecomService 接口的定义文件是：frameworks/base/telecomm/java/com/android/internal/telecom/ITelecomService.aidl，同样在 framework/base 下，它定义了 placeCall、addNewIncomingCall、endCall 和 getCallState 等接口，而 ITelecomService 接口实现的服务程序在什么地方呢？我们可以通过全文搜索关键字"extends ITelecomService"的方式获取其服务的实现，详情如下：

```
$ jgrep "ITelecomService.Stub"
packages/services/Telecomm/src/com/android/server/telecom/TelecomServiceImpl.java:83:
    private final ITelecomService.Stub mBinderImpl = new ITelecomService.Stub() {
```

ITelecomService 的实现逻辑在 packages/services/Telecomm 工程下的 TelecomServiceImpl.java 文件中。TelecomManager 获取 ITelecomService 服务并调用其 placeCall 方法继续传递 intent 发出通话呼叫请求，将涉及第一次跨进程的服务调用。

Dialer 应用中（com.android.dialer 进程内）的拨号流程分析到这里，将完成第一次的跨进程访问，对以上流程进行回顾和总结，详情如下：

- DialpadFragment 提供用户拨号的交互界面
- CallIntentBuilder 创建拨号请求的 intent 对象
- TelecomManager 继续传递拨号请求 intent 对象

Dialer 应用中的拨号流程总结，如图 3-1 所示。

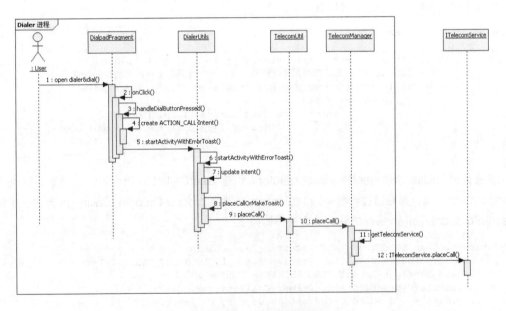

图 3-1　Dialer 应用拨号流程

3.1.4　ITelecomService 接收拨号请求服务

ITelecomService 的接口服务实现逻辑是 TelecomServiceImpl.java，其代码文件的详细路径是：packages/services/Telecomm/src/com/android/server/telecom/TelecomServiceImpl.java，packages/services/Telecomm 是我们跟踪拨号流程涉及的第二个代码库，查看对应的 Android.mk 文件，此代码库将编译出 Telecom.apk Android 应用程序，以后统一称其为 Telecom 应用。

接着查看此服务的定义，打开 packages/services/Telecomm/AndroidManifest.xml 文件，详情

如下：

```
<service android:name=".components.TelecomService"
        android:singleUser="true"
        android:process="system">//指定运行进程为system_server系统进程
    <intent-filter>
        <action android:name="android.telecom.ITelecomService" />
    </intent-filter>
</service>
```

重点关注此服务的运行进程指定为 system，即此服务将运行在 system_server 系统进程空间，而它的唯一 action 是 android.telecom.ITelecomService。

Context.TELECOM_SERVICE 系统服务名"telecom"与服务定义的 Action：android.telecom.ITelecomService 目前还没有对应起来，后面在 Telecom 应用的解析中将重点分析。到此，Dialer 应用的 com.android.dialer 进程提供用户拨号界面并响应用户的拨号请求，把拨号请求包装成 action 为 Intent.ACTION_CALL 的 intent 对象。通过调用 ITelecomService 服务提供的 placeCall 接口，将拨号请求 intent 对象发送给了 Telecom 应用（system_server 进程），完成了第一次跨进程的服务调用，传递的是包括拨号请求相关信息的 intent 对象。

继续关注 TelecomServiceImpl.java 文件中的 placeCall 方法中的逻辑，将响应 Dialer 应用发起的跨进程服务接口调用，其中的关键代码如下：

```
synchronized (mLock) {
    try {
        final Intent intent = new Intent(Intent.ACTION_CALL, handle);
        //创建新的intent对象，并根据接收到的intent信息更新intent对象的属性
        mUserCallIntentProcessorFactory.create(mContext, userHandle)
                .processIntent(
                        intent, callingPackage, isSelfManaged ||
                                (hasCallAppOp && hasCallPermission));
    }
}
```

首先通过 mUserCallIntentProcessorFactory 创建 UserCallIntentProcessor 对象，并执行其 processIntent 方法，然后通过判断 intent 的 action 来调用 processOutgoingCallIntent 方法，继续调用 sendBroadcastToReceiver 方法，此方法详情如下：

```
private boolean sendBroadcastToReceiver(Intent intent) {
    intent.putExtra(CallIntentProcessor.KEY_IS_INCOMING_CALL, false);
    intent.setFlags(Intent.FLAG_RECEIVER_FOREGROUND);
    intent.setClass(mContext, PrimaryCallReceiver.class);
    Log.d(this, "Sending broadcast as user to CallReceiver");
    mContext.sendBroadcastAsUser(intent, UserHandle.SYSTEM);
    return true;
}
```

将发出一个定向广播，由 Telecom 应用中的 PrimaryCallReceiver 对象接收。

同一个应用中为什么要使用广播来传递消息呢？因为 ITelecomService 的服务方法 placeCall，即 TelecomServiceImpl.java 中的 placeCall 方法经过层层调用：sendBroadcastToReceiver 中的逻辑通过广播的方式将同步方法调用转换成异步处理，也就是 ITelecomService 的服务方法 placeCall 快速返回给了 Dialer 应用的调用，而 Telecom 应用中接收到广播后继续处理对应的拨号请求 intent 对象。

跟进 PrimaryCallReceiver 对象的 onReceive 中的处理逻辑，将调用 CallIntentProcessor 类中的 processOutgoingCallIntent 方法，其主要逻辑如下：

```
// 确保广播返回之前发送消息给CallsManager,完成IncallUI通话界面的加载和显示
Call call = callsManager
        .startOutgoingCall(handle, phoneAccountHandle, clientExtras, initiatingUser,
            intent);
if (call != null) {
    sendNewOutgoingCallIntent(context, call, callsManager, intent);
}
```

这里有两个非常关键的处理逻辑 CallsManager.startOutgoingCall 和 sendNewOutgoingCallIntent 调用。跟进 sendNewOutgoingCallIntent 方法，其调用过程是：sendNewOutgoingCallIntent→NewOutgoingCallIntentBroadcaster.processIntent→mCallsManager.placeOutgoingCall。这两个关键的处理逻辑最终是调用了 CallsManager 对象的两个不同方法。

- startOutgoingCall
- placeOutgoingCall

根据代码中的注释，推测一下这两个名字对应的方法究竟会承载什么样的业务呢，startOutgoingCall 将开始拨号前的准备工作，而 placeOutgoingCall 将继续传递拨号请求，实现将拨号请求发送给 BP Modem 处理。

3.1.5　CallsManager 拨号流程处理

1. CallsManager.startOutgoingCall

CallsManager.startOutgoingCall 的主要逻辑是创建、更新和保存 Call 对象，Call 对象的名字非常特殊，其代码是：packages/services/Telecomm/src/com/android/server/telecom/Call.java，由 Telecom 应用中的 com.android.server.telecom.Call 类定义。代码逻辑详情如下：

```
Call startOutgoingCall(Uri handle, PhoneAccountHandle phoneAccountHandle, Bundle extras,
        UserHandle initiatingUser, Intent originalIntent) {
    ......
    // 使用传入的Uri构造Call对象,Call对象增加连接服务时Uri可能会发生变化,但是在大多数情况下Uri
    // 是保持不变的
    if (call == null) {
        //getNextCallId生成可标识唯一Call对象的Id
        call = new Call(getNextCallId(), mContext,
                this,
                mLock,
                mConnectionServiceRepository,
                mContactsAsyncHelper,
                mCallerInfoAsyncQueryFactory,
                mPhoneNumberUtilsAdapter,
                handle,
                null /* gatewayInfo */,
                null /* connectionManagerPhoneAccount */,
                null /* phoneAccountHandle */,
                Call.CALL_DIRECTION_OUTGOING /* callDirection */,
                false /* forceAttachToExistingConnection */,
                false, /* isConference */
                mClockProxy);
        ...... call.XXX
    }
    ......//其他Call类型处理
```

```
        } else if (!mCalls.contains(call)) {
            // 我们有可能重复使用以前保存在mCalls列表中的Call对象,所以需要检查mCalls列表中是否
            // 已经包含了Call对象,可参考reuseOutgoingCall方法

            addCall(call);//保存并触发增加Call对象的通知
        }
        return call;
    }
```

接下来,我们继续查看CallsManager对象addCall方法的处理逻辑。

2. CallsManager.addCall

```
    private void addCall(Call call) {
        call.addListener(this);//Call对象设置Listener为CallsManager对象
        mCalls.add(call);
        ......//更新Call对象mIntentExtras属性

        updateCanAddCall();//更新addCall状态
        // 循环mListeners调用其onCallAdded 方法,目的是快速设置Call对象为前端通话,即第一路通话
        for (CallsManagerListener listener : mListeners) {
            //通过已注册的CallsManagerListener进行回调,通知增加了Call对象
            listener.onCallAdded(call);
        }
    }
```

CallsManager 对象将保存多个 Call 对象到 mCalls 集合中,Call 对象则设置 Listener 对象为 CallsManager,对象之间相互引用。而 CallsManager 对象通过 mListeners 发出 onCallAdded 消息回调。那么 mListeners 究竟是什么呢?摘录出 CallsManager 类的属性定义和构造方法中的关键逻辑,详情如下:

```
//mListeners是一个CallsManagerListener类型的Set集合
private final Set<CallsManagerListener> mListeners = Collections.newSetFromMap(
        new ConcurrentHashMap<CallsManagerListener, Boolean>(16, 0.9f, 1));
//在CallsManager的构造方法中增加的Listeners如下
mListeners.add(mInCallWakeLockController);
mListeners.add(statusBarNotifier);
mListeners.add(mCallLogManager);
mListeners.add(mPhoneStateBroadcaster);
mListeners.add(mInCallController);
mListeners.add(mCallAudioManager);
mListeners.add(missedCallNotifier);
mListeners.add(mHeadsetMediaButton);
mListeners.add(mProximitySensorManager);
```

跟进这九个 Listener 对象,在拨号流程中重点关注 mInCallController 这个注册的 CallsManagerListener 对象,即 com.android.server.telecom.InCallController。

3. InCallController.onCallAdded 消息回调

```
    public void onCallAdded(Call call) {
        if (!isBoundAndConnectedToServices()) {//判断是否已经绑定服务
            Log.i(this, "onCallAdded: %s; not bound or connected.", call);
            // 没有绑定或连接 InCallService 服务
            bindToServices(call);
        } else { ...... }
    }

    public void bindToServices(Call call) {
        if (mInCallServiceConnection == null) {
```

```
        ......
        if (carModeComponentInfo != null &&
                !carModeComponentInfo.getComponentName().equals(
                    mSystemInCallComponentName)) {
            //重点关注创建的 InCallServiceBindingConnection 对象
            carModeInCall = new InCallServiceBindingConnection(carModeComponentInfo);
        }
        //使用 InCallServiceBindingConnection 对象创建 mInCallServiceConnection 对象
        mInCallServiceConnection =
                new CarSwappingInCallServiceConnection(systemInCall, carModeInCall);
    }
    //当前获取的 car mode 是 false 状态,因为是在 Nexus 6P 手机上运行
    mInCallServiceConnection.setCarMode(shouldUseCarModeUI());

    // 执行绑定通话界面 InCallService 服务的真实操作,判断返回结果
    if (mInCallServiceConnection.connect(call) ==
            InCallServiceConnection.CONNECTION_SUCCEEDED) {
        //只有当我们成功连接到通话界面服务后,才执行连接无界面的 IncallService 服务
        connectToNonUiInCallServices(call);
    } else {......}
}
```

需要梳理一下当前几个类的关系,InCallServiceConnection 作为 InCallController 的内部公有类,InCallServiceBindingConnection、EmergencyInCallServiceConnection 和 CarSwappingInCallServiceConnection 这三个类作为 InCallController 的内部私有类,全部继承于 InCallServiceConnection 类。

通过分析 mInCallServiceConnection.connect 的处理逻辑可知,mCurrentConnection.connect(call) 实际调用的是 InCallServiceBindingConnection 对象的 connect 方法,其绑定服务的主要代码逻辑详情如下。

```
@Override
public int connect(Call call) {
    ......
    //这里非常关键,指明了绑定 InCallService.SERVICE_INTERFACE 服务
    Intent intent = new Intent(InCallService.SERVICE_INTERFACE);
    intent.setComponent(mInCallServiceInfo.getComponentName());
    ......//更新 intent 对象扩展属性
    mIsConnected = true;
    if (!mContext.bindServiceAsUser(intent, mServiceConnection,
                Context.BIND_AUTO_CREATE |
                Context.BIND_FOREGROUND_SERVICE,
                UserHandle.CURRENT)) {
        mIsConnected = false;
    }
    ......
    return mIsConnected ? CONNECTION_SUCCEEDED : CONNECTION_FAILED;
}

private final ServiceConnection mServiceConnection = new ServiceConnection() {
    @Override
    public void onServiceConnected(ComponentName name, IBinder service) {
        onConnected(service);
    }

    @Override
    public void onServiceDisconnected(ComponentName name) {
        onDisconnected();
    }
};
```

```java
protected void onConnected(IBinder service) {
    boolean shouldRemainConnected =
            InCallController.this.onConnected(mInCallServiceInfo, service);
    if (!shouldRemainConnected) {
        disconnect();
    }
}
```

上面的代码逻辑体现了绑定服务的全部过程。首先，创建 InCallServiceBindingConnection 对象，创建该对象的同时将同步创建一个 mServiceConnection 对象，此对象为匿名的 ServiceConnection 类型，重写了 onServiceConnected 和 onServiceDisconnected 方法；接着，创建 action 为 InCallService.SERVICE_INTERFACE 的 intent 对象，并更新了 PhoneAccount 和 Call 的一些关键信息；然后，调用 Android 系统的 bindServiceAsUser 方法绑定服务；最后是绑定服务成功以后的收尾工作——onConnected 系统回调，将发起对 InCallController.this.onConnected 的调用，该方法中的主要处理逻辑详情如下：

```java
private boolean onConnected(InCallServiceInfo info, IBinder service) {
    //获取 IInCallService 服务并保存
    IInCallService inCallService = IInCallService.Stub.asInterface(service);
    mInCallServices.put(info, inCallService);

    try {
        //调用服务方法增加 Adapter，InCallAdapter 实现了 IInCallAdapter.aidl 接口
        inCallService.setInCallAdapter(
                new InCallAdapter(
                        mCallsManager,
                        mCallIdMapper,
                        mLock,
                        info.getComponentName().getPackageName()));
    } catch (RemoteException e) {......}
    //将之前保存的 Call 对象通过 inCallService 发送出去
    for (Call call : calls) {
        try {
            //注意这里有 Call 对象的转换，将 com.android.server.telecom.Call 转换为可跨进
            //程传递的对象 android.telecom.ParcelableCall
            inCallService.addCall(ParcelableCallUtils.toParcelableCall(
                    call,
                    true /* includeVideoProvider */,
                    mCallsManager.getPhoneAccountRegistrar(),
                    info.isExternalCallsSupported(),
                    includeRttCall));
        } catch (RemoteException ignored) {......}
    }
    try {
        //通知 Call 相关状态的变化
        inCallService.onCallAudioStateChanged(mCallsManager.getAudioState());
        inCallService.onCanAddCallChanged(mCallsManager.canAddCall());
    } catch (RemoteException ignored) {
    }
    return true;
}
```

拨号流程跟踪到此，Telecom 应用中完成了第一次绑定服务和对应服务的接口调用。对 bindToService 流程的回顾和总结，如图 3-2 所示。

重点关注以下三个接口调用以及调用的时序，同时此服务的服务端接口实现逻辑也将作为后续关注和分析的重点。

- 步骤 8：bindService
- 步骤 13：setInCallAdapter
- 步骤 14：addCall

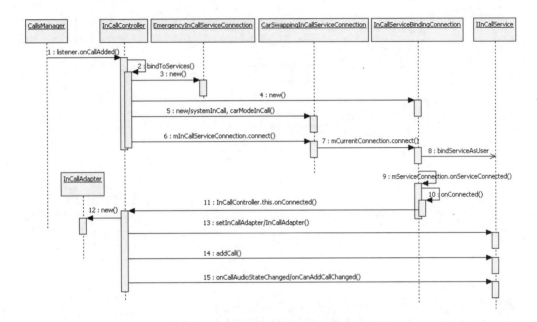

图 3-2　bindToService 流程

这里究竟绑定了什么服务呢？查看文件 frameworks/base/telecomm/java/android/telecom/InCallService.java，其中的 SERVICE_INTERFACE 定义为 "android.telecom.InCallService"，并在 AndroidManifest.xml 文件中查找此关键字，即可找到对应的服务信息，详情如下：

```
public static final String SERVICE_INTERFACE = "android.telecom.InCallService";

$ mangrep android.telecom.InCallService
packages/apps/Dialer/java/com/android/incallui/AndroidManifest.xml:91: <action android:name="android.telecom.InCallService"/>
```

查看 packages/apps/Dialer/java/com/android/incallui/AndroidManifest.xml 中的 Service 定义，详情如下：

```
<service
    android:directBootAware="true"
    android:exported="true"
    android:name="com.android.incallui.InCallServiceImpl"
    android:permission="android.permission.BIND_INCALL_SERVICE">
    ......//省略元数据信息
  <intent-filter>
    <action android:name="android.telecom.InCallService"/>
  </intent-filter>
</service>
```

拨号流程跟踪到这里，我们将再次进入 Dialer 应用中的代码逻辑。InCallController 通过绑定服务的方式，开启拨号流程中的第二次跨进程访问，从 Telecom 应用的 system_server 进程再次回到 Dialer 应用的 com.android.dialer 进程。

在分析 IInCallService 服务之前,我们先总结一下 Telecom 应用中的拨号流程,如图 3-3 所示。

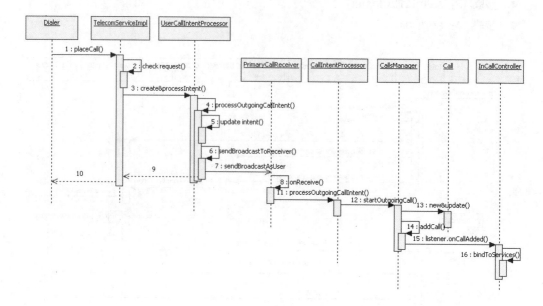

图 3-3　Telecom 中的拨号流程 1

重点关注以下几点。

- 步骤 1：placeCall 是拨号流程中第一次跨进程服务接口调用。
- 步骤 7：sendBroadcast 将同步的服务接口调用转为异步的方式,步骤 9、步骤 10 将立即返回接口调用 Dialer 应用的调用方,而步骤 11 到步骤 16 将继续执行拨号的请求消息传递。
- 步骤 13：创建和更新 Call 对象（此对象的定义在 Telecom 应用中）。
- 步骤 16：bindToService,将发起拨号流程中第二次跨进程服务接口调用。

3.1.6　IInCallService 服务的响应过程

InCallServiceImpl 类继承于 InCallService 类,类代码文件在 packages/apps/Dialer 工程下,而 InCallService 类对应的代码文件则在 framework 下,其服务接口的定义文件为：frameworks/base/telecomm/java/com/android/internal/telecom/IInCallService.aidl,主要定义了 addCall、setInCallAdapter、updateCall 等接口方法。

InCallController 在拨号流程中,首先绑定服务,接着调用服务的 setInCallAdapter、addCall 和 onCallXXXChanged 接口。IInCallService 服务的响应逻辑是什么呢？接下来,我们分别分析 IInCallService 的 onBind、setInCallAdapter 和 addCall 服务接口。

1. onBind 服务被绑定的响应方法

```
public IBinder onBind(Intent intent) {
    ......省略初始化的一些操作
    InCallPresenter.getInstance().onServiceBind();//设置服务绑定状态 mServiceBound 属性
    InCallPresenter.getInstance().maybeStartRevealAnimation(intent);//开启 Activity
    TelecomAdapter.getInstance().setInCallService(this);//保存服务
    ......
    return super.onBind(intent);//调用父类 InCallService 的 onBind 方法
}
```

```
//InCallService 的 onBind 相关方法
public IBinder onBind(Intent intent) {
  return new InCallServiceBinder();
}
// InCallServiceBinder 实现了 IInCallService.aidl 的接口，这些接口通过发送 Handler 消息，将服务
//接收到的服务请求转化为异步处理方式
private final class InCallServiceBinder extends IInCallService.Stub {
@Override
public void setInCallAdapter(IInCallAdapter inCallAdapter) {
  mHandler.obtainMessage(MSG_SET_IN_CALL_ADAPTER, inCallAdapter).sendToTarget();
}
@Override
public void addCall(ParcelableCall call) {
  mHandler.obtainMessage(MSG_ADD_CALL, call).sendToTarget();
}
@Override
public void updateCall(ParcelableCall call) {
  mHandler.obtainMessage(MSG_UPDATE_CALL, call).sendToTarget();
}
......
}
```

mHandler 发送 Handler 消息将 IInCallService 的同步调用转换为异步处理，在后续的 setInCallAdapter 和 addCall 接口响应中继续分析其实现逻辑。

查看 InCallPresenter.getInstance().maybeStartRevealAnimation 方法的逻辑实现，将启动 InCallActivity，即通话界面的核心逻辑如下：

```
final Intent activityIntent =
    InCallActivity.getIntent(mContext, false, true, false /* forFullScreen */);
activityIntent.putExtra(TouchPointManager.TOUCH_POINT, touchPoint);
mContext.startActivity(activityIntent);
```

activityIntent 通过 intent.setClass(context, InCallActivity.class) 的方式指定了启动 InCallActivity，其 AndroidManifest.xml 中的配置信息如下：

```
<activity
    android:directBootAware="true"
    android:excludeFromRecents="true"
    android:exported="false"
    android:label="@string/phoneAppLabel"
    android:launchMode="singleInstance"
    android:name="com.android.incallui.InCallActivity"
    android:resizeableActivity="true"
    android:screenOrientation="nosensor"
    android:taskAffinity="com.android.incallui"
    android:theme="@style/Theme.InCallScreen">
</activity>
```

通过上面 InCallActivity 的定义，重点关注下面几点。

● exported

exported 为 false，说明无法以跨进程的方式启动此 Activity，只能从当前进程启动。

● launchMode

launchMode 为 singleInstance，说明为单例，方便界面管理，减少内存开销。

● name

指定 com.android.incallui.InCallActivity 类对应的 Activity 配置项。

InCallService 的 onBind 流程跟踪和分析总结如图 3-4 所示。

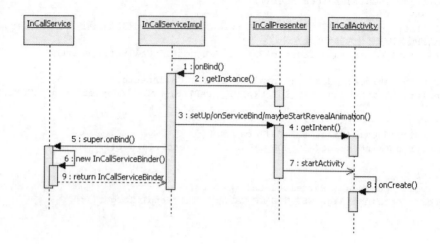

图 3-4　InCallService 类的 onBind 流程

重点关注 InCallPresenter 的创建和初始化操作，即图 3-4 中的步骤 2 和步骤 3；以及步骤 7：调用 startActivity 初始化 InCallActivity 通话界面。

 onBind 只获取了 intent 对象，并没有获取 Call 对象的通话相关信息，可以理解为 InCallActivity 界面的预加载过程；这里做一个猜测，与 Call 对象相关的信息获取或更新应该是通过 IInCallService 服务的接口 addCall 和 updateCall 来完成消息传递的。

2. setInCallAdapter 设置 Adapter

```
//IInCallService 的服务响应方法 setInCallAdapter
public void setInCallAdapter(IInCallAdapter inCallAdapter) {
  //通过 mHandler 发出异步处理的消息 MSG_SET_IN_CALL_ADAPTER，接口立即返回 void
  mHandler.obtainMessage(MSG_SET_IN_CALL_ADAPTER, inCallAdapter).sendToTarget();
}

//mHandler 作为 InCallService 类的匿名内部类对象，异步接收和处理 mHandler 发出的消息
private final Handler mHandler = new Handler(Looper.getMainLooper()) {
@重写
public void handleMessage(Message msg) {
    switch (msg.what) {
        case MSG_SET_IN_CALL_ADAPTER:
            String callingPackage = getApplicationContext().getOpPackageName();
            mPhone = new Phone(
                new InCallAdapter((IInCallAdapter) msg.obj), callingPackage,
                getApplicationContext().getApplicationInfo().targetSdkVersion);
            mPhone.addListener(mPhoneListener);
            onPhoneCreated(mPhone);
            break;
......
```

setInCallAdapter 接口的响应逻辑，主要是创建 Phone 对象和设置 Phone 对象的 Listener 属性。Phone 的类名非常特殊，通过 Android Studio 打开类名，将搜索出来很多结果，此处究竟使用了哪个 Phone 对象呢？通过查看当前类的 import 代码，并没有发现与 import Phone 相关的逻辑，说明 Phone 与 InCallService 是同一个 package，即 frameworks/base/telecomm/java/android/telecom/

Phone.java 代码。

InCallService 类的 setInCallAdapter 流程总结如图 3-5 所示。

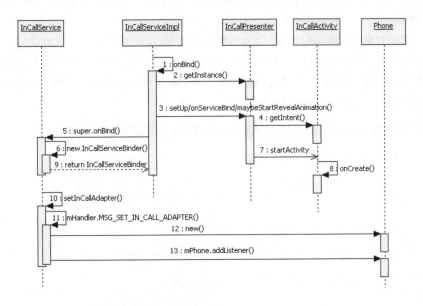

图 3-5　InCallService 类的 onBind 和 setInCallAdapter 流程

重点关注步骤 12 和步骤 13，创建 Phone 对象和增加 Listener 为 InCallService 类的 mPhoneListener 对象，也是拨号流程 Dialer 应用中的第一个 Listener。在创建 Phone 对象之前有个容易被忽略的地方，就是通过 IInCallAdapter Binder 对象创建了 InCallAdapter。

设置 IInCallAdapter Binder 对象有什么作用呢？我们看一下其接口定义文件 frameworks/base/telecomm/java/com/android/internal/telecom/IInCallAdapter.aidl 中的内容，原来它提供了 answerCall、rejectCall、mute、setAudioRoute、playDtmfTone 等控制通话的接口，因此通过 Binder 对象可跨进程访问 Telecom 应用，即 system_server 进程的系统服务相关接口。

3. addCall 增加主动拨号 Outgoing Call

addCall 的方法响应模式与 setInCallAdapter 方法一致，都是通过 mHandler 对象发送 MSG_ADD_CALL 消息，并在其 handleMessage 方法中异步响应，关键代码逻辑如下：

```
// mHandler 响应 MSG_ADD_CALL 消息
case MSG_ADD_CALL:
    mPhone.internalAddCall((ParcelableCall) msg.obj);
//Phone 对象的 internalAddCall 处理逻辑
final void internalAddCall(ParcelableCall parcelableCall) {
    //通过 parcelableCall 相关信息创建 Call 对象
    Call call = new Call(this, parcelableCall.getId(), mInCallAdapter,
            parcelableCall.getState(), mCallingPackage, mTargetSdkVersion);
    mCallByTelecomCallId.put(parcelableCall.getId(), call);
    mCalls.add(call);
    checkCallTree(parcelableCall);
    //通过 parcelableCall 更新当前 Call 对象相关信息
    call.internalUpdate(parcelableCall, mCallByTelecomCallId);
    fireCallAdded(call);//发出 addCall 的消息通知
}
```

（1）在 Telecom 应用中，首先会创建 Call 对象，Dialer 应用中也会创建 Call 对象，但这两个 Call 对象的定义是不同的。Telecom 中的 Call 对象是当前应用中的定义 packages/services/Telecomm/src/com/android/server/telecom/Call.java，而 Dialer 应用中 Call 对象的定义是 frameworks/base/telecomm/java/android/telecom/Call.java。

（2）Call 对象的创建与转换。从 Telecom 应用中创建 com.android.server.telecom.Call，并通过此对象创建跨进程传递的 android.telecom.parcelableCall 对象（支持序列化和反序列化，因此可以跨进程传递此对象），而 Dialer 应用中是接收到 parcelableCall 对象后，通过此对象相关信息创建 android.telecom.Call 对象。

继续分析 fireCallAdded(call)方法，关键业务逻辑和流程详情如下：

```
private void fireCallAdded(Call call) {
    for (Listener listener : mListeners) {
        //InCallService 在处理 MSG_SET_IN_CALL_ADAPTER 消息时设置的 Listener
        listener.onCallAdded(this, call);//传递 Call 对象
    }
}
// InCallService.mPhoneListener 的 onCallAdded 响应
private Phone.Listener mPhoneListener = new Phone.Listener() {
@Override
public void onCallAdded(Phone phone, Call call) {
    InCallService.this.onCallAdded(call);//InCallServiceImpl 重写了此方法,Java 多态将调
                                         //用子类对应的重写方法
}
......//省略 onCallAdded、onCallRemoved、onAudioStateChanged 等接口实现,都是基于
//InCallService.this.onXXX 的调用
}
```

接着，我们将连续执行两次 onCallAdded 调用：

```
InCallPresenter.getInstance().onCallAdded(call);
mCallList.onCallAdded(mContext, call, latencyReport);
```

进入 CallList 类的 onCallAdded 方法，其中 dialerCallListener.onDialerCallUpdate()调用与当前拨号流程有关。dialerCallListener 为 CallList 类的私有内部类 DialerCallListenerImpl 对象，其核心逻辑如下：

```
@Override
public void onDialerCallUpdate() {
  onUpdateCall(mCall);//更新 Call 信息
  notifyGenericListeners();//调用 CallList 类的 notifyGenericListeners 方法
}
//CallList 类的 notifyGenericListeners 方法
private void notifyGenericListeners() {
  for (Listener listener : mListeners) {//拨号流程 Dialer 应用中的第二个 Listener
    //InCallPresenter.setUp 方法中增加的 Listener mCallList.addListener(this)
    listener.onCallListChange(this);
  }
}
//InCallPresenter.onCallListChange 中的核心逻辑
for (InCallStateListener listener : mListeners) {//拨号流程 Dialer 应用中的第三个 Listener
  listener.onStateChange(oldState, mInCallState, callList);
}
```

在 Dialer 通话界面展示的流程跟踪过程中，已经涉及三个 Listener，可见其逻辑的复杂度，很容易就跟踪错误。InCallStateListener 是一个 interface 接口，它定义了一个方法 onStateChange，在

Dialer 工程下找到了八个实现此接口的类，详情如下：
- ConferenceManagerPresenter
- VideoCallPresenter
- CallButtonPresenter
- CallCardPresenter
- ProximitySensor
- DialpadPresenter
- VideoPauseController
- StatusBarNotifier

这些类通过名字就能看出是与通话界面相关的，特别是 DialpadPresenter、CallCardPresenter、CallButtonPresenter 这几个类名，在后面解析通话界面的章节我们再进行详细解析。拨号流程跟踪到此，我们仅需要掌握其中的展示和更新通话界面即可。InCallService 类的 addCall 流程的分析和总结，如图 3-6 所示。

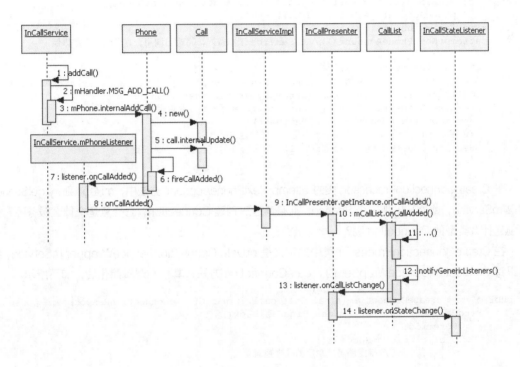

图 3-6 InCallService 类的 addCall 流程

重点关注三个 Listener 的消息回调，分别是步骤 7、步骤 13 和步骤 14，而这些步骤中，都将传递 android.telecom.Call 对象，最终交给 InCallStateListener 对象进行通话界面中与通话相关信息的展示和更新。

再次回顾拨号流程，Dialer 应用提供用户拨号操作界面，用户输入电话号码发起拨号请求，到 Telecom 应用接收拨号请求，完成第一次跨进程服务接口调用，Telecom 应用创建 com.android.server.telecom.Call 对象，然后通过此对象信息创建可跨进程传递的 android.telecom.parcelableCall 对象，完成第二次跨进程服务接口调用；回到 Dialer 应用的 com.android.dialer 进程中的 IInCallService 服务响应，通过接收到的 parcelableCall 对象信息创建本地的 android.telecom.Call 对象，最后根据此对象更新和显示 InCallActivity 通话界面相关信息。

3.1.7 继续分析 CallsManager.placeOutgoingCall

我们再次回到 Telecom 应用的拨号流程中，CallsManager 分别调用 startOutgoingCall 和 placeOutgoingCall。startOutgoingCall 方法将通过绑定服务和调用其服务接口，启动和更新 Dialer 应用中的 InCallActivity，展示出通话界面；但拨号请求并未发送到 BP Modem 处理。我们继续分析 placeOutgoingCall 的处理逻辑，其关键的调用过程是：sendNewOutgoingCallIntent→NewOutgoingCallIntentBroadcaster. processIntent→mCallsManager.placeOutgoingCall→call.startCreateConnection→CreateConnectionProcessor.process→attemptNextPhoneAccount 等一系列调用，这个流程相对简单，请读者自己去跟踪和分析。我们重点关注一下 CreateConnectionProcessor 类的 attemptNextPhoneAccount 方法，其关键逻辑详情如下。

```
//又出现一个 Service
mService = mRepository.getService(phoneAccount.getComponentName(),
        phoneAccount.getUserHandle());
if (mService == null) {
    attemptNextPhoneAccount();//如 Service 未准备好，将递归调用，直到超过次数限制
} else {
    mConnectionAttempt++;//计数
    mCall.setConnectionManagerPhoneAccount(attempt.connectionManagerPhoneAccount);
    mCall.setTargetPhoneAccount(attempt.targetPhoneAccount);
    mCall.setConnectionService(mService);//更新 Call 对象相关信息
    setTimeoutIfNeeded(mService, attempt);
    //创建 Connection, 我们推测一下会是什么样的连接呢？
    mService.createConnection(mCall, this);
}
```

在 CreateConnectionProcessor 类的 attemptNextPhoneAccount 方法中，mRepository.getService 获取 mService，此处又出现一个 Service，而且通过它创建 Connection，它究竟是个什么服务呢？又将创建什么类型的 Connection 呢？

在 CreateConnectionProcessor 类中的定义是 private ConnectionServiceWrapper mService，继续跟踪 ConnectionServiceWrapper 的 createConnection 方法，其核心逻辑简化后，如下所示。

```
public void createConnection(final Call call, final CreateConnectionResponse response) {
    BindCallback callback = new BindCallback() {
        @Override
        public void onSuccess() {
            ......//绑定服务成功后的处理逻辑
        }
        @Override
        public void onFailure() {
            ...... //绑定服务失败后的处理逻辑
        }
    };
    mBinder.bind(callback, call);//又是一处绑定服务操作
}
void bind(BindCallback callback, Call call) {
    mCallbacks.add(callback);
    if (mServiceConnection == null) {
        Intent serviceIntent = new Intent(mServiceAction).setComponent(mComponentName);
        ServiceConnection connection = new ServiceBinderConnection(call);
        final int bindingFlags = Context.BIND_AUTO_CREATE |
```

```
                        Context.BIND_FOREGROUND_SERVICE;
        ......//开始绑定服务
        isBound = mContext.bindService(serviceIntent, connection, bindingFlags);
        ......
    } else {......}
}
```

mBinder 是 ConnectionServiceWrapper 的父类 ServiceBinder 的内部类 Binder2 类型对象,在创建 ConnectionServiceWrapper 对象时,mBinder 被同步创建;mServiceAction 指定了将要绑定的服务,它在 ConnectionServiceWrapper 构造方法中调用了 super(ConnectionService.SERVICE_INTERFACE, ...),即将 mServiceAction 属性初始化为 ConnectionService.SERVICE_INTERFACE。ConnectionService 的代码文件在 frameworks/base 下,即 frameworks/base/telecomm/java/android/telecom/ConnectionService.java 中。

同样在 Android 源码中查找 AndroidManifest.xml 文件中对 service 的定义,在 packages/services/Telephony 中找到了对应的定义,详情如下。

```
public static final String SERVICE_INTERFACE = "android.telecom.ConnectionService";
<service
        android:singleUser="true"
        android:name="com.android.services.telephony.TelephonyConnectionService"
        android:label="@string/pstn_connection_service_label"
android:permission="android.permission.BIND_TELECOM_CONNECTION_SERVICE" >
    <intent-filter>
        <action android:name="android.telecom.ConnectionService" />
    </intent-filter>
</service>
```

拨号流程在 Telecom 应用中将发起第二次绑定服务的跨进程服务访问,绑定的服务对象在 packages/services/Telephony 代码库中,这是我们涉及的第三个代码库。查看对应的 Android.mk 文件,此代码库将编译出 TeleService.apk Android 应用程序,以后统一称其为 TeleService 应用。我们先不急着分析 TeleService 应用的业务逻辑,继续分析 Telecom 应用中绑定服务成功后的处理逻辑,将在 ServiceBinderConnection 对象的 onServiceConnected 方法中响应,其代码逻辑详情如下。

```
private final class ServiceBinderConnection implements ServiceConnection {
    @Override
    public void onServiceConnected(ComponentName componentName, IBinder binder) {
        setBinder(binder);//保存绑定服务成功后的 binder 对象
        handleSuccessfulConnection();//bind Service 成功后的处理
    }
    @Override
    public void onServiceDisconnected(ComponentName componentName) {
        handleDisconnect();
    }
}
```

setBinder 方法是在 ConnectionServiceWrapper 的父类 ServiceBinder 中实现的方法,其中关键的逻辑是调用 setServiceInterface 方法保存 binder 对象,而 setServiceInterface 方法在 ServiceBinder 类中定义为 abstract(抽象)方法,最终在子类 ConnectionServiceWrapper 中实现了该方法。这是一个非常典型的模板方法设计模式的实现,请读者试着去扩展此设计模式。应用在此处有什么作用?能给我们的代码带来什么优势?

```
@Override//ConnectionServiceWrapper 实现父类抽象接口 setServiceInterface
protected void setServiceInterface(IBinder binder) {
```

```
            mServiceInterface = IConnectionService.Stub.asInterface(binder);//保存 binder 对象
            addConnectionServiceAdapter(mAdapter);//服务调用增加 apapter 对象
        }

mServiceInterface.addConnectionServiceAdapter(adapter, Log.getExternalSession());

// mAdapter 作为 ConnectionServiceWrapper 对象的成员属性
private final Adapter mAdapter = new Adapter();
// Adapter 作为 ConnectionServiceWrapper 类的私有内部类,实现了
IConnectionServiceAdapter.aidl 接口
private final class Adapter extends IConnectionServiceAdapter.Stub {
    ......setActive、setRinging、setDialing、setConnectionCapabilities
}
```

addConnectionServiceAdapter 方法将通过 mServiceInterface 调用 addConnectionServiceAdapter 接口增加 IConnectionServiceAdapter,为成功绑定服务后的第一次跨进程调用。

这里的流程是不是和绑定 IInCallService 服务的流程非常相似呢?
首先绑定服务,然后调用 addConnectionServiceAdapter 增加 Adapter 绑定对象,最后还有没有类似 addCall 的操作呢?请读者先自行思考这个问题。

IConnectionServiceAdapter 接口的定义文件是:frameworks/base/telecomm/java/com/android/internal/telecom/IConnectionServiceAdapter.aidl,主要定义了 setActive、setRinging、setDialing、setConnectionCapabilities 等接口方法。通过接口名,我们可以判断出当前绑定的服务"android.telecom.ConnectionService"将通过此 Adapter 接口调用去更新 Telecom 中的通话状态。

到此,ServiceBinderConnection 类的 onServiceConnected 方法中的 setBinder 已经分析完成,我们继续查看 handleSuccessfulConnection 中的处理逻辑。

```
private void handleSuccessfulConnection() {
    for (BindCallback callback : mCallbacks) {
        callback.onSuccess();
    }
    mCallbacks.clear();
}
//在 ConnectionServiceWrapper 的 createConnection 方法中创建的 callback 对象响应 onSuccess 方
//法调用

//通过通话相关信息创建 ConnectionRequest 支持序列化和反序列化对象
ConnectionRequest connectionRequest = new ConnectionRequest.Builder()
            .setAccountHandle(call.getTargetPhoneAccount())
            .setAddress(call.getHandle())
            .setExtras(extras)
            .setVideoState(call.getVideoState())
            .setTelecomCallId(callId)//关键点
            .setShouldShowIncomingCallUi(
                    !mCallsManager.shouldShowSystemIncomingCallUi(call))
            .setRttPipeFromInCall(call.getInCallToCsRttPipeForCs())
            .setRttPipeToInCall(call.getCsToInCallRttPipeForCs())
            .build();//建造者模式的应用,创建对象

try {
    mServiceInterface.createConnection(//第二次跨进程访问
            call.getConnectionManagerPhoneAccount(),
            callId,
            connectionRequest,
            call.shouldAttachToExistingConnection(),
            call.isUnknown(),
```

```
                  Log.getExternalSession());
} catch (RemoteException e) {......}
```

拨号流程中，Telecom 第二次绑定服务与第一次绑定服务的处理过程非常相似，都分三步走：
- bind Service
- addConnectionServiceAdapter
- createConnection

第二次绑定服务的服务对象为：SERVICE_INTERFACE，即"android.telecom.ConnectionService"，其关键流程总结如图 3-7 所示。

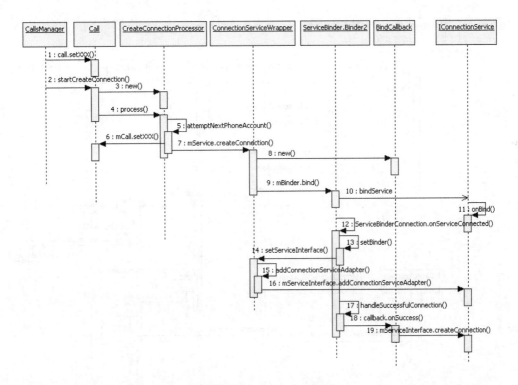

图 3-7 createConnection 流程

图 3-7 中，重点关注以下三个关键步骤：
- 绑定服务

步骤 10：创建 intent 的 Action 有一个比较隐含的设置，在 ConnectionServiceWrapper 类的构造方法中调用了 super 构造方法，从而设置了绑定服务的对象为 ConnectionService.SERVICE_INTERFACE。

- 第一次跨进程服务接口调用 addConnectionServiceAdapter

步骤 16：将传递实现 IConnectionServiceAdapter.aidl 接口 Stub 的跨进程访问 binder 对象。

- 第二次跨进程服务接口调用 createConnection

步骤 19：通过 Call 对象拨号请求相关信息创建 ConnectionRequest 对象，传递给 packages/services/Telephony 中对应的服务。

拨号流程中，Telecom 应用第一次跨进程服务调用，将与 Call 对象相关的拨号请求信息传递给了 Dialer 应用，去加载和展现通话界面；那么第二次跨进程服务调用，Call 拨号请求相关信息转换成了 ConnectionRequest 对象并传递给了 TeleService 应用。我们在这里做一个推测，TeleService 将接收到的 ConnectionRequest 请求相关信息传递给 BP Modem 来发起电话拨号请求。

3.1.8　Telecom 应用拨号流程回顾与总结

拨号流程跟踪到这一步，已经涉及三个应用的消息传递 Dialer、Telecom 和 TeleService。可见流程复杂而漫长，消息类型又相近，比如，两次服务的绑定过程，几个 Call 对象的创建、转换和传递，各种 Listener 消息回调等。

下面，我们对 Telecom 中的拨号流程进行回顾和总结，详情如图 3-8 所示。

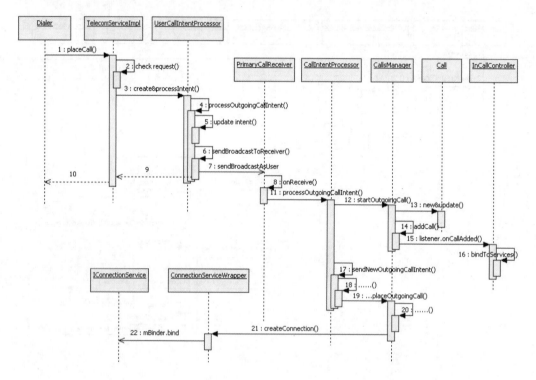

图 3-8　Telecom 中的拨号流程 2

我们重点关注以下几点。

- 异步处理

步骤 7 将 TelecomServiceImpl 服务请求从同步转换成了异步方式处理，步骤 9、步骤 10 直接返回给了 Dialer 应用。

- 第一次绑定服务

CallsManager.startOutgoingCall 方法调用将绑定 Dialer 应用中的 IInCallService 服务，并调用服务提供的 setInCallAdapter 和 addCall 方法，加载和更新 InCallActivity 通话界面，需重点关注 InCallController 中的服务处理机制。

- 第二次绑定服务

CallsManager.placeOutgoingCall 方法调用将绑定 TeleService 应用中的 TelephonyConnectionService 服务，并调用服务提供的 addConnectionServiceAdapter 和 createConnection 方法，继续传递拨号请求。

因此，Telecom 应用 CallsManager 对象的 startOutgoingCall 和 placeOutgoingCall 方法，两次绑定不同的服务，并且过程也非常相似，分三步走，总结如下：

① bind Service
② setInCallAdapter/addConnectionServiceAdapter
③ addCall/createConnection

3.1.9　IConnectionService 服务的响应过程

根据 AndroidManifest.xml 中对 android.telecom.ConnectionService 服务的定义，其服务的 Java 类为 com.android.services.telephony.TelephonyConnectionService，继承自 android.telecom.ConnectionService 抽象类。在 frameworks/base 工程下，代码文件为 frameworks/base/telecomm/java/android/telecom/ConnectionService.java，其服务响应的关键逻辑简化后的代码框架详情如下。

```java
public abstract class ConnectionService extends Service {
    private final IBinder mBinder = new IConnectionService.Stub() {
            ......addConnectionServiceAdapter、createConnection、answer、hold
    }

    @Override
    public final IBinder onBind(Intent intent) {
         return mBinder;
    }
}
```

frameworks/base/telecomm/java/com/android/internal/telecom/IConnectionService.aidl 文件作为 IConnectionService 服务的接口定义，主要定义了 addConnectionServiceAdapter、createConnection、answer、hold 等接口。通过这些接口的名字，可以知道此服务主要提供了 Call 状态管理的接口供 Telecom 应用调用，比如接听电话、保持呼叫、挂断电话等。

Telecom 应用的 ConnectionServiceWrapper 对象在拨号流程中，首先绑定服务，接着调用服务的 addConnectionServiceAdapter 和 createConnection 接口。TelephonyConnectionService 服务的响应逻辑是什么呢？接下来，我们分别从 IConnectionService 的 onBind、addConnectionServiceAdapter 和 createConnection 服务接口加以解析。

1. onBind 服务被绑定的响应方法

TelephonyConnectionService 继承于 ConnectionService 类，并未重写父类的 onBind 方法。onBind 逻辑简单，返回了 IConnectionService.Stub 类型的 mBinder 对象。

2. addConnectionServiceAdapter 设置 Adapter

```java
//使用 Handler 的异步消息处理机制，将服务调用的同步方式转为异步方式处理，addConnectionServiceAdapter
//服务接口将立即返回
@Override
public void addConnectionServiceAdapter(IConnectionServiceAdapter adapter,
        Session.Info sessionInfo) {
    mHandler.obtainMessage(MSG_ADD_CONNECTION_SERVICE_ADAPTER, args).
                sendToTarget();
}
//查看 mHandler 对消息 MSG_ADD_CONNECTION_SERVICE_ADAPTER 的响应逻辑
case MSG_ADD_CONNECTION_SERVICE_ADAPTER: {
    IConnectionServiceAdapter adapter = (IConnectionServiceAdapter) args.arg1;
```

```
        mAdapter.addAdapter(adapter);//mAdapter 再嵌套一层 Adapter
        onAdapterAttached();//验证 Adapter 是否可用
}
```

addConnectionServiceAdapter 接口的响应逻辑相对比较简单，我们只需重点关注使用 Handler 的异步消息处理机制，将服务调用的同步方式转为异步处理即可，与 IInCallService 的处理机制相同。

3. createConnection 继续发送拨号请求

ConnectionService 服务的接口 createConnection 的响应逻辑仍然是通过 mHandler 将同步调用转为异步处理。mHandler 发出 MSG_CREATE_CONNECTION 消息，并在 handleMessage 中响应此方法，再调用父类的 createConnection 方法，此方法的关键逻辑如下：

```
private void createConnection(
        final PhoneAccountHandle callManagerAccount,
        final String callId,
        final ConnectionRequest request,
        boolean isIncoming,
        boolean isUnknown) {
    //将调用 onCreateOutgoingConnection 创建 Connection 对象，注意与 ConnectionService 的
    //区别 Connection connection = isUnknown ?
            onCreateUnknownConnection(callManagerAccount, request)
            : isIncoming ? onCreateIncomingConnection(callManagerAccount, request)
            : onCreateOutgoingConnection(callManagerAccount, request);
    }
    ......//跨进程调用 Telecom 服务接口，通过 Connection 已经成功创建
    mAdapter.handleCreateConnectionComplete(
            callId,
            request,
            new ParcelableConnection(
                    request.getAccountHandle(),
                    connection.getState(),
                    connection.getConnectionCapabilities(),
                    connection.getConnectionProperties(),
                    connection.getSupportedAudioRoutes(),
                    connection.getAddress(),
                    connection.getAddressPresentation(),
                    ......//省略其他 Connection 信息
                    createIdList(connection.getConferenceables()),
                    connection.getExtras()));
}
```

上面的处理逻辑主要有两个：利用 onCreateXXXConnection 创建 Connection 对象和通过 mAdapter 传递过来的 Binder 对象进行 handleCreateConnectionComplete 接口回调。首先看看 Connection 对象的创建过程，TelephonyConnectionService 重写了父类 ConnectionService 的 onCreateOutgoingConnection 方法，其关键逻辑如下：

```
......//首先是对 Emergency Call 紧急呼叫进行判断和处理
//这里同样得到 Phone 对象，还记得 Dialer 中的 Phone 对象吗？它们有什么区别
final Phone phone = getPhoneForAccount(request.getAccountHandle(), isEmergencyNumber);
Connection resultConnection = getTelephonyConnection(request, numberToDial,
        isEmergencyNumber, handle, phone);
//如果失败，最终的连接将不是 TelephonyConnection
//所以将不能打电话
if(resultConnection instanceof TelephonyConnection) {
    placeOutgoingConnection((TelephonyConnection) resultConnection, phone, request);
}
return resultConnection;
```

```
//继续跟进 placeOutgoingConnection 的处理逻辑
com.android.internal.telephony.Connection originalConnection = null;
try {
    if (phone != null) {
        originalConnection = phone.dial(number, null, videoState, extras);
    }
} catch (CallStateException e) {//拨号失败的处理逻辑,设置 DisconnectCause
    connection.setDisconnected(DisconnectCauseUtil.toTelecomDisconnectCause(
            cause, e.getMessage()));
    return;
}

if (originalConnection == null) {//拨号失败的处理逻辑,设置 DisconnectCause
    connection.setDisconnected(DisconnectCauseUtil.toTelecomDisconnectCause(
            telephonyDisconnectCause, "Connection is null"));
} else {//设置 Connection
    connection.setOriginalConnection(originalConnection);
}
```

重点跟进 phone 对象以及 phone.dial 方法的调用,原来 phone 是 com.android.internal.telephony. GsmCdmaPhone 类型对象,其代码为 frameworks/opt/telephony/src/java/com/android/internal/ telephony/GsmCdmaPhone.java。frameworks/opt/telephony/是我们涉及的第四个代码库,查看对应的 Android.mk 文件,此代码库将编译出 telephony-common.jar,我们以后统一称其为 Telephony。

继续跟进 phone 对象的 dial 方法,可以发现 dial→dialInternal→mCT.dial 的调用过程,mCT 即 GsmCdmaCallTracker,其 dial 方法的关键代码逻辑如下。

```
public synchronized Connection dial(String dialString, int clirMode, UUSInfo uusInfo,
                                    Bundle intentExtras)
        throws CallStateException {
    ......//又出现一个 Connection 类型
    mPendingMO = new GsmCdmaConnection(mPhone,
                    checkForTestEmergencyNumber(dialString),
            this, mForegroundCall, isEmergencyCall);
    //继续调用 dial 拨号请求
    mCi.dial(mPendingMO.getAddress(), clirMode, uusInfo, obtainCompleteMessage());
    ......
    //更新通话状态,并发出通知
    updatePhoneState();
    mPhone.notifyPreciseCallStateChanged();
    return mPendingMO;
}
```

这里的 mCi 即 RIL 对象,其 Java 代码是 frameworks/opt/telephony/src/java/com/android/internal/ telephony/RIL.java,这里将发出 RIL 的拨号请求。到此,我们跟踪拨号流程已经到了 HAL(硬件抽象层),在这一层不同的芯片厂家将完成不同的实现,比如高通平台将 RIL 请求转为 QMI 消息与 Modem 交互,MTK 平台则采用 AT 命令的方式与 Modem 交互。

TeleService 应用中的拨号流程已经分析完成,总结如图 3-9 所示。

图 3-9 中所示的 TeleService 拨号流程,我们重点关注以下步骤。

● 异步处理

步骤 6 和步骤 10 通过 Handler 消息处理将同步转为异步。

● RIL 拨号请求

将拨号请求消息转换为 RIL 拨号请求,对应步骤 15、步骤 17 和步骤 18。步骤 18 在 RIL Java 对象中处理,将发出 RIL 请求,交给 HAL 层的 rild 进程处理。

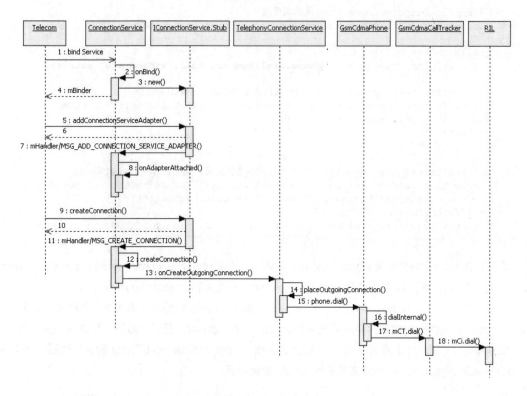

图 3-9 TeleService 拨号流程

我们既然采用 Nexus 6P 手机作为调试的工具，再次发起拨号请求，同步查看手机在 RIL 处理的日志，详情如下：

```
// GsmCdmaPhone 发起 CS call
D/GsmCdmaPhone( 1109): [GsmCdmaPhone] Trying (non-IMS) CS call
//获取的国家编码 CN 即中国，紧急呼救号码 110、119、112 等
D/PhoneNumberUtils( 1109): slotId:0 subId:1 country:CN emergencyNumbers:
 110,119,120,112,911,122,999
D/RILJ ( 1109): [3952]> DIAL [SUB0]//RIL Java 对象发起 DIAL RIL 请求
D/GsmCdmaCallTracker( 1109): [GsmCdmaCallTracker] update phone state, old=IDLE new=O
FFHOOK  //更新 phone 状态为摘机状态
//QMI 消息处理，这部分逻辑为厂家定制，未开源
I/RILQ (  604): (0/604):RIL[0][event] qcril_qmi_voice_all_call_status_ind_hdlr: call
state CC IN PROGRESS for conn id 1
I/RILQ (  604): (0/604):RIL[0][event] qcril_qmi_voice_all_call_status_ind_hdlr: RILVIMS:
update 1 calls[uncached qmi call id 1, call state 4]
I/RILQ (  604): (0/604):RIL[0][event] qcril_qmi_voice_all_call_status_ind_hdlr: call
state ORIGINATING for conn id 1
//RIL Java 对象接收到 DIAL 的返回消息
D/RILJ ( 1109): [3952]< DIAL   [SUB0]
```

通过上面的日志可以分析出几个关键信息。

● 进程信息

GsmCdmaPhone、RIL 等代码均运行在 PID 为 1109 的进程 com.android.phone 空间；QMI 相关消息处理运行在 PID 为 604 的进程 rild 空间。

● RIL 消息配对

DIAL 拨号请求的 RIL 信息编号为 3952，在请求和接收返回消息时将通过此消息编号进行匹配，

">"是 RIL 请求消息下发,"<"是接收到返回消息。
- QMI 消息

qcril_qmi_XXX 为与消息相关的内容,604 号进程即代表在 rild 进程中运行,在 Android 源码中未找到对应的开源代码。

3.1.10　TelecomAdapter 接收消息回调

ConnectionServiceWrapper.Adapter 将接收 TeleService 应用的接口回调,其中将通过 this 调用 ConnectionServiceWrapper 对象的 handleCreateConnectionComplete 方法,接着是 mPendingResponses 属性对象的 handleCreateConnectionSuccess 方法调用,即 CreateConnectionProcessor 对象,最后是 mCallResponse.handleCreateConnectionSuccess 对象,即 Call 对象的 handleCreateConnectionSuccess 方法响应 TeleService 应用的接口回调,其中的处理逻辑详情如下。

```
public void handleCreateConnectionSuccess(
        CallIdMapper idMapper,
        ParcelableConnection connection) {
    ......//根据 connection 对象使用 setXXX 方法设置 Call 对象的对应属性

    for (String id : connection.getConferenceableConnectionIds()) {
        mConferenceableCalls.add(idMapper.getCall(id));
    }
    switch (mCallDirection) {
        case CALL_DIRECTION_INCOMING: //来电流程处理
            for (Listener l : mListeners) {
                l.onSuccessfulIncomingCall(this);
            }
            break;
        case CALL_DIRECTION_OUTGOING://拨号流程处理
            for (Listener l : mListeners) {
                l.onSuccessfulOutgoingCall(this,
                        getStateFromConnectionState(connection.getState()));
            }
            break;
        ......
    }
}
```

我们重点关注拨号流程,而与来电相关的处理逻辑暂不解析。但是,也请读者思考一下,这里的来电流程将承载什么业务逻辑?

Listener 究竟是什么呢?Call 类中有 Listener 的接口定义,同时也定义了 ListenerBase 抽象类,它实现了 Listener 接口。ListenerBase 抽象类实现了 Listener 接口的所有方法,并且这些方法都是空实现,没有具体逻辑。ListenerBase 抽象类有三个子类,分别是:

- CallsManager
- InCallController 匿名内部类对象 mCallListener
- IncomingCallNotifier 匿名内部类对象 mCallListener

这三个类中,仅有 CallsManager 重写了父类 ListenerBase 的 onSuccessfulOutgoingCall 方法,因此,我们接着查看 CallsManager 对象的 onSuccessfulOutgoingCall 方法逻辑,修改 Call 对象的状态时,经过 markCallAsDialing→setCallState 的调用,又找到 Listener 的消息回调,详情如下:

```
for (CallsManagerListener listener : mListeners) {
    listener.onCallStateChanged(call, oldState, newState);
}
```

CallsManagerListener 在 CallsManager 类中定义了接口，而 CallsManagerListenerBase 类实现了其接口，并且这些接口基本上都是空实现，它有 12 个子类：CallLogManager、HeadsetMediaButton、IncomingCallNotifier 和 InCallController 等。通过前面的流程分析，还记得 InCallController 与 Dialer 的交互吗？它们将启动和更新通话界面。难道这里要更改通话界面的通话状态为拨号中？我们先做一个推论。

继续对 CallController 的 onCallStateChanged 调用 updateCall 方法，将 Call 信息传递给 Dialer，具体逻辑详情如下。

```
private void updateCall(Call call, boolean videoProviderChanged, boolean rttInfoChanged) {
    //通过已经修改状态的 Call 对象构造跨进程传递的 ParcelableCall 对象
    ParcelableCall parcelableCall = ParcelableCallUtils.toParcelableCall(
            call,
            videoProviderChanged /* includeVideoProvider */,
            mCallsManager.getPhoneAccountRegistrar(),
            info.isExternalCallsSupported(),
            rttInfoChanged && info.equals(mInCallServiceConnection.getInfo())));
    try {
        inCallService.updateCall(parcelableCall);
    } catch (RemoteException ignored) {
    }
}
```

mAdapter 的接口回调是将当前呼出的电话状态进行更新，更新为 dialing，即正在拨号的状态，最终会调用 IInCallService 的接口去更新通话界面，其流程总结如图 3-10 所示。

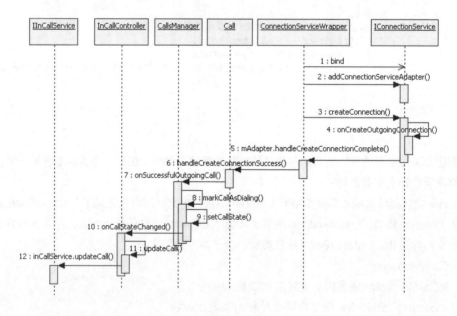

图 3-10 通话状态更新流程

3.1.11 拨号流程总结

在拨号流程中，需要把拨号请求发送到 RIL，我们跟踪到了五个代码库和三个 Android 系统应用

Dialer、Telecom、TeleService，需要重点掌握以下几点。

- 拨号入口 DialpadFragment

如何找到 DialpadFragment 是拨号界面的过程，还请读者认真思考和总结，找到属于自己的查找和调试代码的方法。

Dialer 应用中的拨号流程总结可参考图 3-1，将发起第一次的跨进程服务接口调用，即 TelecomManager 的 placeCall 方法调用。

- 第一次跨进程访问

拨号流程中的第一次跨进程访问，将从 Dialer 应用访问到 Telecom 应用中的 ITelecomService 服务接口，重点参考 framework/base 代码库下的 TelecomManager.java 代码中的处理逻辑。

- Telecom 应用第一次绑定服务

Telecom 应用中 InCallController 对象的 bindToServices 方法将绑定 Dialer 应用中的 IInCallService 服务，并调用该服务提供的 setInCallAdapter 和 addCall 等方法，Dialer 应用中将展示和更新 InCallActivity 进行通话界面的显示。图 3-2 总结了 Telecom 应用第一次绑定服务的核心流程。

- Telecom 应用第二次绑定服务

Telecom 应用中 ConnectionServiceWrapper 对象的 createConnection 方法将绑定 TeleService 应用中的 IConnectionService 服务，并调用该服务提供的 addConnectionServiceAdapter 和 createConnection 等方法，TeleService 应用将通过 RIL 对象发出拨号的 RIL 请求。图 3-7 总结了 Telecom 应用第二次绑定服务的核心流程。

- Adapter 第一次回调

TeleService 应用发出拨号请求后，将通过 Adapter Binder 对象跨进程调用 ConnectionServiceWrapper 的 mAdapter 对象中的服务接口，再通过 Telecom 中的处理，最终调用 Dialer 应用中 IInCallService 服务的 updateCall 接口来更新通话状态。图 3-10 总结了 TeleService 应用调用 Telecom 应用 Adapter 来回调消息以更新 Call 状态的核心流程。

- Telecom 应用中的拨号流程

CallsManager 对象的 startOutgoingCall 和 placeOutgoingCall 方法，分别进行两次绑定服务的操作，并且两次绑定过程非常相似，都是分三步走，绑定服务、setInCallAdapter/addConnectionServiceAdapter 和 addCall/createConnection，其中的核心流程可参考图 3-8。

- Dialer 和 TeleService 应用中对应服务的响应

IInCallService 和 IConnectionService 分别是 Dialer 应用和 TeleService 应用中提供的服务，重点关注这两个服务提供的接口的处理方式是通过 Handler 发送消息，将同步的服务调用接口转换成异步的消息处理。

3.2 来电流程分析

手机的通话功能，最常用的应用场景就是发起拨号和接收来电。上一节已经非常详细地分析和总结了本地主动拨号的流程，本节将继续分析和学习手机被动接收到来电请求后如何展示来电界面的核心流程。

本地主动拨号流程可以理解为手机操作者主动发起的操作请求，主要流程是从上层到下层发送拨号请求的过程。那么，手机被动接收来电的流程呢？可以理解为与主动拨号流程正好相反的过程，手机 BP Modem 侧接收到网络端的来电请求，消息从 Modem 发给 RIL，RIL 再发给 TeleService 应用，然后再传递给 Telecom 应用，最终 Dialer 应用接收到来电请求，进行来电响铃（可选震动）和展示来电界面，通知手机用户有新的来电了。

来电流程分析该从什么地方着手呢？根据拨号流程的分析经验，选择 Android HAL 层中的 RIL 开始分析，因为 Modem 接收到来电后，最先将消息通知给 RIL。

3.2.1 分析 radio 来电日志

在 Nexus 6P 手机中插入 SIM 电话卡，并给此电话卡打电话，同时查看手机的 radio 日志，手机接收到来电请求的 RIL 消息，汇总如下：

```
I/RILQ  (604): (0/604):RIL[0][event] qcril_qmi_voice_all_call_status_ind_hdlr:
call state INCOMING for conn id 1
I/RILQ  (604): (0/604):RIL[0][event] qcril_qmi_voice_voip_create_call_info_entry: Created
call info entry 0x7a49883000 with call android id 1, qmi id 1, media id 7
D/RILJ  (1109): [UNSL]< UNSOL_RESPONSE_CALL_STATE_CHANGED [SUB0]
D/RILJ  (1109): [4185]> GET_CURRENT_CALLS [SUB0]
D/RILJ  (1109): [4185]< GET_CURRENT_CALLS
 {[id=1,INCOMING,toa=129,norm,mt,0,voc,noevp,,cli=1,,1] } [SUB0]
D/GsmCdmaCallTracker( 1109): [GsmCdmaCallTracker] pendingMo=null,
 dc=id=1,INCOMING,toa=129,norm,mt,0,voc,noevp,,cli=1,,1
D/GsmCdmaCallTracker( 1109): [GsmCdmaCallTracker] update phone state, old=IDLE new=RINGING
D/Phone  (1109): Event EVENT_CALL_RING_CONTINUE Received state=RINGING
D/Phone  (1109): Sending notifyIncomingRing
```

我们重点查看日志中三个 INCOMING 的 Call 状态。首先，是 PID 为 604 的进程 RILQ 消息，QMI 接收到了 INCOMING 的 Call 状态；接着，是 PID 为 1109 的 com.android.phone 进程中 RILJ 接收到 UNSOL_RESPONSE_CALL_STATE_CHANGED，即 Call 状态变化的 RIL 上报消息；紧跟着，RILJ 发起 GET_CURRENT_CALLS 查询当前 Call 列表编号 4185 的 RIL 请求，RIL 返回了 id 为 1、状态为 INCOMING 的 Call；然后，将 INCOMING 的 Call 交给 GsmCdmaCallTracker 处理，更改 phone 状态为 RINGING，即响铃状态；最后，Phone 对象发出 INCOMING 的消息通知。

以上是对 Android Radio 日志的分析过程和结果，接下来，我们将查看代码来验证我们的分析结果是否正确。

3.2.2 UNSOL_RESPONSE_CALL_STATE_CHANGED 消息处理

在 Java 代码中查找关键字 "UNSOL_RESPONSE_CALL_STATE_CHANGED"，发现 RIL.java 文件中的 responseToString 方法将 RIL 请求编号转换为字符串。反推此方法，将发现当前类的 unsljLog 方法，继续反推发现 RadioIndication 类有 17 个方法调用 RIL 对象的 unsljLog 方法打印消息，处理此 RIL 消息的代码逻辑详情如下：

```java
public void callStateChanged(int indicationType) {
    mRil.processIndication(indicationType);
    mRil.unsljLog(RIL_UNSOL_RESPONSE_CALL_STATE_CHANGED);
    mRil.mCallStateRegistrants.notifyRegistrants();
}
```

上面的代码逻辑有两个问题：RadioIndication 是一个什么对象？其 callStateChanged 方法是由谁调用的？

第 3 章 深入解析通话流程

```
//又是一个服务对象
public class RadioIndication extends IRadioIndication.Stub
```

RIL.java 文件将对其进行引用，并有一个 mRadioIndication 属性对象，在 RIL 的构造方法中同步创建此对象。

```
private IRadio getRadioProxy(Message result) {
try {
    mRadioProxy = IRadio.getService(HIDL_SERVICE_NAME[mPhoneId == null ? 0 : mPhoneId]);
    if (mRadioProxy != null) {
        mRadioProxy.linkToDeath(mRadioProxyDeathRecipient,
            mRadioProxyCookie.incrementAndGet());
        mRadioProxy.setResponseFunctions(mRadioResponse, mRadioIndication);
    } else {
        riljLoge("getRadioProxy: mRadioProxy == null");
    }
} catch (RemoteException | RuntimeException e) {......}
return mRadioProxy;
}
```

通过 IRadio 服务调用 setResponseFunctions 接口，将设置跨进程的 RadioIndication 对象，此处将开启再一次的跨进程服务调用。通过相关代码的查找，发现是 Native 层的 Android 服务，留待 RIL 中重点解析。

继续查看 mRil.mCallStateRegistrants.notifyRegistrants 调用，mRil 即 RIL Java 对象，其 mCallStateRegistrants 属性以及相关更新的方法在 RIL 的父类 BaseCommands 中定义和实现，详情如下。

```
protected RegistrantList mCallStateRegistrants = new RegistrantList();
```

有关 RegistrantList 的消息处理下一节将做扩展讲解，其中涉及一个非常著名的设计模式 Observer（观察者模式）。

3.2.3 扩展 RegistrantList 消息处理机制

RegistrantList 消息处理机制包括两个重要的 Java 类：RegistrantList.java 和 Registrant.java。其中有一个重要的关键字就是 "registrant"，中文意思为 "登记者"。其实这里采用了 GoF 中非常著名的设计模式——观察者模式，其类图如图 3-11 所示。

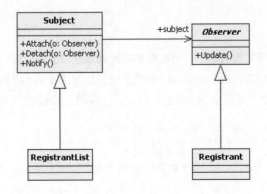

图 3-11 RegistrantList 中使用的观察者模式

什么是观察者模式？不清楚的读者可查阅 GoF 提出的 23 个常用设计模式相关资料。Android 源码中使用大量的设计模式，如单例模式、命令模式、模板方法等，理解这些常用的设计模式，对深入学习 Android 源码将有很大的帮助，同时也能提高设计和编程能力。

RegistrantList 使用的观察者模式中，RegistrantList 为通知者，Registrant 为观察者。RegistrantList 作为通知者，负责对通知者的增加（add/addUnique）、删除（remove），并且能够发出通知（notifyRegistrants）；而 Registrant 作为观察者，由其 internalNotifyRegistrants 方法负责响应通知者发出的 notifyRegistrants 通知。

这里仍然使用 mCallStateRegistrants 的 RegistrantList 对象作为学习实例。在 BaseCommands.java 源代码中，mCallStateRegistrants 作为众多 RegistrantList 类型属性之一，它的处理方法主要有两个：

第一个，观察者对对象列表的管理，对其进行注册和取消，其代码为：

```java
//根据 Handler 对象、消息处理类型、Object 数据对象生成观察者的 Observer 对象，并增加到通知者
//的 RegistrantList 数组列表中
@Override
public void registerForCallStateChanged(Handler h, int what, Object obj) {
    Registrant r = new Registrant (h, what, obj);
    mCallStateRegistrants.add(r);
}

//根据 Handler 对象取消对应的观察者对象 Registrant，将从通知者的 RegistrantList 数组列表中删除对应
//的观察者对象
@Override
public void unregisterForCallStateChanged(Handler h) {
    mCallStateRegistrants.remove(h);
}
```

第二个，通知者发起对所有观察者的通知，其方法为 notifyRegistrants，它会调用内部方法 internalNotifyRegistrants 以轮询的方式对每个观察者发出通知。internalNotifyRegistrants 方法的代码逻辑如下。

```java
ArrayList registrants = new ArrayList(); // registrants 作为 ArrayList，保存了所有注册的观
                                         //察者对象，即 Registrant 对象
//加了同步锁 synchronized
private synchronized void internalNotifyRegistrants(Object result, Throwable exception) {
    int i = 0;
    for(int s = this.registrants.size(); i < s; ++i) {//循环 registrants ArrayList
        Registrant r = (Registrant)this.registrants.get(i); //获取每个 Registrant 对象
        r.internalNotifyRegistrant(result, exception); //向每个 Registrant 对象发出通知
    }
}
```

通知者的处理逻辑已经非常清楚了，那么观察者的处理逻辑又如何呢？通过前面的学习，知道观察者有两个重要的方法：构造方法 Registrant 和响应通知的方法 internalNotifyRegistrant。

Registrant 的构造方法主要是由 Handler、what、Object 三个组合成 Registrant 对象。

```java
public Registrant(Handler h, int what, Object obj) {
    this.refH = new WeakReference(h);//弱引用 Handler
    this.what = what; //Handler 消息类型编号
    this.userObj = obj; //Object 数据对象，用于封装传递的数据
}
```

接着，继续查看 internalNotifyRegistrant 响应通知的方法，这里使用基本的 Handler 消息处理机

制。将 Handler 对象传递给其他线程，并在其他线程中创建消息对象和发起消息通知，最后在 Handler 定义的 handleMessage 方法中接收响应消息，详情如下。

```
void internalNotifyRegistrant (Object result, Throwable exception) {
    Handler h = getHandler();//获取 Handler 对象
    if (h == null) {//异常判断和处理
        clear();
    } else {
        Message msg = Message.obtain();//创建消息对象
        msg.what = what; //设置消息类型编号
        msg.obj = new AsyncResult(userObj, result, exception);//Object 封装数据
        h.sendMessage(msg);//发送 Handler 消息，将同步的消息通知转为异步消息处理
    }
}
```

至此，结合 Handler 消息处理机制，我们应该已经掌握了 RegistrantList 消息处理机制。那么，在前一节的 mCallStateRegistrants.notifyRegistrants 方法调用后，会有哪些 Registrant 对象进行响应呢？你有解决思路吗？具体思路如下步骤所示。

● 查找 RegistrantList 对象注册观察者的方法

在 Android 源码中一般为 registerForXXX 方法，此方法中调用 RegistrantList 对象的 add/addUnique 等来注册 Registrant 对象的方法。

● 通过 registerForXXX 方法反推其调用者

若要查找 registerForXXX 方法的调用方，重点关注其调用的形参，特别是第一个参数，多数情况为 this 或 mHandler，而第二个参数将明确在 Handler 中响应什么消息类型。

● Handler 的 handleMessage 用于响应 RegistrantList 对象发出的消息通知。

3.2.4 GsmCdmaCallTracker 消息处理

通过上一节提供的方法，我们能定位到 mCallStateRegistrants.notifyRegistrants 发出通知后，仅一个地方可响应此消息通知，即 GsmCdmaCallTracker 类的 handleMessage 方法，其代码处理逻辑详情如下：

```
public GsmCdmaCallTracker (GsmCdmaPhone phone) {
    this.mPhone = phone;
    mCi = phone.mCi;
    // GsmCdmaCallTracker 构造方法中注册了 this 和 EVENT_CALL_STATE_CHANGE
    mCi.registerForCallStateChanged(this, EVENT_CALL_STATE_CHANGE, null);
    mCi.registerForOn(this, EVENT_RADIO_AVAILABLE, null);
    mCi.registerForNotAvailable(this, EVENT_RADIO_NOT_AVAILABLE, null);
    ......
}
// handleMessage 响应 EVENT_CALL_STATE_CHANGE 消息
public void handleMessage(Message msg) {
    case EVENT_CALL_STATE_CHANGE:
        pollCallsWhenSafe();//调用 pollCallsWhenSafe，将进入父类 CallTracker 对应的方法
}
//继承父类 CallTracker 的 pollCallsWhenSafe 方法
protected void pollCallsWhenSafe() {
    mNeedsPoll = true;
    if (checkNoOperationsPending()) {
        mLastRelevantPoll = obtainMessage(EVENT_POLL_CALLS_RESULT);
        mCi.getCurrentCalls(mLastRelevantPoll);
    }
}
```

出现一个新的 Handler 消息 EVENT_POLL_CALLS_RESULT，handleMessage 方法中对 EVENT_POLL_CALLS_RESULT 消息类型的处理将调用 handlePollCalls 方法，该方法的代码很多，与来电通知相关的信息是：mPhone.notifyNewRingingConnection(newRinging)的调用。

 还记得 Dialer 应用中的 Phone 对象吗？此处也有一个 Phone 对象，Dialer 应用中的 Phone 对象对应的 Java 程序是 frameworks/base/telecomm/java/android/telecom/Phone.java，而此处 TeleService 应用中的 Phone 对象对应的 Java 程序是 frameworks/opt/telephony/src/java/com/android/internal/telephony/Phone.java，请大家注意区分。

GsmCdmaPhone 类实现了 Phone 抽象对象，查看其 notifyNewRingingConnection 方法，将调用父类的相关方法 super.notifyNewRingingConnection (c)，其核心逻辑如下：

```
AsyncResult ar = new AsyncResult(null, cn, null);
mNewRingingConnectionRegistrants.notifyRegistrants(ar);
```

继续跟踪 mNewRingingConnectionRegistrants 的注册方法，发现 PstnIncomingCallNotifier 类的构造方法中的逻辑会注册相关消息，调用 registerForNotifications 方法完成 Phone 对象 NewRinging 的消息回调，其核心处理逻辑详情如下。

```
mPhone.registerForNewRingingConnection(mHandler, EVENT_NEW_RINGING_CONNECTION, null);
```

Phone 对象发出 NewRinging 通知时，PstnIncomingCallNotifier 的 mHandler 属性对象的 handleMessage 方法将接收和响应 EVENT_NEW_RINGING_CONNECTION 类型的消息，连续调用 handleNewRingingConnection→sendIncomingCallIntent 方法，最后调用 TelecomManager.from(mPhone.getContext()).addNewIncomingCall(handle, extras)。

还记得分析拨号流程时 TelecomManager 对象的 placeCall 方法吗？同样，addNewIncomingCall 也将完成一次跨进程的接口调用。在分析 addNewIncomingCall 接口调用之前，我们首先对接收到来电的流程做一次总结，详情如图 3-12 所示。

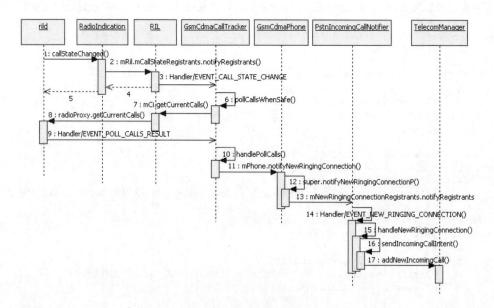

图 3-12　来电流程 1

我们重点关注以下四点。

- RadioIndication 响应 rild 的回调 callStateChanged，对应步骤 1 到步骤 3 的处理逻辑。
- RIL 与 GsmCdmaCallTracker 交互查询 Call List，对应步骤 6 到步骤 10。
- 两次 RegistrantList 消息通知，对应步骤 2 和步骤 13。
- TelecomManager 的 addNewIncomingCall 调用，将进行接收来电消息的第一次跨进程调用。

3.2.5　ITelecomService 处理来电消息

TelecomServiceImpl.addNewIncomingCall 方法将接收到来电消息，其关键处理逻辑如下。

```
Intent intent = new Intent(TelecomManager.ACTION_INCOMING_CALL);
intent.putExtra(TelecomManager.EXTRA_PHONE_ACCOUNT_HANDLE,
        phoneAccountHandle);
intent.putExtra(CallIntentProcessor.KEY_IS_INCOMING_CALL, true);
if (extras != null) {
    extras.setDefusable(true);
    intent.putExtra(TelecomManager.EXTRA_INCOMING_CALL_EXTRAS, extras);
}
mCallIntentProcessorAdapter.processIncomingCallIntent(
        mCallsManager, intent);
```

首先创建 Action 为 TelecomManager.ACTION_INCOMING_CALL 的 intent 对象，并更新相关的属性，接着调用 mCallIntentProcessorAdapter 对象的 processIncomingCallIntent 方法。是不是感觉似曾相识？原来在拨号流程中，此对象调用了 processOutgoingCallIntent 方法来分别调用 CallsManager 的 startOutgoingCall 和 placeOutgoingCall 方法启动通话界面和继续发出拨号请求。

而在来电流程中，processIncomingCallIntent 方法同样会调用 CallsManager 对象的 processIncomingCallIntent 方法，其核心处理逻辑详情如下。

```
callsManager.processIncomingCallIntent(phoneAccountHandle, clientExtras);

void processIncomingCallIntent(PhoneAccountHandle phoneAccountHandle, Bundle extras) {
    Call call = new Call(
            getNextCallId(),
            mContext,
            this,
            mLock,
            ......
            mClockProxy);

    ......//更新 Call Info
    if (!isHandoverAllowed || (call.isSelfManaged() && !isIncomingCallPermitted(call,
            call.getTargetPhoneAccount())))  {//失败处理
        notifyCreateConnectionFailed(phoneAccountHandle, call);
    } else {
        //将创建与 Telephony 的连接，进行再一次跨进程访问
        call.startCreateConnection(mPhoneAccountRegistrar);
    }
}
```

同样是 CallsManager 中的处理逻辑，我们来找一下来电处理逻辑与拨号处理逻辑的异同。

拨号流程中，CallsManager.startOutgoingCall 创建 Call 对象并调用 addCall 去绑定 Dialer 应用的 IInCallService 服务，来展示通话界面；而 processIncomingCallIntent 方法中并没有调用 addCall 来展示来电通知界面。

拨号流程中的CallsManager.placeOutgoingCall方法，通过调用Call对象的startCreateConnection与Telephony方法创建Connection，而来电流程中的processIncomingCallIntent方法，同样调用Call对象的startCreateConnection方法，只不过拨号流程和来电流程使用的Call对象属性有所不同。

Nexus 6P手机收到来电请求，到目前为止我们分析的流程还没有展现来电界面，那么来电流程中，通话界面何时加载和展现呢？回想下拨号请求流程中，Telecom中mAdapter的回调处理逻辑。我们跟踪到Call对象的handleCreateConnectionSuccess时，有个逻辑分支就是用来处理IncomingCall来电信息的，其关键逻辑详情如下。

```
for (Listener l : mListeners) {
    l.onSuccessfulIncomingCall(this);
}
```

查找ListenerBase抽象类的三个子类，发现仅有CallsManager子类重写了父类ListenerBase的onSuccessfulIncomingCall方法，其中最关键的逻辑如下。

```
new IncomingCallFilter(mContext, this, incomingCall, mLock,
        mTimeoutsAdapter, filters).performFiltering();
```

通过名字可知，其就是对来电进行过滤操作。继续分析IncomingCallFilter类的performFiltering方法，发现最关键的信息是Handler启动线程异步处理逻辑中有一个对mListener.onCallFilteringComplete的回调，还有一个Listener消息调用，它是CallFilterResultCallback的接口定义，而CallsManager实现了其接口，再次进入CallsManager对象的onCallFilteringComplete方法，其中的处理逻辑根据Filter的过滤结果将决定调用rejectCallAndLog或addCall。

注意

太多的Listener消息回调、Handler消息发送和接收以及RegistrantList消息注册和发出通知，读者是不是有点晕呢，会将这些消息弄混淆？
（1）我们可以通过打印trace日志来验证分析的流程是否正确。
（2）后面的章节将重点对一些关键的点展开分析，以区分Listener和Handler。

我们的手机在没有设置任何过滤的配置信息时，IncomingCallFilter将通过的检查发起对addCall的调用，将来电的Call对象信息进行第二次跨进程服务调用，发送给Dialer应用的IInCallService服务，服务接收到Call对象后，将展示通话界面为接收到来电界面。

在CallsManager类的onSuccessfulIncomingCall和addCall方法中加入以下打印trace调用的堆栈日志代码，代码和打印的日志详情如下。

```
android.util.Log.e("XXX", "", new RuntimeException());
```

```
CallsManager onSuccessfulIncomingCall
com.android.server.telecom.CallsManager.onSuccessfulIncomingCall(CallsManager.java:447)
com.android.server.telecom.Call.handleCreateConnectionSuccess(Call.java:1525)
com.android.server.telecom.CreateConnectionProcessor.handleCreateConnectionSuccess
(CreateConnectionProcessor.java:398)
com.android.server.telecom.ConnectionServiceWrapper.handleCreateConnectionComplete
(ConnectionServiceWrapper.java:1362)
com.android.server.telecom.ConnectionServiceWrapper.-wrap0(Unknown Source:0)
com.android.server.telecom.ConnectionServiceWrapper$Adapter.handleCreateConnection
Complete(ConnectionServiceWrapper.java:78)
com.android.internal.telecom.IConnectionServiceAdapter$Stub.onTransact(Iconnection
ServiceAdapter.java:78)
android.os.Binder.execTransact(Binder.java:697)
```

```
addCall
com.android.server.telecom.CallsManager.addCall(CallsManager.java:2135)
com.android.server.telecom.CallsManager.onCallFilteringComplete(CallsManager.java:491)
com.android.server.telecom.callfiltering.IncomingCallFilter$2.loggedRun(IncomingCall
Filter.java:99)
```

通过来电过程中打印出来的 Telecom 应用中调用的堆栈信息，可以得出和之前的分析完全匹配的结论。请读者阅读和跟踪对应的调用堆栈信息，以理解来电相关流程，并解惑心中不确定的流程。

Telecom 应用中响应 addNewIncomingCall 服务接口调用，对接收到来电的业务处理逻辑进行回顾和总结，如图 3-13 所示。

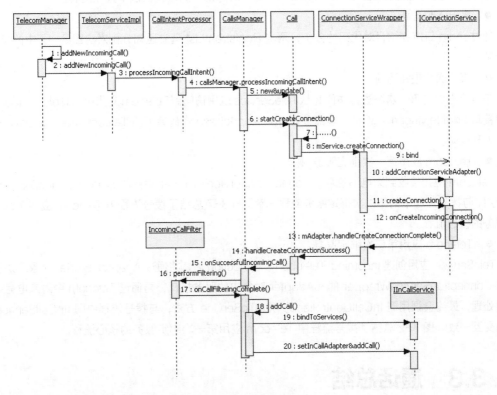

图 3-13　来电流程 2

我们重点关注以下四点。

- 第一次跨进程服务接口调用

TelecomManager 的 addNewIncomingCall 接口调用对应图 3-13 中的步骤 2，将调用 Telecom 应用中的 ITelecomService 服务接口，完成从 TeleService 应用到 Telecom 应用的跨进程服务接口调用。

- Telecom 应用第一次绑定服务

Telecom 应用中 Call 对象的 createConnection 调用，将绑定 TeleService 应用中的 IConnectionService 服务，与拨号流程中 Telecom 应用的第二次绑定服务处理逻辑相同，唯一不同的是 Call 对象的属性有所区别。

- Telecom 应用第二次绑定服务

通过 TeleService 对 Adapter 接口的回调，再经过 IncomingCallFilter 的来电过滤，最终绑定 Dialer

应用中的 IInCallService 服务，与拨号流程中 Telecom 应用的第一次绑定服务处理逻辑相同，只是展示通话界面为来电状态。

- IncomingCallFilter

作为 Telecom 应用独立的业务，下一章将重点解析其实现机制。

3.2.6 来电流程总结

来电流程总结如下：首先是 RIL 接收到消息，再分别传递给 TeleService、Telecom 和 Dialer 三个系统应用。我们重点掌握图 3-12 和图 3-13 两个时序图中的流程及以下几点内容。

- 来电入口分析

通过 radio 日志并结合对代码的分析，两次 RegistrantList 消息通知分别对应图 3-12 中的步骤 2 和步骤 13。

- 第一次跨进程访问

来电流程中的第一次跨进程访问将从 TeleService 应用访问到 Telecom 应用中的 ITelecomService 服务接口 addNewIncomingCall，重点参考 framework/base 代码库下的 TelecomManager.java 代码中的处理逻辑。

- Telecom 应用第一次绑定服务

Telecom 应用接收到来电消息后，ConnectionServiceWrapper 对象的 createConnection 方法将被调用，与拨号流程中的第二次绑定服务流程一致。图 3-7 总结了拨号流程中 Telecom 应用第二次绑定服务的核心流程。

- Telecom 应用第二次绑定服务

TeleService 应用创建 Incoming TelephonyConnection 后，将通过 Adapter Binder 对象跨进程地调用 ConnectionServiceWrapper 的 mAdapter 对象中的服务接口，再通过 Telecom 中对来电号码的过滤处理，最终会调用到 InCallController 的 bindToService 方法，与拨号流程中同 IInCallService 服务的交互一致。图 3-2 总结了拨号流程中 Telecom 应用第一次绑定服务的核心流程。

3.3 通话总结

通过前面两个小节对主动拨号和被动接收来电流程的分析，发现通话流程的实现非常漫长和复杂，包括各种服务绑定、服务接口调用、Listener 消息回调、Handler 消息发送和处理。本节首先汇总一下通话的相关代码信息，然后再对一些与通话相关的核心流程进行汇总和总结，比如通话控制命令的下发、底层通话状态改变的消息上报等。

3.3.1 通话关键代码汇总

对主动拨号和被动来电流程的分析过程中，共涉及五个代码库，分别是：

- packages/apps/Dialer
- packages/services/Telecomm
- packages/services/Telephony

- frameworks/base/telecomm
- frameworks/opt/telephony

具体代码库的详情如表 3-1 和表 3-2 所示。

表 3-1 通话关键代码

代码库	Android.mk 核心配置	备注
packages/apps/Dialer	LOCAL_PACKAGE_NAME := Dialer include $(BUILD_PACKAGE)	编译出 Dialer.apk 应用
packages/services/Telecomm	LOCAL_JAVA_LIBRARIES := telephony-common LOCAL_PACKAGE_NAME := Telecom LOCAL_CERTIFICATE := platform $(BUILD_PACKAGE)	编译出 Telecom.apk 应用,并使用平台签名
packages/services/Telephony	LOCAL_JAVA_LIBRARIES := telephony-common LOCAL_PACKAGE_NAME := TeleService LOCAL_CERTIFICATE := platform $(BUILD_PACKAGE)	编译出 TeleService.apk 应用,并使用平台签名
frameworks/base/telecomm		编译出 framework.jar
frameworks/opt/telephony	LOCAL_MODULE := telephony-common include $(BUILD_JAVA_LIBRARY)	编译出 telephony-common.jar

表 3-2 通话相关代码名称统一约定及运行进程汇总

代码库	名称	进程
packages/apps/Dialer	Dialer	com.android.dialer
packages/services/Telecomm	Telecom	system_server
packages/services/Telephony	TeleService	com.android.phone
frameworks/base/telecomm	framework	
frameworks/opt/telephony	Telephony	system_server&com.android.phone

表 3-1 和表 3-2 中涉及的代码,有以下几个重要内容需要我们继续分析。

1. Telecom 应用运行空间

Telecom 应用为什么运行在 system_server 进程空间？查看 Telecom 应用的配置文件 AndroidManifest.xml,可以发现以下关键信息。

```
android:sharedUserId="android.uid.system"//指定该应用将使用system系统用户

<application android:label="@string/telecommAppLabel"
        android:icon="@mipmap/ic_launcher_phone"
        android:allowBackup="false"
        android:supportsRtl="true"
        android:process="system"//指定该应用将在system_server进程运行
        android:usesCleartextTraffic="false"
        android:defaultToDeviceProtectedStorage="true"
        android:directBootAware="true">
```

> 这样 Telecom 应用就可以使用系统用户和运行在系统进程里了吗？其实还需要一个非常关键的信息，那就是系统签名 LOCAL_CERTIFICATE := platform。没有此签名，系统将认为此应用是非法的，会列入黑名单不予启动，这也是 Android 系统的安全机制之一。

2. TeleService 应用运行空间

查看 TeleService 应用的配置文件 AndroidManifest.xml，可以发现以下关键信息。

```
<manifest xmlns:android="http://schemas.android.com/apk/res/android"
          xmlns:androidprv="http://schemas.android.com/apk/prv/res/android"
          package="com.android.phone"//进程名为默认的 com.android.phone
          coreApp="true"
          android:sharedUserId="android.uid.phone"//使用 radio 用户
          android:sharedUserLabel="@string/phoneAppLabel"
```

> 这样 TeleService 应用就可以使用系统用户和运行在系统进程里了吗？其实还需要一个非常关键的信息，那就是系统签名 LOCAL_CERTIFICATE := platform，没有此签名，系统将认为该应用是非法的，不予启动，这也是 Android 系统的安全机制之一。

3. Telephony jar 包的依赖关系

Telecom 应用和 TeleService 应用在 Android.mk 文件中明确了对 telephony-common.jar 的依赖关系，因此在这两个应用中可以方便地创建和使用 Telephony 中的类和对象。

3.3.2 通话状态更新消息上报流程

有了拨号流程和来电流程分析和跟踪的经验，就不再跟踪和分析通话状态更新消息的上报对应的 Android 源码了。

拨号成功后，对方接听了此路通话，那么通话界面将更新当前通话为通话中的状态，并开始通话计时，可以理解为 Modem→RIL→TeleService→Telecom→Dialer，一层一层上报通话状态为"通话中"的消息处理和发送过程，总结其关键流程如图 3-14 所示。

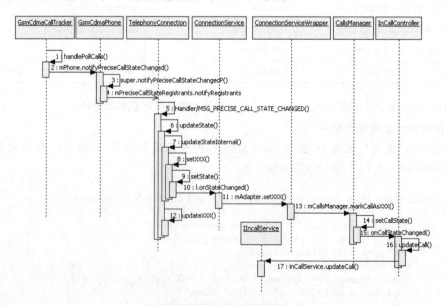

图 3-14 通话状态更新消息上报流程

我们需要重点关注以下五点内容。

- 三个应用的 Call 信息传递

TeleService 应用首先接收到通话状态更新的消息，通过 Telecom 的 Adapter 服务设置不同的通话状态（步骤 11）；接着 Telecom 应用更新 Call 状态（步骤 13 和步骤 14）；最后 Telecom 调用 IInCallService 的 updateCall 接口更新 Call 状态（步骤 17）。

- RegistrantList 消息处理

步骤 2 和步骤 3：在 GsmCdmaPhone 对象发出 RegistrantList 消息通知后，在 TelephonyConnection 对象的 mHandler 匿名内部类对象的 handleMessage 中响应 MSG_PRECISE_CALL_STATE_CHANGED 类型的 Handler 消息，该 Handler 消息的注册入口在 TelephonyConnection 抽象类的两个子类 GsmConnection 和 CdmaConnection 的 setOriginalConnection 方法中实现注册 MSG_PRECISE_CALL_STATE_CHANGED 类型的 Registrant，并在 GsmConnection 的构造方法中调用 setOriginalConnection 接口进行消息注册的初始化操作；而 TelephonyConnection 对象在 TelephonyConnectionService 类的 onCreateOutgoingConnection 和 onCreateIncomingConnection 方法中创建。

- TelephonyConnection 对象的 Listener 注册

步骤 10：通过 Listener 对象的 onStateChanged 进行消息回调，那么 Listener 对象是什么？又是在什么地方注册的？是在 ConnectionService 的 createConnection 方法中，首先创建 TelephonyConnection 对象，然后调用 addConnection 方法，设置当前类的私有内部类 Connection.Listener 对象 mConnectionListener 为 TelephonyConnection 对象的 Listener。

- IConnectionServiceAdapter 接口汇总

IConnectionServiceAdapter 的 Stub 接口实现在 Telecom 应用 ConnectionServiceWrapper 类的私有内部类 Adapter 中，它主要由 setActive、setRinging、setDialing、setAudioRoute 等设置 Call 相关状态信息的接口，以及 onConnectionEvent、onRttInitiationSuccess、onRemoteRttRequest 等消息通知接口构成。

- IInCallService 接口汇总

IInCallService 的 Stub 接口实现在 framework/base 下的 InCallService 抽象类的私有内部类 InCallServiceBinder 中，它主要由 setInCallAdapter、addCall、updateCall 等增加和更新 Call 对象相关的接口，以及 onConnectionEvent、onCallAudioStateChanged 等消息通知接口构成。而在 Dialer 应用中，InCallServiceImpl 继承了抽象类 InCallService。

3.3.3 控制通话消息下发流程

我们在通话界面若想更改当前通话状态，比如挂断/接听当前接收来电，挂断/保持当前通话等操作，可以理解为是控制通话消息下发的过程，从 Dialer→Telecom→TeleService→RIL→Modem，通话控制消息一层一层的下发，最终交给 Modem 处理具体的通话控制。通话状态更改完成后，再将通话状态变更的消息进行一层一层的上报，最后交给 Dialer 应用对通话界面进行更新和显示。

控制通话消息下发流程与通话状态更新消息上报流程正好是两个相反的信息传递和交互过程，这两个流程的组合就完成了通话所需的流程。

控制通话消息下发流程，如图 3-15 所示。图中的流程以将当前接通的电话更改为"通话保持"状态为实例，我们需要重点关注以下三点。

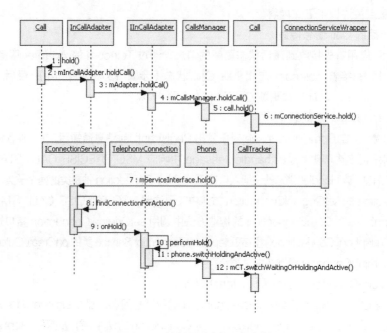

图 3-15　控制通话消息下发流程

- 三个应用的控制消息传递

Dialer 应用展示的通话界面或来电界面均有控制通话状态请求的界面控件，通过滑动或是点击相关的控件，将触发通话状态控制，调用 android.telecom.Call 对象的 hold 方法，如步骤 1 所示；在步骤 3 的 Dialer 应用中，调用 IInCallAdapter 的 holdCall 服务接口完成第一次的跨进程服务接口调用，进入 Telecom 应用。Telecom 应用首先更新 Call 状态，此处的 Call 对象为 Telecom 应用内部定义的类 com.android.server.telecom.Call，步骤 7 调用 IConnectionService 的 hold 服务接口，完成第二次的跨进程服务接口调用，进入到 TeleService 应用。TeleService 应用中经过层层方法调用，由 CallTracker 对象进行 Call 的 hold 操作，并将请求发给 RIL 对象，发出对应的 RIL 请求。

- IInCallAdapter 接口汇总

IInCallAdapter 的 Stub 接口实现在 Telecom 应用的 InCallAdapter 类中，它主要由 answerCall、rejectCall、playDtmfTone、mergeConference 等接口构成；InCallAdapter 对象则在 InCallController 对象绑定 IInCallService 成功后创建。

- IConnectionService 接口汇总

IConnectionService 的 Stub 接口实现在 frameworks/base 下的 ConnectionService 抽象类的匿名内部类中，mBinder 为其对象，它主要由 addConnectionServiceAdapter、createConnection 等创建 TelephonyConnection 接口，以及 answer、reject、hold、playDtmfTone 等控制通话状态的接口构成。而在 TeleService 应用中，TelephonyConnectionService 继承了抽象类 TelephonyConnection。

3.4　建立 Android 通话模型

通过前面三个小节的学习和探索，我们已经知道 Android Telephony 中的 Dialer、Telecom 和 TeleService 三个系统应用主要负责的业务及运行进程空间，并详细解析和总结了主动拨号流程和被动接收来电流程，另外，总结了通话状态更新消息上报和控制通话状态消息下发两个流程。

结合这四个通话的核心流程，下面将对 Android Telephony 的通话功能进行抽象，通过抽象出模型，可以站在更高的层次去认识和理解 Android Telephony 通话功能，如图 3-16 所示。

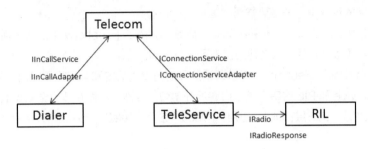

图 3-16 Android 通话模型

我们需要重点掌握以下几点。
- 系统的分层

Dialer、Telecom 和 TeleService 三大应用可理解为：Dialer 应用是普通的 Android App 应用，其运行进程的用户信息和进程信息，也能说明此问题；Telecom 应用运行在 system_server 进程上，其进程用户名为 system 系统用户，说明它是运行在 Android Framework 框架层；TeleService 应用运行的进程名是 com.android.phone，用户名是 radio，承载着 Telephony Call 协议栈，同样可以认为它运行在 Android Framework 框架层；最后是 RIL，它运行在 HAL（硬件抽象层）。

- 交互方式

Dialer、Telecom、TeleService 和 RIL 都是通过服务进行交互的。在图 3-16 中，它们之间有箭头连接的都是通过 Service 跨进程的接口调用实现的。Dialer 与 TeleService 之间没有直接的消息传递，要通过 Telecom 进行消息中转，Telecom 与 RIL 之间同样没有直接的消息传递，要通过 TeleService 进行消息中转。

通过服务进行跨进程接口调用实现消息的传递，服务接口调用本身就是同步的接口调用，在 Service 端的实现将转换为异步的方式处理，待消息处理完成后，再使用回调的接口传递消息处理的结果。

- 分解通话相关流程

根据消息的传递方向，可分成两大类。

第一类控制通话消息下发流程：应用层通过框架层向 RIL 发起通话管理和控制相关 RIL 请求，RIL 转换成对应的消息发送给 Modem 执行，其中包括拨号、接听电话、拒接电话、保持、恢复通话等；

第二类为通话状态更新消息上报流程：RIL 接收到 Modem 的通话状态变化通知，通过框架层向应用层发起通话状态变化通知，包括来电、电话接通、进入通话中等。

本 章 小 结

本章根据通话的四个主要流程并结合代码进行了详细的解析和总结：
- 主动拨号流程

- 被动接收来电流程
- 本地主动控制通话状态流程
- 通话状态变更消息上报流程

这些通话相关流程之间有什么关系或者规律呢？通过建立 Android 通话模型可以回答这个问题，在学习 Telephony 的过程中，一定要掌握和理解这个模型。

另外，本章总结出的这四个流程并不能完全覆盖 Android Telephony 中的所有通话功能，如三方通话、会议电话、紧急呼救等，感兴趣的读者可自行学习。

要深入认识 Android Telephony，仅从通话流程的角度去研究和学习是不够的，还需要对 Dialer、Telecom、TeleService 和 RIL 这四个核心内容展开学习。当然，后面章节也会陆续展开，分析和讲解 Android 通话模块中涉及的三层：Dialer 应用、Telecom 应用、TeleService 应用，Telephony Frameworks 层和 RIL。

第 4 章

详解 Telecom

学习目标

- 掌握 Telecom 加载入口及分析方法。
- 总结和演进 Telecom 交互模型。
- 掌握 Listener 消息回调机制。
- 扩展学习 CallsManager。

上一章我们对拨号流程和来电流程做了详细的解析和总结，并抽象出 Android Telephony 通话模型，让我们能站在更高地层次全面地认识和理解通话的架构和业务实现。本章开始将对通话中重要的业务模块进行解析和学习。为什么首先选择 Telecom 呢？因为在 Android Telephony 通话模型中它作为 Dialer 和 TeleService 的消息中转站，消息的处理任务非常繁重和复杂，并且它作为拨号和来电消息处理的关键入口，承上启下，因此重要性不言而喻。

- 呈上

相对 Dialer 应用而言，发送 Call 状态变化消息给 IInCallService 服务，并接收 Dialer 应用发出的 Call 状态的控制消息。

- 启下

相对 TeleService 应用而言，Telecom 应用继续传递接收到 Dialer 应用发出的 Call 状态的控制消息给 IConnectionService 服务，并接收 TeleService 应用发出的 Call 状态变化消息传递给 Dialer 应用。

4.1 Telecom 应用加载入口

同样，在解析 Telecom 应用之前，首先需要确定 Telecom 应用的加载入口和加载时机。查看 Telecom 应用对应的 AndroidManifest.xml 应用的配置文件，看看是否能找到一些蛛丝马迹？但并未发现应用定义的 application 程序入口。

下一步我们该从哪里入手呢？还记得上一章对拨号流程和来电流程的解析和总结过程中，业务流程进入 Telecom 应用的入口吗？这个入口就是 TelecomManager 类的 placeCall 和 addNewIncomingCall

方法。因此，我们首先解析 TelecomManager 类的关键实现机制，作为突破 Telecom 应用的加载入口。

4.1.1 TelecomManager 类核心逻辑分析

上一章分析拨号流程的过程中，在 TelecomManager 类获取 ITelecomService 服务的过程中，Context.TELECOM_SERVICE 系统服务名 "telecom" 与服务定义的 Action：android.telecom.ITelecomService 并没有对应和关联起来。本节将 TelecomManager 对象的获取和 ITelecomService 服务的加载过程这两个方面作为 Telecom 应用的解析入手点。

在拨号流程和来电流程中，最先获取 TelecomManager 对象，在 TelecomManager 类的 from 方法中定义了获取 TelecomManager 对象的方法，代码详情如下：

```java
public static TelecomManager from(Context context) {
    return (TelecomManager) context.getSystemService(Context.TELECOM_SERVICE);
}
```

在 Android 源码的其他地方获取 TelecomManager 对象，是将其类中 from 方法中的逻辑拷贝出来直接使用，如 TelecomUtil 类中的 getTelecomManager 方法。

TelecomManager 是一个普通的 Java 类，怎么能与 ITelecomService 服务关联起来呢？我们发现当前类中的 getTelecomService 方法可以获取 Service 服务的 Binder 对象，代码逻辑详情如下：

```java
private ITelecomService getTelecomService() {
    if (mTelecomServiceOverride != null) {
        return mTelecomServiceOverride;
    }
    return ITelecomService.Stub.asInterface(ServiceManager.getService(
                Context.TELECOM_SERVICE));
}
```

关键字 "Context.TELECOM_SERVICE" 在 TelecomManager 类中第二次出现了，通过其对应的 Java 文件，可以发现以下详情：

```java
frameworks/base/core/java/android/app/SystemServiceRegistry.java
static {
registerService(Context.TELECOM_SERVICE, TelecomManager.class,
        new CachedServiceFetcher<TelecomManager>() {
    @Override
    public TelecomManager createService(ContextImpl ctx) {
        //创建并保存 TelecomManager 对象
        return new TelecomManager(ctx.getOuterContext());
}});
}

frameworks/base/services/core/java/com/android/server/telecom/TelecomLoaderService.java
//service 为 ITelecomService 服务的 Binder 对象
ServiceManager.addService(Context.TELECOM_SERVICE, service);
```

TelecomManager 类中有两个非常关键的方法 from 和 getTelecomService，都用到了 Context.TELECOM_SERVICE 字符串常量 "telecom"，需要大家仔细区分。

1．对象的创建和获取

SystemServiceRegistry 类中的构造方法是私有的，也就是说没有 SystemServiceRegistry 对象的存在，在它的静态代码块中将创建和保存 TelecomManager 对象；此对象的创建时机在 system_server 进程的启动过程中。

TelecomManager 类中的 from 方法，将使用 Context 对象的 getSystemService 获取 SystemServiceRegistry 中保存的 TelecomManager 普通 Java 对象。

2. ITelecomService 系统服务

继续分析 TelecomLoaderService 类中的逻辑，并加入 trace 日志的打印，详情如下：

```
com.android.server.telecom.TelecomLoaderService.connectToTelecom(TelecomLoaderService.
java:177)
com.android.server.telecom.TelecomLoaderService.onBootPhase(TelecomLoaderService.java:172)
com.android.server.SystemServiceManager.startBootPhase(SystemServiceManager.java:153)
……//省略部分匿名类的调用过程
com.android.server.am.ActivityManagerService.systemReady(ActivityManagerService.java:14252)
com.android.server.SystemServer.startOtherServices(SystemServer.java:1672)
com.android.server.SystemServer.run(SystemServer.java:391)
com.android.server.SystemServer.main(SystemServer.java:267)
java.lang.reflect.Method.invoke(Native Method)
com.android.internal.os.RuntimeInit$MethodAndArgsCaller.run(RuntimeInit.java:438)
com.android.internal.os.ZygoteInit.main(ZygoteInit.java:787)
```

原来，在 Android 系统启动过程中，SystemServer 加载时将启动 ITelecomService 系统服务，而 TelecomLoaderService 类中的 connectToTelecom 方法调用，将以绑定服务的方式绑定 ITelecomService 服务，在绑定服务成功的回调接口中，会将 Binder 服务对象添加到 ServiceManager 中，其服务名为 "telecom"，关键代码逻辑详情如下：

```
//ITelecomService 服务信息与 Telecom 应用的配置信息匹配
private static final String SERVICE_ACTION = "com.android.ITelecomService";
private class TelecomServiceConnection implements ServiceConnection {
    @Override
    public void onServiceConnected(ComponentName name, IBinder service) {
        //添加到系统服务中，其服务名为 "telecom"
        ServiceManager.addService(Context.TELECOM_SERVICE, service);
    }
    ……
}
//绑定 Telecom 应用中的 ITelecomService 服务
private void connectToTelecom() {
    TelecomServiceConnection serviceConnection = new TelecomServiceConnection();
    Intent intent = new Intent(SERVICE_ACTION);
    intent.setComponent(SERVICE_COMPONENT);
    int flags = Context.BIND_IMPORTANT | Context.BIND_FOREGROUND_SERVICE
            | Context.BIND_AUTO_CREATE;
    if (mContext.bindServiceAsUser(intent, serviceConnection, flags, UserHandle.SYSTEM)) {
        mServiceConnection = serviceConnection;
    }
}
```

TelecomManager 对象和 telecom 系统服务均是在手机启动过程中，SystemServer 加载时同步创建的。我们需要在这里做一下区分：

- TelecomManager 类的 from 方法通过 Context 获取的是 TelecomManager 对象。
- TelecomManager 类的 getTelecomService 方法通过 ServiceManager 获取的是 ITelecomService 服务对象。

> Context.TELECOM_SERVICE 为 "telecom" 的系统服务与 Action 为 "android.telecom.ITelecomService" 的 Telecom 应用中的服务，已经成功匹配。
> com.android.server.telecom.components.TelecomService 类作为 Telecom 应用的加载入口，在 SystemServer 系统启动过程中加载。

4.1.2 Telecom 应用代码汇总

在讲解 ITelecomService 服务的加载过程之前，我们先来看看 Telecom 应用的代码相关信息，其核心代码详情如图 4-1 所示。

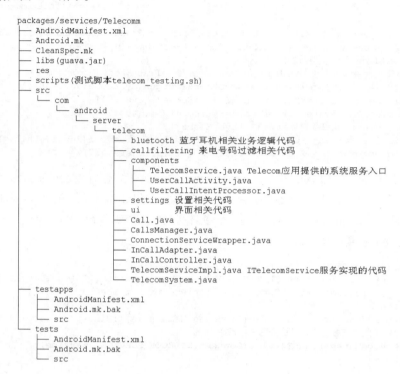

图 4-1 Telecom 应用关键代码结构

我们需重点关注以下几点。

- 代码库

再次明确代码库 packages/services/Telecomm，注意 Telecomm 有两个 m 字符，而其编译出的应用文件名为 Telecom.apk。

- 系统签名

packages/services/Telecomm 代码库根据 Android.mk 编译脚本，将编译出 Telecom.apk Android 应用文件，并使用平台签名，可以保障获取到 system 用户权限并运行在 system_server 系统进程空间。

- Java 程序包名

Telecom 应用统一使用了 com.android.server.telecom 包名，此包名下包括了上一章解析通话流程时比较重要的类，如 Call、CallsManager、ConnectionServiceWrapper 和 TelecomServiceImpl 等。

另外，此包名下还有五个子包名：bluetooth、callfiltering、components、settings 和 ui。

- Test 工程

有两个测试工程 testapps 和 tests，它们都有对应的 Android.mk 和 AndroidManifest.xml，为了更加方便地编译和调试，已经将 Android.mk 文件改成 Android.mk.bak 文件，让我们单独编译当前模块时，不必再编译测试相关的工程。

4.1.3　ITelecomService 的 onBind 过程

通过 AndroidManifest.xml 应用配置文件中对 android.telecom.ITelecomService 服务的配置，可以找到 com.android.server.telecom.components.TelecomService 类，它就是 Telecom 应用的加载入口。根据上一小节的分析，此服务将在 SystemServer 系统启动过程中被加载。

TelecomService 类继承于系统的抽象 Service 类并实现了 onBind 方法，相关的代码逻辑详情如下：

```java
public class TelecomService extends Service implements TelecomSystem.Component {
    @Override
    public IBinder onBind(Intent intent) {
        initializeTelecomSystem(this);
        synchronized (getTelecomSystem().getLock()) {
            return getTelecomSystem().getTelecomServiceImpl().getBinder();
        }
    }

    static void initializeTelecomSystem(Context context) {
        if (TelecomSystem.getInstance() == null) {
            ......
            TelecomSystem.setInstance(//设置 TelecomSystem 实例对象
                new TelecomSystem(
                    context,
                    ......//省略匿名类的对象创建
                );
        }
    }
    @Override
    public TelecomSystem getTelecomSystem() {
        return TelecomSystem.getInstance();//获取 TelecomSystem 的单例对象
    }
}
```

TelecomService 类 onBind 方法的逻辑比较简单，有两个非常重要的内容。
（1）创建 TelecomSystem 对象，初始化 Telecom 应用中的核心对象。
（2）获取 Binder 对象并返回。
总结 Telecom 应用的加载流程，详情如图 4-2 所示。
我们重点关注以下几点。
● 创建匿名对象
通过匿名类创建对应的匿名对象，如 MissedCallNotifierImpl.MissedCallNotifierImplFactory、CallerInfoAsyncQueryFactory 和 ClockProxy 等对象，对应步骤 3。
● TelecomSystem 对象的创建过程
从步骤 5 到步骤 14 为 TelecomSystem 对象的创建过程，其中非常关键的是 CallsManager 和 TelecomServiceImpl 对象的创建。
CallsManager 对象创建时将同步创建 CallsManagerListener 对象，并注册 Listener 消息通知。本章将重点解析和区分 Telecom 应用中的核心 Listener 消息处理机制。
TelecomServiceImpl 对象创建的同时，也将同步创建 ITelecomService.Stub 的匿名类对象。
步骤 4、步骤 8、步骤 11、步骤 13 反映出 TelecomSystem 对象的创建过程是同步的 Java 方法调用，并未加入异步的处理机制。

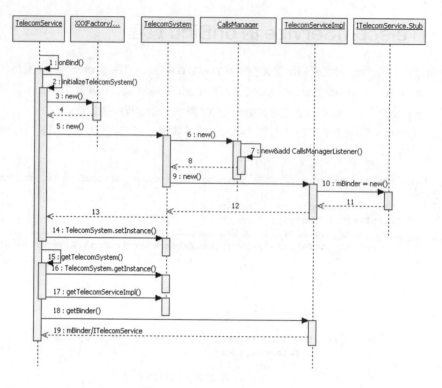

图 4-2　Telecom 应用加载流程

- Binder 对象的获取

从步骤 15 到步骤 19，TelecomService 类获取的 Binder 对象是 TelecomServiceImpl 对象的内部 ITelecomService.Stub 匿名对象，它们之间没有继承关系，千万不要被类名给迷惑了。先获取单例的 TelecomSystem 对象，再获取 TelecomServiceImpl 对象，最终获取到 ITelecomService.Stub 类型的 Binder 对象。

上面是对 TelecomSystem 初始化的流程讲解，那么这些核心类之间的关系究竟如何？总结如图 4-3 所示。

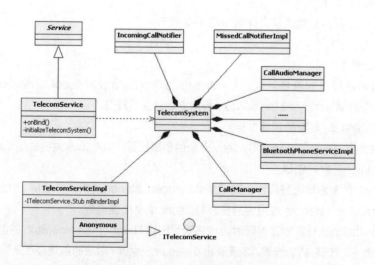

图 4-3　Telecom 应用加载核心类

我们重点关注以下几点。

- Telecom 加载入口

Telecom 应用的加载入口是 TelecomService 类的 onBind 方法，它是一个 Service 类型，并且在 AndroidManifest.xml 中明确定义了它是一个服务。

- TelecomService 和 TelecomServiceImpl 的关系

TelecomService 服务通过 TelecomServiceImpl 的 mBinderImpl 属性对象，承载了 ITelecomService 服务，它们之间本没有继承关系，通过 TelecomSystem 才有了一定的依赖关系。

- TelecomSystem 的组合关系

本例中使用组合来表示它们之间的强依赖关系，请读者注意区分聚合和组合。

4.1.4　第二个拨号入口

通过观察 Telecom 应用的 AndroidManifest.xml 配置文件，发现其中还有一个重要的 Activity 配置信息，详情如下。

```
<activity android:name=".components.UserCallActivity"
        android:label="@string/userCallActivityLabel"
        android:theme="@style/Theme.Telecomm.Transparent"
        android:permission="android.permission.CALL_PHONE"
        android:excludeFromRecents="true"
        android:process=":ui">
  <intent-filter>
        <action android:name="android.intent.action.CALL" />
        <category android:name="android.intent.category.DEFAULT" />
        <data android:scheme="tel" />
  </intent-filter>
  ......//省略 intent-filter
</activity>
```

".components.UserCallActivity·"的 Activity 定义，该配置文件中有 PrivilegedCallActivity 和 EmergencyCallActivity 两个别名 Activity 定义。

老的 Android 版本中，Telecom 应用通过此 Activity 接收拨号 Intent 请求。上一章解析 Dialer 应用的拨号流程中，有一个分支就是使用 context.startActivity 发出拨号请求的 Intent 对象，此 Activity 响应 onCreate 系统调用，从而接收到拨号请求的 intent 对象。

注意　Intent.ACTION_CALL 与 Telecom 应用的 AndroidManifest.xml 配置文件，对 UserCallActivity 的 Action："android.intent.action.CALL"已经成功匹配。

4.2　Telecom 交互模型

上一章解析拨号流程和来电流程的过程中，跟踪到 frameworks/base/telecomm 代码库中的部分代码，最典型的就是 TelecomManager Java 类、跨进程服务调用的 aidl 接口定义文件，以及 InCallService 和 ConnectionService 等 Java 类的定义。

4.2.1 汇总 frameworks/base/telecomm 代码

在解析 Telecom 应用的交互模型之前，我们首先查看 frameworks/base/telecomm 代码库中的代码信息，详情如图 4-4 所示。

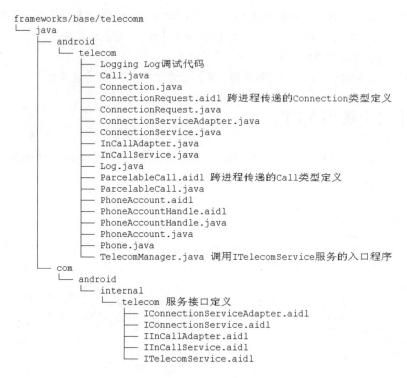

图 4-4　frameworks/base/telecomm 代码结构

frameworks/base/telecomm 代码库主要由两个包构成：android.telecom 和 com.android.internal.telecom，它们没有独立的 Android.mk 编译脚本，将被编译进 framework.jar 中。

通话流程中涉及的所有服务接口定义文件，全部在 com.android.internal.telecom 包下。图 4-4 中列出我们即将接触到的五个 aidl 接口定义文件。

在 android.telecom 包下，有我们非常熟悉的 TelecomManager.java 文件，是调用 ITelecomService 服务的入口程序；并且此包下定义了跨进程接口调用传递的对象，比如 ParcelableCall、ConnectionRequest 和 PhoneAccountHandle 等 aidl 文件，以及相应支持序列化和反序列化的 Java 对象文件；最后还包括了一些 Call、Phone、InCallService、InCallAdapter、ConnectionService、ConnectionServiceAdapter 等服务绑定过程中的关键代码。

为什么将 InCallService 和 ConnectionService 等 Java 程序放在 frameworks/base/telecomm 中，这里涉及一个非常有用的设计模式——模板方法。也就是在 Android Framework 中定义模板，在具体的应用实现类中重写模板的方法。

 关注 com.android.internal.telecom 包名中的 internal，此包下由 aidl 文件编译出的 Java 接口定义程序，并不提供到 Android SDK 中。

4.2.2 绑定 IInCallService 机制

在来电流程和拨号流程中，Telecom 应用有两次绑定服务操作，绑定 IInCallService 的过程将与 Dialer 应用中的服务交互，最终展示和更新通话界面，其中关键的三个步骤如下。
- 绑定服务
- setInCallAdapter
- addCall

上一章中我们已经对 Telecom 应用绑定 IInCallService 的流程做了详细的总结，如图 3-2 InCallController bindToService 所示。本节将从代码的角度进行横向的扩展学习和总结，详情如图 4-5 所示。

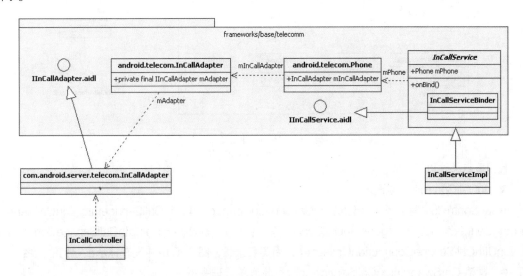

图 4-5　Telecom 应用绑定 IInCallService 核心类图

我们重点关注以下四点。
- frameworks/base/telecomm 包下的代码

IInCallService.aidl 和 IInCallAdapter.aidl 接口定义文件，以及我们非常关注的 InCallService、android.telecom.Phone 和 android.telecom.InCallAdapter 等类。
- IInCallService.aidl 和 IInCallAdapter.aidl 接口实现

com.android.server.telecom.InCallAdapter 实现了 IInCallAdapter.aidl 接口；android.telecom.InCallService 抽象类的私有内部类 InCallServiceBinder 继承实现了 IInCallService.aidl 接口。
- 两个 InCallAdapter

android.telecom.InCallAdapter 在 frameworks/base/telecomm 包下定义，它是一个普通的 Java 类，代理了 IInCallAdapter mAdapter 对象的所有操作；com.android.server.telecom.InCallAdapter 正是 IInCallAdapter 服务接口的实现。图 4-5 中它们之间的依赖关系可以理解为 android.telecom.InCallAdapter 对象通过 IInCallAdapter mAdapter 对象发起跨进程的服务接口调用。
- 区分运行空间

图 4-5 中列出的类将运行在两个进程空间：Dialer 和 Telecom 应用的进程空间。

左下角的两个类 com.android.server.telecom.InCallAdapter 和 InCallController 对象运行在 Telecom

应用进程空间。Telecom 应用提供 IInCallAdapter 服务，InCallController 对象绑定 IInCallService 时保存了 IInCallService Binder 对象，从而使用此对象可以调用 addCall/updateCall 等接口，跨进程访问 Dialer 应用提供的 IInCallService 服务；图 4-5 中除左下角 com.android.server.telecom.InCallAdapter 和 InCallController 两个类，其他类都运行在 Dialer 应用的进程空间。

Telecom 应用主动发起的绑定 IInCallService 服务，其入口是 InCallController 类的 onCallAdded 方法；在绑定 IInCallService 的过程中涉及的几个关键类，如图 4-6 所示。

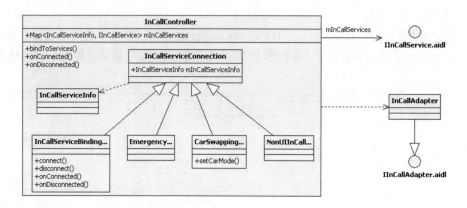

图 4-6　Telecom 应用 InCallController 核心类图

我们重点关注以下几点。

- InCallController 内部类

InCallController 有六个内部类：InCallServiceConnection、InCallServiceInfo、InCallServiceBindingConnection、EmergencyInCallServiceConnection、CarSwappingInCallServiceConnection 和 NonUIInCallServiceConnectionCollection，因类名太长，图 4-6 中并没有标示出它们的全名。

- 理清 CarSwappingInCallServiceConnection 的代理关系

通过调用 setCarMode 方法设置 mIsCarMode 和 mCurrentConnection 属性，在调用 connect 或 disconnect 方法时则调用 mCurrentConnection 对象的对应方法，而 mCurrentConnection 是 InCallServiceConnection 对象类型，即 InCallServiceBindingConnection 对象。

- 内部调用

InCallServiceBindingConnection 对象的 connect 方法绑定 IInCallService——内部匿名 mServiceConnection 对象的 onServiceConnected 方法，将响应绑定成功后的系统回调，最后通过 InCallController.this.onConnected 方式调用主类的 onConnected 方法。

- InCallController 关键属性和方法

mInCallServices 属性保存 IInCallService 的 Binder 对象列表，onCallAdded 和 bindToServices 方法是绑定 IInCallService 的入口，onConnected 方法是响应绑定成功后 setInCallAdapter、addCall 服务接口的调用入口。

4.2.3　绑定 IConnectionService 机制

Telecom 通话流程中有两次绑定服务的关键操作，上一节从代码的角度讲述了绑定 IInCallService 的核心机制，本节将继续分析绑定 IConnectionService，它将与 TeleService 应用中的服务交互，发出通话控制消息或是接收通话状态更新的消息；其中关键的三个步骤如下。
- bind Service
- addConnectionServiceAdapter
- createConnection

上一章我们已经对 Telecom 应用绑定 IConnectionService 的流程做了详细的总结，如图 3-7 中 createConnection 所示，本节将从代码的角度进行横向的扩展学习和总结，详情如图 4-7 所示。

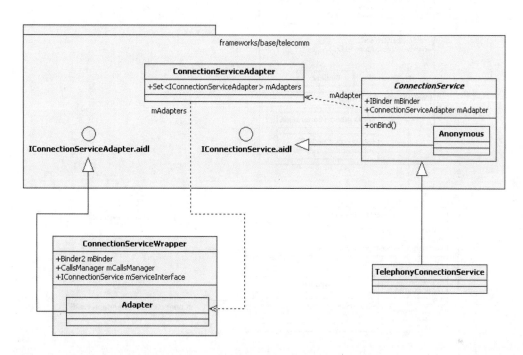

图 4-7 Telecom 应用绑定 IConnectionService 核心类图

重点关注以下三点：
- frameworks/base/telecomm 包下的代码

IConnectionService.aidl 和 IConnectionServiceAdapter.aidl 接口定义文件，以及我们非常关注的 ConnectionService 和 ConnectionServiceAdapter 两个类。

- ConnectionService.aidl 和 IConnectionServiceAdapter.aidl 接口实现

抽象类 ConnectionService 的匿名内部类实现了 ConnectionService.aidl 接口，而 ConnectionService 的 mBinder 属性对象将同步创建该匿名内部类对象。

在 Telecom 应用中，ConnectionServiceWrapper 类的私有内部类 Adapter 继承实现了 ConnectionService.aidl。

- 区分运行空间

图 4-7 中列出的类运行在两个进程空间：Telecom 和 TeleService 应用进程空间；最左边的一个类 ConnectionServiceWrapper 对象运行在 Telecom 应用进程空间，由 Telecom 应用提供 IConnectionServiceAdapter 服务，而 TeleService 应用中的 TelephonyConnectionService 对象保存了 IConnectionServiceAdapter Binder 对象，使用此对象可以调用 setActive/setOnHold 等接口，跨进程调用 Telecom 应用提供的 IConnectionServiceAdapter 服务，发送通话状态变化的消息。

除了 ConnectionServiceWrapper 类，图 4-7 中其他类均运行在 TeleService 应用的进程空间。

Telecom 应用主动发起的绑定 IConnectionService 调用过程的入口是 ConnectionServiceWrapper 类的 createConnection 方法；在绑定 IConnectionService 的过程中涉及的几个关键类，如图 4-8 所示。

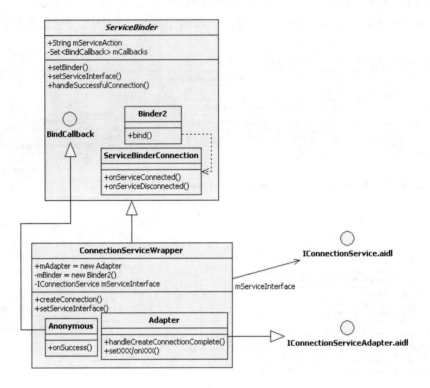

图 4-8　Telecom 应用 ConnectionServiceWrapper 核心类图

重点关注以下四点。
- ConnectionServiceWrapper 内部类及继承关系

ConnectionServiceWrapper 是一个普通的 Java 类型，继承了抽象的 ServiceBinder 类，但是在 Telecom 应用中，此对象是通过 ConnectionServiceRepository 的 getService 方法获取，其名称会有点误导，需要我们特别小心。

ServiceBinder 有五个内部接口或内部类，图中并没有完全展示，仅列出了 BindCallback、Binder2 和 ServiceBinderConnection。ConnectionServiceWrapper 有一个内部类 Adapter，它实现了 IConnectionServiceAdapter 服务接口，并且在方法调用过程中临时创建三个 BindCallback 的匿名对象。

- Binder2

Binder2 对象负责绑定 IConnectionService 的所有处理逻辑，作为 ConnectionServiceWrapper 对象的 mBinder 属性，与 ConnectionServiceWrapper 对象同步创建。

ServiceBinderConnection 对象的 onServiceConnected 方法将响应绑定 IConnectionService 成功后的回调，而回调过程将调用主类对象的 setBinder 和 handleSuccessfulConnection 两个方法。

- 绑定服务回调逻辑

setBinder 和 handleSuccessfulConnection 两个方法的实现都在 ServiceBinder 类中。setBinder

方法将保存绑定服务成功后的 Binder 对象，调用子类中实现的 setServiceInterface 方法，最终调用服务的 addConnectionServiceAdapter 接口；而 handleSuccessfulConnection 方法则通过 BindCallback 对象进行回调，在创建 Connection 的过程中调用服务的 createConnection 方法。

- ConnectionServiceWrapper 关键属性和方法

mServiceAction 属性确定绑定服务的 Action；mServiceInterface 属性保存 IConnectionService 的 Binder 对象；mAdapter 属性对象是 IConnectionServiceAdapter 服务 Binder 对象；

createConnection 方法是绑定 IConnectionService 的程序入口，onServiceConnected、setServiceInterface、onSuccess 等不同对象的方法是响应绑定服务成功后，调用服务 addConnectionServiceAdapter、createConnection 接口的入口。

4.2.4 演进 Telecom 交互模型

我们对 Telecom 和 frameworks/base/telecomm 的代码进行了总结和分析，并且横向扩展了 Telecom 在通话流程中两次绑定服务的核心代码。是时候推出 Telecom 交互模型了。

还记得图 3-16 的 Android 通话模型吗？Telecom 位于 Dialer 和 TeleService 两个应用之间，通过四个服务接口的相互调用，作为消息传递和处理的中转站。下面对此模型进一步扩展，详情如图 4-9 所示。

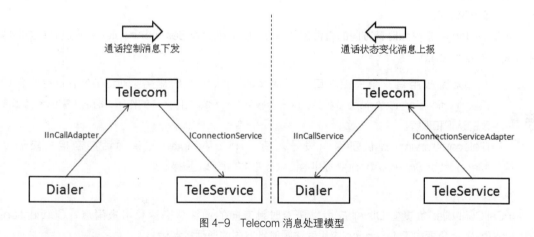

图 4-9 Telecom 消息处理模型

重点关注以下两点。

- 通话控制消息下发

左边的模型说明了通话控制消息下发的过程，Dialer 应用接收用户操作后，通过 IInCallAdapter 服务接口调用发出通话控制消息；Telecom 接收通话控制消息，经过处理后通过 IConnectionService 服务接口，将通话控制消息发送给 TeleService 应用。

- 通话状态变化消息上报

右边的模型说明了通话状态变化消息上报的过程，TeleService 应用接收 RIL 消息后，通过 IConnectionServiceAdapter 服务接口调用发出通话状态变化消息；Telecom 接收通话状态变化消息，经过处理后通过 IInCallService 服务接口，将通话状态变化消息发送给 Dialer 应用。

两个消息的发起方都是 IXXXAdapter 接口，最终的消息接收方都是 IXXXService，归并图 4-9 所示 Telecom 消息处理模型，如图 4-10 所示。

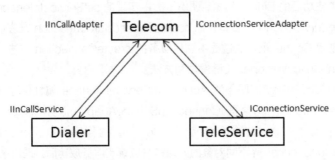

图 4-10　Telecom 交互模型

重点关注以下五点。
- 消息顺时针方向传递

消息顺时针方向传递为通话控制消息下发的流程。
- 消息逆时针方向传递

消息逆时针方向传递为通话状态变化消息上报的流程。
- 服务承载

Telecom 应用承载 IInCallAdapter 和 IConnectionServiceAdapter 两个 Adapter 服务；而 Dialer 和 TeleService 应用各自承载着 IInCallService 和 IConnectionService 两个 Service 服务。
- 接收消息

IInCallAdapter 接收顺时针方向的通话控制消息；IConnectionServiceAdapter 接收逆时针方向的通话状态变化上报消息。

注意

Telecom 应用有三个关键入口：ITelecomService 系统服务、IInCallAdapter 服务和 IConnectionServiceAdapter 服务。这三个服务究竟承载着什么样的作用，相信大家已经有了直观的认识和理解。

ITelecomService 提供了两个拨号入口：IInCallAdapter 提供通话控制相关接口，IConnectionServiceAdapter 提供通话状态变化的消息上报接口。

- 发送消息

InCallController 负责向 Dialer 应用发送逆时针方向的通话状态变化上报消息，ConnectionServiceWrapper 负责向 TeleService 应用发送顺时针方向的通话控制消息。

4.3　核心 Listener 回调消息处理

在解析拨号流程和来电流程的过程中，遇见了各式各样的 Listener 消息回调。本节将重点解析 Telecom 中的 Listener 消息处理机制，学完本节，将非常容易区分这些 Listener 消息。

4.3.1　CallsManagerListener

在 TelecomSystem 初始化过程中创建 CallsManager 对象时，将同步创建 CallsManagerListener 对象，并注册 Listener 消息通知，查看 CallsManager 类的构造方法，其中的关键逻辑详情如下：

```
private final Set<CallsManagerListener> mListeners = Collections.newSetFromMap(
            new ConcurrentHashMap<CallsManagerListener, Boolean>(16, 0.9f, 1));

mListeners.add(mInCallWakeLockController);
mListeners.add(statusBarNotifier);
mListeners.add(mCallLogManager);
mListeners.add(mPhoneStateBroadcaster);
mListeners.add(mInCallController);
mListeners.add(mCallAudioManager);
mListeners.add(missedCallNotifier);
mListeners.add(mHeadsetMediaButton);
mListeners.add(mProximitySensorManager);
```

在 Telecom 应用加载的初始化过程中，将创建 CallsManagerListener 对象，并增加到 CallsManager 对象的 mListeners 列表中；而通话的相关状态或属性发生改变时，CallsManager 将遍历 mListeners 列表，进行 onXXX 的消息回调，其代码逻辑总结如下：

```
for (CallsManagerListener listener : mListeners) {
    listener.onXXX();
}
```

CallsManagerListener 接口的继承关系，总结如图 4-11 所示。

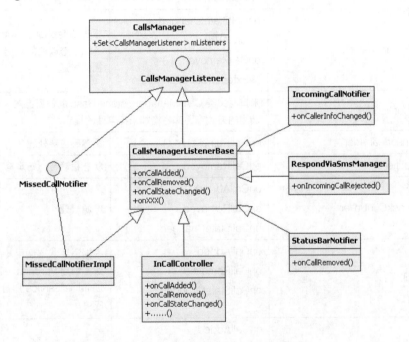

图 4-11　CallsManagerListener 继承关系类图

我们重点关注以下几点。

● CallsManagerListener 接口定义

CallsManagerListener 接口定义在 CallsManager 类中，CallsManagerListenerBase 类实现了此接口的所有方法，而且这些方法都没有任何的业务逻辑代码。在面向对象编程中，使用多态的一种典型方式是子类可根据业务需要选择父类中的方法进行重写。

● CallsManagerListenerBase 子类

Telecom 应用中 CallsManagerListenerBase 总共有 12 个子类，图 4-8 中仅展示了比较关键和常

用的 5 个子类。CallsManagerListenerBase 所有子类的作用及重写的方法，总结如表 4-1 所示。

表 4-1　CallsManagerListenerBase 子类汇总表

类名	重写父类方法	关键业务
IncomingCallNotifier	onCallAdded onCallRemoved onCallStateChanged	手机状态栏显示来电信息，并附加了接听和拒接操作入口
RespondViaSmsManager	onIncomingCallRejected	拒接来电后，发送短信
CallAudioManager	onCallStateChanged onCallAdded onCallRemoved ……	根据通话状态设置 Audio 音频策略
HeadsetMediaButton	onCallAdded onCallRemoved onExternalCallChanged	耳机按钮控制事件的响应，根据通话状态进行接听或是挂断电话操作
CallLogManager	onCallStateChanged	记录通话日志
PhoneStateBroadcaster	onCallStateChanged onCallAdded onCallRemoved onExternalCallChanged	根据 telephony.registry 服务发起通话状态改变的注册消息回调和广播的发送
MissedCallNotifierImpl	未重写父类 CallsManagerListenerBase 的任何方法，通过对象方法调用的方式在通知栏增加未接电话的提示	
ProximitySensorManager	onCallRemoved	距离感应器处理
StatusBarNotifier	onCallRemoved	状态栏静音、扬声器状态同步
InCallWakeLockController	onCallAdded onCallRemoved onCallStateChanged	电源管理
InCallController	onCallAdded onCallRemoved onCallStateChanged ……	根据通话状态调用 updateCall 方法，创建 ParcelableCall 对象，通过 IInCallService 服务传递给 Dialer 应用
Anonymous in mCallsManagerListener in BluetoothPhoneServiceImpl	onCallAdded onCallRemoved onCallStateChanged ……	蓝牙耳机的交互

CallsManagerListener 的 12 个子类承载了通话的扩展功能，比如通话日志、Audio 音频策略、状态栏的信息同步等。

4.3.2　Call.Listener

拨号流程或是来电流程中，都会创建 com.android.server.telecom.Call 对象，此类中定义了 Listener

接口，主要有以下几个方法：
- onSuccessfulOutgoingCall
- onFailedOutgoingCall
- onSuccessfulIncomingCall
- onFailedIncomingCall
- onXXXChanged

这些接口方法都是以 onXXX 开头的，并且都传递 Call 对象的引用，注意与 CallsManagerListener 接口的区别；当前 Call 类中的内部抽象类 ListenerBase 实现了 Listener 接口的所有方法，与 CallsManagerListenerBase 类的实现机制相同，ListenerBase 实现的所有方法没有具体的代码逻辑。

Call.Listener 接口的继承关系，总结如图 4-12 所示。

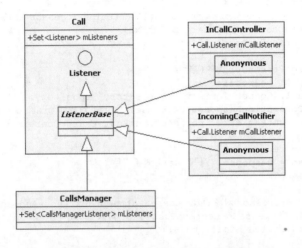

图 4-12　Call.Listener 接口继承关系类图

我们重点关注以下四点。
- Call. Listener 接口定义

Call. Listener 接口在 Call 类中定义，ListenerBase 抽象类实现了此接口的所有方法，并且这些方法都没有任何的业务逻辑代码，与 CallsManagerListenerBase 类的设计思想是一致的。
- ListenerBase 子类

Telecom 应用中 ListenerBase 总共有三个子类：CallsManager 和 InCallController、IncomingCallNotifier 两个匿名内部类。
- Call 对象的 mListeners

Call 对象在 CallsManager 中创建，同时调用 call.addListener(this)逻辑，将 CallsManager 对象作为 Call 对象的 mListeners 中的一员；

InCallController 和 IncomingCallNotifier 两个匿名内部类对象在 TelecomSystem 创建的过程中将同步创建，对象名均为：mCallListener；InCallController 在绑定 IInCallService 成功后调用 addCall 方法的过程中，调用 call.addListener(mCallListener)逻辑，将 mCallListener 添加到 Call 对象的 mListeners 中；同理，IncomingCallNotifier 在 Call 变化的消息回调中，将 mCallListener 添加到 Call 对象的 mListeners 中。
- CallsManager

CallsManager 作为 Call.Listener 接口的子类，由 Call 对象触发 mListeners Call 对象变化的消息

回调，CallsManager 对象将通过自己的 mListeners，继续发出 Call 对象变化的消息回调，而这一次的消息回调将接收并处理 12 个对象；因此，可以将 CallsManager 看作 Call 对象变化 Listener 消息回调的消息中转站，将 Call.Listener 和 CallsManagerListener 这两个 Listener 紧密联系在一起。

4.3.3　CreateConnectionResponse

不论拨号流程还是来电流程，Telecom 在 Call 对象创建完成后，都将调用其 startCreateConnection 方法最终完成绑定 IConnectionService 服务相关的操作；在此过程中将涉及 CreateConnectionResponse 接口对象的创建和传递过程，代码框架总结如下。

```
//Call 对象的 startCreateConnection 方法
void startCreateConnection(PhoneAccountRegistrar phoneAccountRegistrar) {
    //传递了两个 this，即 Call 对象
    mCreateConnectionProcessor = new CreateConnectionProcessor(this, mRepository, this,
        phoneAccountRegistrar, mContext);
    mCreateConnectionProcessor.process();
}
// CreateConnectionProcessor 类的构造方法
public CreateConnectionProcessor(
        Call call, ConnectionServiceRepository repository,
        CreateConnectionResponse response,//Call 对象即 CreateConnectionResponse
        PhoneAccountRegistrar phoneAccountRegistrar, Context context) {
    mCall = call;
    mRepository = repository;
    mCallResponse = response;//保存 Call 对象引用
    ......
}
//CreateConnectionProcessor 对象的 process-> attemptNextPhoneAccount 最终完成以下调用逻辑，
//仍然关注 this 参数，即 CreateConnectionProcessor 对象
mService.createConnection(mCall, this);

//ConnectionServiceWrapper 类的 createConnection 方法，CreateConnectionProcessor 对象也是
//CreateConnectionResponse 接口对象
public void createConnection(final Call call, final CreateConnectionResponse response) {
    BindCallback callback = new BindCallback() {
        @Override
        public void onSuccess() {
            String callId = mCallIdMapper.getCallId(call);
            mPendingResponses.put(callId, response);//保存到 mPendingResponses 列表
```

CreateConnectionResponse 接口定义了两个方法：handleCreateConnectionSuccess 和 handleCreateConnectionFailure，它总共有两个子类：Call 和 CreateConnectionProcessor，Call 和 CreateConnectionProcessor 都是 CreateConnectionResponse 接口对象。

通过上面的代码框架了解到，CreateConnectionProcessor 的 mCallResponse 属性是 Call 对象，ConnectionServiceWrapper 的 mPendingResponses 将保存 CreateConnectionProcessor 对象列表。

4.3.4　总结 Listener 消息

通过前面的学习了解到，在 Telecom 应用中主要处理两种消息类型：顺时针方向下发的通话控制消息和逆时针方向上报的通话状态变化消息。而 Listener 消息回调承载着上报消息的业务处理逻辑，其应用场景是 ConnectionServiceWrapper 的 Adapter 服务对象接收到 TeleService 应用的接口调用，通知当前 Connection 和 Call 的状态或属性发生的变化，再经过一系列的 Listener 消息回调处理，最

终由 InCallController 创建 ParcelableCall 对象，使用 IInCallService 服务接口调用发送给 Dialer 应用。

可见这些 Listener 消息处理在 Telecom 应用中的重要性，继续对 CallsManagerListener、Call.Listener 和 CreateConnectionResponse 三个核心消息回调的分析结果进行汇总和总结，形成 Telecom 应用中消息回调的全貌，如图 4-13 所示。

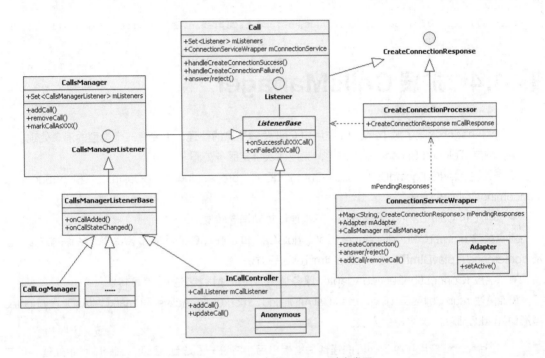

图 4-13　Telecom 主要 Listener 消息类图

重点关注以下几点。

● 接收上报消息的入口

IConnectionServiceAdapter 服务即 Adapter 对象。

● 第一条消息回调通道

CreateConnectionResponse 消息回调的过程是第一条消息回调通道，也就是在 Connection 相关的接口调用过程中，IConnectionServiceAdapter 服务即 Adapter 对象接收到上报消息，通过消息回调将消息发送到 Call 对象，再通过 Call.Listener 对象进行消息回调 CallsManager 对象和 InCallController 的内部匿名对象。

● 第二条消息回调通道

Adapter 对象接收到上报消息，绕过了 Call 对象的相关消息处理过程，直接使用 Connection-ServiceWrapper 对象的 mCallsManager 属性调用 CallsManager 对象的方法，再通过 CallsManagerListener 对象进行 Listener 消息回调，最后交给 InCallController 处理。

两条消息回调通道的调用过程，都会调用 CallsManager 对象的 setCallState 方法更新 Telecom 应用中的 Call 属性，详细的调用过程总结如下：

```
//第一条消息回调通道的调用过程
ConnectionServiceWrapper.mAdapter.handleCreateConnectionComplete
-> mPendingResponses(CreateConnectionProcessor).handleCreateConnectionSuccess
-> mCallResponse(Call).handleCreateConnectionSuccess
```

```
-> mListeners(CallsManager).onSuccessfulXXXCall->setCallState

//第二条消息回调通道的调用过程
ConnectionServiceWrapper.mAdapter.setActive
-> mCallsManager.markCallAsActive->setCallState

//CallsManager 对象的调用过程
CallsManager.setCallState
-> InCallController.onCallStateChanged->updateCall->inCallService.updateCall
```

4.4 扩展 CallsManager

CallsManager 承载了对 Telecom 应用中 Call 关键消息的处理，主要有三个类型的消息处理。

- 响应 ITelecomService 服务调用，完成两次绑定服务处理。

重点关注 startOutgoingCall、placeOutgoingCall、processIncomingCallIntent 和 onSuccessfulIncomingCall 四个方法。

- 响应 IInCallAdapter 服务调用，完成通话控制消息转发。

重点关注 answerCall、disconnectCall、holdCall 和 unholdCall 等形如 XXXCall 名称的方法，以及 conference、playDtmfTone 和 stopDtmfTone 等方法。

- 响应 IConnectionServiceAdapter 服务调用，完成通话状态变化消息转发。

重点关注 markCallAsRinging、markCallAsDialing、markCallAsActive 和 markCallAsDisconnected 等形如 markCallAsXXX 名称的方法。

Telecom 应用接收到外界的通话关键信息后，将统一汇总到 CallsManager 中处理，可与 Telecom 交互模型结合起来，进一步完善 Telecom 交互模型，如图 4-14 所示。

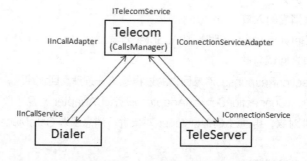

图 4-14 Telecom 交互模型

4.4.1 记录通话日志

CallLogManager 类负责记录通话日志，它重写了 CallsManagerListener 父类的 onCallStateChanged 方法。在响应 CallsManager Listener 消息回调时，通过判断通话的发起方和通话断开的 DisconnectCause 将通话日志分为呼出、呼入、拒绝和未接四类。onCallStateChanged 的关键处理逻辑详情如下。

```
@Override//重写父类CallsManagerListener方法
public void onCallStateChanged(Call call, int oldState, int newState) {
    int disconnectCause = call.getDisconnectCause().getCode();//获取通话断开原因
    boolean isNewlyDisconnected = //通话断开
            newState == CallState.DISCONNECTED || newState == CallState.ABORTED;
    //四个主要的通话日志记录类型
    if (!call.isIncoming()) {
        type = Calls.OUTGOING_TYPE;
    } else if (disconnectCause == DisconnectCause.MISSED) {
        type = Calls.MISSED_TYPE;
    } else if (disconnectCause == DisconnectCause.REJECTED) {
        type = Calls.REJECTED_TYPE;
    } else {
        type = Calls.INCOMING_TYPE;
    }
    logCall(call, type, showNotification);
}
```

logCall 的调用逻辑相对比较简单，先将通话日志信息封装成 AddCallArgs 对象，再通过 LogCallAsyncTask 在后台执行 android.provider.CallLog 对象的 addCall 方法，完成通话日志 SQLite 数据库的写入操作。

4.4.2 耳机 Hook 事件

HeadsetMediaButton 类负责监听耳机 Hook 按键的事件，在事件消息的回调响应过程中，可接收到耳机 Hook 按键的长按或短按事件，将其交给 CallsManager 对象的 onMediaButton 方法处理，包括接听电话、拒接电话和通话静音三个操作。

HeadsetMediaButton 的匿名内部类对象 mSessionCallback 注册为 Android 系统 MediaSession 的 Call Back 对象，在主类 HeadsetMediaButton 的 onCallAdded、onCallRemoved 和 onExternalCallChanged 方法的 Listener 回调响应过程中，通过 Handler 消息 MSG_MEDIA_SESSION_SET_ACTIVE 的处理过程设置 MediaSession 的激活状态。简单地说，在通话的过程中需要激活 MediaSession 来接收耳机 Hook 按键的事件；通话断开以后则需要激活 MediaSession，不再接收耳机 Hook 按键的事件。

4.4.3 通知栏信息同步

1. IncomingCallNotifier

IncomingCallNotifier 类负责在手机状态栏显示或隐藏来电信息，通过 onCallAdded、onCallRemoved、onCallStateChanged 方法的消息回调，在 updateIncomingCall 方法中实现显示或隐藏来电信息，关键逻辑详情如下。

```
//创建 Notification.Builder 对象并发出通知,在通知栏显示 builder 对象中的内容
private void showIncomingCallNotification(Call call) {
    Notification.Builder builder = getNotificationBuilder(call,
            mCallsManagerProxy.getActiveCall());
    mNotificationManager.notify(NOTIFICATION_TAG,
            NOTIFICATION_INCOMING_CALL, builder.build());
}
//发出 cancel 通知,在通知栏不再显示来电相关信息
private void hideIncomingCallNotification() {
    mNotificationManager.cancel(NOTIFICATION_TAG,
            NOTIFICATION_INCOMING_CALL);
}
```

getNotificationBuilder 方法在创建 Notification.Builder 对象的逻辑中，由 Builder 对象增加两个 Action，提供用户接听来电和拒接来电操作，而对应操作的 Intent 对象分别是 answerIntent 和 rejectIntent。

2. StatusBarNotifier

StatusBarNotifier 类负责通知栏中通话静音和扬声器状态的同步显示，仅重写了父类 CallsManagerListener 的 onCallRemoved 方法，通话断开的时候调用 notifyMute 或 notifySpeakerphone 来取消通知栏通话静音和扬声器的信息显示。

```
@Override//重写父类方法
public void onCallRemoved(Call call) {
    if (!mCallsManager.hasAnyCalls()) {
        notifyMute(false);//取消静音通知栏图标显示
        notifySpeakerphone(false);//取消扬声器通知栏图标显示
    }
}
public void notifyMute(boolean isMuted) {
    if (isMuted) {
        mStatusBarManager.setIcon(//增加静音通知栏图标显示
            SLOT_MUTE,
            android.R.drawable.stat_notify_call_mute,
            0,  /* iconLevel */
            mContext.getString(R.string.accessibility_call_muted));
    } else {//删除静音通知栏图标显示
        mStatusBarManager.removeIcon(SLOT_MUTE);
    }
}
```

StatusBarNotifier 类响应 onCallRemoved 消息回调只能取消静音或扬声器通知栏图标的显示。在我们打开静音或是扬声器时，CallAudioRouteStateMachine 将通过 StatusBarNotifier 对象直接调用 notifyMute 或 notifySpeakerphone 方法，在通知栏上显示对应的信息。

本 章 小 结

本章详细总结和进一步讲解 Telecom 交互模型以及 Listener 消息回调机制。

- Telecom 消息入口

ITelecomService、IInCallAdapter 和 IConnectionServiceAdapter 三个服务作为三种不同类型的消息的入口。

- Telecom 消息出口

InCallController 和 ConnectionServiceWrapper 两次绑定服务操作作为 Telecom 应用消息的出口。

- 下发顺时针、上报逆时针消息机制

Telecom 应用通过消息入口和出口的五个服务，承载两种类型的消息：控制通话下发的顺时针方向消息处理和控制上报通话状态变化的逆时针方法消息处理，如图 4-13 所示。

- Listener 消息处理机制

Call 和 CallsManager 对象作为 Listener 回调消息的交换中心，控制图 4-12 中的 Listener 对象关系。

- CallsManagerListener

12 个 CallsManagerListener 对象接收到 CallsManager mListeners 消息回调，判断当前 Call 的属性和状态，将实现通话日志、通知栏信息同步、电源管理、耳机交互等通话扩展功能。

第 5 章
详解 TeleService

学习目标

- 解析 TeleService 系统应用的加载过程。
- 掌握 Telephony Phone 模型设计及核心消息处理方式。
- 理解 PhoneAccount 的作用。
- 掌握 TeleService 系统应用提供的 Service 服务。
- 区分各种 Connection 对象及对应的消息处理。
- 掌握 slotId、phoneId、subId 的关系与区别。

上一章对 Telecom 系统应用进行了详细的分析和总结，其中最核心的内容是对通话业务交互模型的总结；在通话业务的交互模型中，TeleService 负责与 RILJ 对象进行交互，完成通话管理，处理两种类型的消息：

- 顺时针方向的通话管理和控制消息

IConnectionService 服务接收 Telecom 应用中转的通话管理和控制请求，将请求提交给 Telephony 模型处理，最终与 RILJ 对象进行交互，完成通话管理和控制。

- 逆时针方向的通话变化消息上报

Telephony 业务模型与 RILJ 对象交互，通过 Telecom 应用设置的 Adapter 跨进程 Binder 对象，发出通话信息或状态发生改变的跨进程消息上报。

5.1 加载过程分析

解析 TeleService 系统应用的加载过程之前，首先查看一下 TeleService 的代码结构，详情如图 5-1 所示。

针对图 5-1 所示的 TeleService 系统应用代码结构，重点关注以下几点。

- TeleService 系统应用的依赖关系

查看 TeleService 系统应用的编译文件 Android.mk 依赖 telephony-common.jar 包，即 Telephony

业务模型的实现。

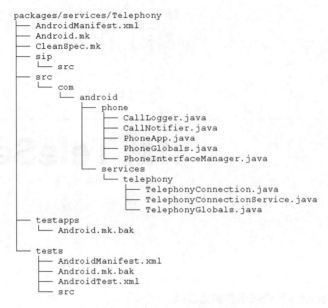

图 5-1　TeleService 代码结构

- 包路径

TeleService 系统应用主要有两个包。

com.android.phone：TeleService 系统应用基础包，主要包括 phoneApp、PhoneInterfaceManager 等 Java 代码文件。

com.android.services.telephony：与 IonnectionService 服务实现相关的 Java 代码，如 TelephonyConnection、TelephonyConnectionService 等 Java 代码文件。

5.1.1　应用基本信息

查看 TeleService 系统应用的 AndroidManifest.xml 配置文件，其中定义了此应用的加载方式和加载入口，配置信息详情如下。

```
<manifest xmlns:android="http://schemas.android.com/apk/res/android"
    xmlns:androidprv="http://schemas.android.com/apk/prv/res/android"
    package="com.android.phone"
    coreApp="true"
    android:sharedUserId="android.uid.phone"
    android:sharedUserLabel="@string/phoneAppLabel"
>
<application android:name="PhoneApp"
    android:persistent="true"
    android:label="@string/phoneAppLabel"
    ......
```

package="com.android.phone"唯一标识应用程序及基础包名，TeleService 系统应用的运行进程名也是 com.android.phone。

android:sharedUserId="android.uid.phone"设置 TeleService 系统应用进程时使用的用户是 android.uid.phone，即 radio 用户，读者可通过 adb shell ps 命令查看并验证 com.android.phone 进

程的相关信息。

android:persistent 属性设置为 true，定义此 Application 应用常驻内存。如果进程异常退出或被人为"杀掉"，Android 系统机制会将此应用重新唤醒。读者可尝试使用 adb kill pid 的方式杀掉 com.android.phone 进程，来验证此进程是否被系统重新唤醒并完成初始化操作。

android:name 属性设定为 PhoneApp，再结合 package 属性定义，可知 TeleService 系统应用加载入口的 Java 类是 com.android.phone.PhoneApp。

TeleService 系统应用加载入口的逻辑相对简单，详情如下。

```
public class PhoneApp extends Application {//继承系统 Application 类
    @Override//重写父类的 onCreate 方法
    public void onCreate() {
        //非常关键，除系统调用，其他情况下不加载
        if (UserHandle.myUserId() == 0) {
            // 系统用户运行 PhoneApp，将启动和加载全局的电话属性和状态
            mPhoneGlobals = new PhoneGlobals(this);
            mPhoneGlobals.onCreate();

            mTelephonyGlobals = new TelephonyGlobals(this);
            mTelephonyGlobals.onCreate();
        }
    }
}
```

PhoneApp 类的 onCreate 方法中最关键的是 UserHandle.myUserId() == 0 的条件判断，系统加载 TeleService 系统应用的 com.android.phone 进程时，方能发起两个 onCreate 方法调用，以避免其他应用或进程非法加载 TeleService 系统应用。

汇总 TeleService 应用的基本信息，重点关注以下三点。

- TeleService 是常驻内存的系统级应用，在开机过程中或进程异常退出时，Android 系统主动完成应用加载。
- TeleService 系统应用的运行进程为 com.android.phone，其用户名为 radio。
- 加载入口 Java 类：com.android.phone.PhoneApp，继承于 Application 类，重写了 onCreate 方法。onCreate 方法是 TeleService 系统应用的加载入口。

5.1.2 PhoneGlobals.onCreate

继续学习 TeleService 系统应用的加载逻辑，有关 PhoneGlobals.onCreate 方法的调用，简化后的主要代码逻辑详情如下。

```
public void onCreate() {
    if (mCM == null) {
        //初始化 Telephony 框架
        PhoneFactory.makeDefaultPhones(this);//创建 Phone 对象加载 Telephony 模型
        //TeleService 中的 CallManager，注意与 Telecom 中的 CallsManager 加以区分
        mCM = CallManager.getInstance();
        for (Phone phone : PhoneFactory.getPhones()) {//循环创建的 Phone 对象
            mCM.registerPhone(phone);//注册消息
        }
        //ITelephony 服务初始化和发布
        phoneMgr = PhoneInterfaceManager.init(this, PhoneFactory.getDefaultPhone());
    }
}
```

有关 PhoneGlobals 对象的 onCreate 方法中的核心逻辑，总结出以下四个关键点。
- 创建 Phone 对象

调用 PhoneFactory 类的静态方法 makeDefaultPhones 创建 GsmCdmaPhone 对象，将同步创建和初始化 Telephony 业务模型的核心对象，比如 RILJ 和各种 Tracker 对象。

为区分 HAL 层与 Telephony Framework 层与 RIL 相关的信息，com.android.phone 进程空间的 com.android.internal.telephony.RIL 对象，本书统一约定为 RILJ。

- CallManager 的消息注册

CallManager 向所有创建的 GsmCdmaPhone 对象注册相关消息（双卡或多卡）。

- 初始化 ITelephony 服务

通过 PhoneInterfaceManager 的 init 方法创建 ITelephony.Stub 类型的 Binder 对象，再通过 ServiceManager.addService 的调用增加系统服务 ITelephony。

- 其他操作

包括 TelephonyDebugService、NotificationMgr 等初始化操作，并注册了两个广播接收器 mReceiver 和 mSipReceiver。

PhoneFactory、CallManager 等代码在 frameworks/opt/telephony 代码库中，由此代码库可编译出 telephony-common.jar，最终运行在 TeleService 系统应用的进程空间，即 com.android.phone 进程，实现了 TeleService 与 Telephony 业务模型无缝结合。
PhoneFactory.makeDefaultPhones 将加载以 GsmCdmaPhone 对象为核心的 Telephony 业务模型。

5.1.3　TelephonyGlobals.onCreate

接下来查看 TeleService 系统应用加载逻辑中的第二个 onCreate 方法调用，即 TelephonyGlobals.onCreate，其代码逻辑详情如下。

```
public void onCreate() {
    //支持多 SIM 卡设备
    Phone[] phones = PhoneFactory.getPhones();
    for (Phone phone : phones) {//初始化 TTY
        mTtyManagers.add(new TtyManager(mContext, phone));
    }

    有关 TelecomAccountRegistry.getInstance(mContext).setupOnBoot();//加载 PhoneAccount
}
```

有关 TelephonyGlobals 对象的 onCreate 方法中的核心逻辑，总结出以下两个关键点。
- 初始化 TTY

Text Telephones (TTY)即聋哑人电话，在手机插入专用设备后支持收发文本，但需网络支持。
- 加载 PhoneAcount

5.2　Telephony Phone

TeleService 系统应用的加载过程也可以理解为 Telephony 业务模型的加载过程，即以创建

GsmCdmaPhone 对象为中心，同步创建 GsmCdmaCallTracker、ServiceStateTracker、DcTracker 和 RILJ 等关键业务对象的过程，同时向 RILJ 对象注册 Handler Message 回调消息。总结其关键流程如图 5-2 所示。

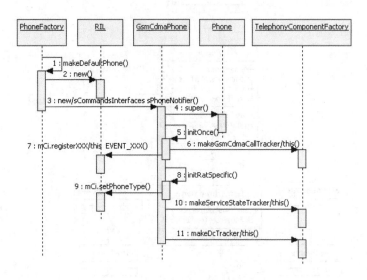

图 5-2　Telephony 业务模型加载流程

图 5-2 所示 Telephony 业务模型的加载流程，重点关注以下几点。

● 加载入口

PhoneFactory.makeDefaultPhones 作为 Telephony 业务模型的加载入口。

● GsmCdmaPhone 对象创建

步骤 3 中创建 GsmCdmaPhone 对象，将传入对 RILJ、sPhoneNotifier、TelephonyComponentFactory 等对象的引用，以及 Phone 类型的 GSM 或 CDMA。

● phoneId 的隐藏逻辑

步骤 2 和步骤 3 是在 for 循环中创建 RILJ 和 Phone 对象并传入循环下标值作为 phoneId 的参数。比如双卡双待 numPhones 取值 2，将创建两个 RILJ 对象和两个 Phone 对象，phoneId 为 0 代表 SIM 卡卡槽 1 对应的 Telephony 业务模型，phoneId 为 1 代表 SIM 卡卡槽 2 对应的 Telephony 业务模型。

● GsmCdmaPhone 对象的消息注册

步骤 7 中调用 registerXXX 方法向 RILJ 注册 RegistrantList 消息，关注一下 this 对象和 EVENT_XXX 消息类别，也就是向 RILJ 注册 GsmCdmaPhone 对象的 Handler Message 回调消息。

● 创建 Tracker 对象

步骤 6、步骤 10 和步骤 11 在 GsmCdmaPhone 构造方法中，通过 TelephonyComponentFactory 类的 makeXXXTracker 方法，创建非常关键的三个对象：GsmCdmaCallTracker、ServiceStateTracker、DcTracker，它们将分别承载 Telephony 业务模型中非常重要的三个业务能力：Voice Call 语音通话、ServiceState 网络服务和 Data Call 移动数据业务。

5.2.1　GsmCdmaPhone

GsmCdmaPhone 对象作为 Telephony 业务模型中的关键对象、中心对象，其 Java 类的定义和继承关系，总结如图 5-3 所示。

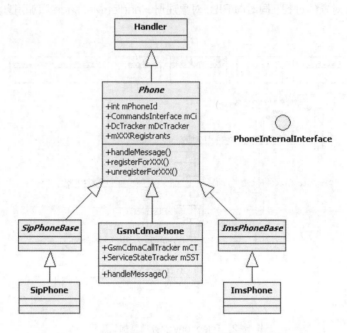

图 5-3 Telephony Phone 对象继承关系

- GsmCdmaPhone 对象本质

Phone 抽象类实现 PhoneInternalInterface 接口并继承 Handler 类，通过重写 handleMessage 方法实现消息的异步处理。

- Phone 抽象类的三个子类

SipPhone 负责 Sip 网络电话业务

GsmCdmaPhone 承载 CS（Circuit Switch，电路交换）域的电信业务

ImsPhone 承载高清语音通话业务

- phoneId

phoneId 对应 SIM 卡卡槽承载的 Telephony 业务模型。

- 关键属性对象

mCi 是 RILJ 对象引用，mDcTracker 是 DcTracker 对象引用，mCT 是 GsmCdmaCallTracker 对象引用，mSST 是 ServiceStateTracker 对象引用。

注意

Phone 对象的继承关系在 Android 8.1 版本中有了非常大的变化，相比 Android 4.0 版本的 Phone 对象，首先减少了 PhoneProxy 代理、PhoneBase 等中间类，其次将 GSMPhone 和 CDMAPhone 两个 Phone 类型合并为一个 GsmCdmaPhone 类型。

汇总 GsmCdmaPhone 对象的关键属性，如表 5-1 所示。

表 5-1 GsmCdmaPhone 对象关键属性

属性	类型	说明
mRilVersion	int	RIL 版本号
mImei	String	IMEI 串号

续表

属性	类型	说明
mImeiSv	String	IMEI 串号的软件版本
mVmNumber	String	语音信箱（Voice Mail）
mCi	RIL	RIL 的 Java 服务对象
mSST	ServiceStateTracker	服务状态跟踪者
mCT	GsmCdmaCallTracker	通话跟踪者
mDcTracker	DcTracker	移动数据跟踪者
mXXXRegistrants	RegistrantList	14 个 Observer 列表
mNotifier	PhoneNotifier	通知 Phone 状态变化

汇总 GsmCdmaPhone 对象的关键方法，如表 5-2 所示。

表 5-2　GsmCdmaPhone 对象关键方法

分类	方法/接口	说明
消息处理	registerForXXX	Handler 消息注册
	unregisterForXXX	取消 Handler 消息注册
	notifyXXX	发出 Handler 消息通知
控制和管理接口	dial\acceptCall\rejectCall…	拨号、接听来电、拒接来电等通话控制相关接口
	setRadioPower\updateServiceLocation\disableLocationUpdates	开关飞行模式、设置位置服务
	setDataRoamingEnabled\setDataEnabled	漫游移动数据设置接口和移动数据开关接口
获取信息接口	getForegroundCall\getBackgroundCall\getRingingCall	获取 Call 通话管理对象
	getServiceState\getCellLocation	获取 ServiceState 驻网服务管理对象和小区信息
	getDataActivityState\getDataRoamingEnabled\getDataEnabled\getActiveApnTypes…	获取移动数据状态、APN 等信息

5.2.2　Composition（组合）关系

GsmCdmaPhone 类的构造方法可体现它与 mCi、mCT、mSST、mDcTracker 等对象之间的关系，它们之间具有强组合（Composition）的逻辑关系。请注意区分 Composition（组合）与 Aggregation（聚合）在语义上的区别。

GsmCdmaPhone 类的构造方法的代码逻辑详情如下。

```
public GsmCdmaPhone(Context context, CommandsInterface ci, PhoneNotifier notifier,
            boolean unitTestMode, int phoneId, int precisePhoneType,
            TelephonyComponentFactory telephonyComponentFactory) {
    super(precisePhoneType == PhoneConstants.PHONE_TYPE_GSM ? "GSM" : "CDMA",
        notifier, context, ci, unitTestMode, phoneId, telephonyComponentFactory);
    initOnce(ci);
    mSST = mTelephonyComponentFactory.makeServiceStateTracker(this, this.mCi);
```

```java
    mDcTracker = mTelephonyComponentFactory.makeDcTracker(this);
    ......
}
protected Phone(String name, PhoneNotifier notifier, Context context, CommandsInterface ci,
                boolean unitTestMode, int phoneId,
                TelephonyComponentFactory telephonyComponentFactory) {
    mPhoneId = phoneId;
    mName = name;
    mNotifier = notifier;
    mCi = ci;
    ......
}
private void initOnce(CommandsInterface ci) {
    mCT = mTelephonyComponentFactory.makeGsmCdmaCallTracker(this);
    ......
}
```

上面的代码需注意以下两点：

- RILJ 和 PhoneNotifier 对象的创建在 PhoneFactory 中完成，GsmCdmaPhone 的构造方法中则通过 supper 调用，将这两个对象的引用传递给父类 Phone 的构造方法，进行 mCi 和 mNotifier 属性的初始化。
- mCT、mSST、mDcTracker 等对象均在 GsmCdmaPhone 的构造方法中同步完成创建，调用这些类的构造方法时传入了 this，即 GsmCdmaPhone 对象。

GsmCdmaPhone 类与其他关键类的组合关系如图 5-4 所示。

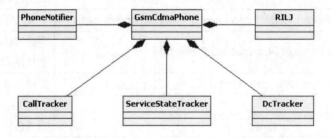

图 5-4　GsmCdmaPhone 类与其他关键类的组合关系

5.2.3　Facade Pattern

表 5-2 汇总的 GsmCdmaPhone 对象的关键方法主要有两类：控制管理接口和信息查询接口，这些接口实现逻辑有一个规律，详情如下：

mCi.doXXX/mCi.getXXX

mCT.doXXX/mCT.getXXX

mSST.doXXX/mSST.getXXX

mDcTracker.doXXX/mDcTracker.getXXX

非常典型的 Facade Pattern（门面设计模式）的应用。GsmCdmaPhone 为 Facade（门面），而 GsmCdmaCallTracker、ServiceStateTracker、DcTracker、RILJ 都作为内部的子系统，总结 Facade Pattern 的设计类图如图 5-5 所示。

第 5 章 详解 TeleService

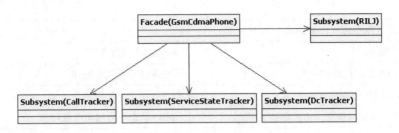

图 5-5 Facade Pattern 设计类图

5.2.4 Handler 消息处理机制

GsmCdmaPhone 类的父类是 Phone 抽象类，它不仅实现了 PhoneInternalInterface 接口，而且继承了 Handler 类，是一个自定义的 Handler 消息处理类。它的 Handler 消息处理机制是如何运行的呢？GsmCdmaPhone 类中的 Handler 消息处理机制可分成以下三个大类：

- 基本 Handler 消息注册和响应机制。
- RegistrantList 封装的 Handler 消息运行机制。
- 创建的 Message 对象作为 RILJ 对象回调入口。

1．基本 Handler 消息注册和响应机制

GsmCdmaPhone 类中基本的 Handler 消息注册和响应机制包括两方面的内容：

- 调用 mCi.registerForXXX 方法，向 RILJ 对象注册单个的 Handler 消息
- Handler 对象 handleMessage 接收并响应 Message 消息

在 GsmCdmaPhone 类的构造方法的调用过程中，即加载 Telephony 业务模型的过程中，以调用 mCi.registerForXXX(this, what, null)的方式向 RILJ 对象发起消息注册。在 GsmCdmaPhone 和 Phone 类中均重写了父类的 handleMessage 方法，从而响应 RILJ 对象发出的 Handler 消息回调通知。

2．RegistrantList 封装的 Handler 消息运行机制

首先，展开介绍 GsmCdmaPhone 类中使用 RegistrantList 实现的 Handler 消息处理机制。

第 3 章已经详细分析了 RegistrantList 封装的 Handler 消息处理机制的实现，这里主要讲解 GsmCdmaPhone 对象使用 RegistrantList 实现的业务逻辑。

在 Phone 抽象类中，一共定义了 14 个 RegistrantList 对象。针对这 14 个 RegistrantList 对象，分别实现了 registerForXXX 和 unregisterForXXX 方法来完成多个 Handler 消息的注册和取消注册，如表 5-3 所示。

表 5-3 Phone 抽象类 RegistrantList 对象汇总

RegistrantList 列表	说明
mPreciseCallStateRegistrants	通话状态变化消息通知
mHandoverRegistrants	SRVCC 通话切换消息通知
mNewRingingConnectionRegistrants	接收到新来电请求的消息通知
mIncomingRingRegistrants	来电响铃消息通知
mDisconnectRegistrants	通话连接断开消息通知
mServiceStateRegistrants	服务状态变化消息通知
mMmiCompleteRegistrants	MMI 执行完毕消息通知

续表

RegistrantList 列表	说明
mMmiRegistrants	执行 MMI 消息通知
mUnknownConnectionRegistrants	出现未知连接消息通知
mSuppServiceFailedRegistrants	附加服务请求失败消息通知
mRadioOffOrNotAvailableRegistrants	Radio 状态不可用消息通知
mSimRecordsLoadedRegistrants	SIM 卡加载完成消息通知
mVideoCapabilityChangedRegistrants	视频通话能力变化消息通知
mEmergencyCallToggledRegistrants	Emergency call/callback 消息通知

RegistrantList 对象需要正常运转 Handler 消息处理，不仅需要提供 registerForXXX 和 unregisterForXXX 方法完成 Handler 消息的注册和取消注册，还需要一个重要的方法，那就是 notifyXXX 方法来发出多个注册的 Handler 通知。

在 GsmCdmaPhone 对象中，这些 RegistrantList 对象发出消息通知的方法的实现逻辑是在什么地方完成的呢？答案就是在 Phone 抽象类和 GsmCdmaPhone 类中均有不同程度的实现。可以找到其中的一些规律：notifyPreciseCallStateChangedP、notifyDisconnectP、notifyServiceStateChangedP、notifyUnknownConnectionP 和 notifyNewRingingConnectionP 共五个 Handler 消息通知方法，它们的名称最后都有一个大写的 P 字母作为方法名称的结尾，P 是 Parent 的首字母。这些方法调用时，都是由其子类调用其父类的 super.notifyXXXP 对应的方法。

以 mPreciseCallStateRegistrants 对象封装 Handler 消息处理机制为例，在 Phone 抽象类中实现 RegistrantList 对象通知方法的详情如下：

```
protected void notifyPreciseCallStateChangedP() {
    AsyncResult ar = new AsyncResult(null, this, null);
    mPreciseCallStateRegistrants.notifyRegistrants(ar);
    mNotifier.notifyPreciseCallState(this);
}
```

GsmCdmaPhone 类中提供的 RegistrantList 对象通知方法，原来是调用父类的 notifyXXX 方法来实现消息的通知。

```
public void notifyPreciseCallStateChanged() {
    /*包作用域范围内，我们习惯调用父类的方法*/
    super.notifyPreciseCallStateChangedP();
}
```

剩余四个 RegistrantList 对象的 notifyXXX 实现方法在 Phone 和它的子类 GsmCdmaPhone 中实现，它们有一个共同点，那就是 notifyXXX 通知消息的调用是在 GsmCdmaPhone 对象提供的通信管理和控制的方法中，调用 notifyXXX 方法来发出 Handler 消息的通知。比如，MMI 拨号请求、输入 PIN 码、请求 USSD 码等一系列主动请求的处理逻辑。

3. 创建的 Message 对象作为 RILJ 对象回调入口

GsmCdmaPhone 对象在与 RILJ 对象的交互过程中创建 Message 对象，作为 RILJ 对象的回调入口。使用这种方式不需要向 RILJ 注册 Hanlder 消息，其生命周期很短，仅在一次交互过程中有效，决定了这种交互方式的灵活性。

GsmCdmaPhone 对象提供的方法中有一些处理逻辑。首先，创建基于 GsmCdmaPhone 对象的 Message 对象，然后将此对象作为参数调用 mCi 对象的方法；其次，RILJ 对象处理完成后，通过

Message 对象进行回调；最后，在 GsmCdmaPhone 对象的 handleMessage 方法中接收和响应 Message 对象发出的 Handler 消息。这种 Handler 消息处理方式的代码逻辑详情如下。

```
//在 GsmCdmaPhone 类的 handleMessage 方法中摘录部分代码
//手机开机或者取消飞行模式后会打开 Radio 无线通信模块，便能接收到 RIL 发出的此消息
case EVENT_RADIO_AVAILABLE: {
    handleRadioAvailable();
}
break;
case EVENT_GET_BASEBAND_VERSION_DONE://接收 RILJ 对象的 Handler 消息回调
    ar = (AsyncResult)msg.obj;
    TelephonyManager.from(mContext).setBasebandVersionForPhone(getPhoneId(),
            (String)ar.result);//获取并记录基带软件版本号

private void handleRadioAvailable() {
    //创建 Message 对象获取基带软件版本号
    mCi.getBasebandVersion(obtainMessage(EVENT_GET_BASEBAND_VERSION_DONE));
    ......
}
```

上面的代码逻辑中，单次 Handler 消息处理的实现机制非常灵活和简单，其特点是：无需提前完成 Handler 消息注册，而是根据业务需要主动向 RILJ 对象发起请求，Message 对象的生命周期仅存在于一次交互过程中。

注意

RILJ 对象响应 GsmCdmaPhone 对象发出的请求后，通过 Message 对象完成回调。RILJ 类中的实现机制将在后续的章节中详细讲解。

5.3 扩展 PhoneAccount

在第 2 章解析通话流程的过程中，经常见到 PhoneAccount、PhoneAccountHandle 等字样。那么 PhoneAccount 在通话流程中究竟起到什么样的作用呢？继续分析 TeleService 系统应用加载过程中调用的 TelephonyGlobals.onCreate 方法，来完成 PhoneAccount 初始化操作，即 TelecomAccountRegistry.getInstance(mContext).setupOnBoot()涉及的业务逻辑。

5.3.1 PhoneAccount 初始化过程

TelecomAccountRegistry 完成 PhoneAccount 初始化业务逻辑包括两个关键步骤。
- 创建 TelecomAccountRegistry 对象
- 调用 setupOnBoot 方法

1. TelecomAccountRegistry 对象的创建过程

TelecomAccountRegistry 类提供了静态同步 getInstance 方法，创建并获取 TelecomAccountRegistry 对象，典型的单例模式的实现方式。TelecomAccountRegistry 的构造方法中通过 Context 获取了三个 Manager 对象：TelecomManager、TelephonyManager 和 SubscriptionManager。

注意

在 TelecomAccountRegistry 对象的创建过程中，会同步创建 mUserSwitchedReceiver、mOnSubscriptionsChangedListener、mPhoneStateListener 三个内部匿名对象。

2. setupOnBoot 方法调用

```
void setupOnBoot() {
    //注册 Subscription 变化的回调 Listener
    SubscriptionManager.from(mContext).addOnSubscriptionsChangedListener(
            mOnSubscriptionsChangedListener);
    //注册 Phone State 变化的回调 Listener
    mTelephonyManager.listen(mPhoneStateListener,
            PhoneStateListener.LISTEN_SERVICE_STATE);
    //注册用户切换广播
    mContext.registerReceiver(mUserSwitchedReceiver,
            new IntentFilter(Intent.ACTION_USER_SWITCHED));
}
```

setupOnBoot 方法注册了两个 Listener 回调和一个广播接收器，使用了三个内部匿名对象：mOnSubscriptionsChangedListener、mPhoneStateListener 和 mUserSwitchedReceiver 作为监听回调的响应。

总结这三个对象的响应逻辑，它们全部发起了 tearDownAccounts 和 setupAccounts 的方法调用，重点查看 PhoneStateListener 的回调响应逻辑，详情如下。

```
private final PhoneStateListener mPhoneStateListener = new PhoneStateListener() {
    @Override
    public void onServiceStateChanged(ServiceState serviceState) {
        int newState = serviceState.getState();//获取当前最新服务状态
        if (newState == ServiceState.STATE_IN_SERVICE && mServiceState != newState) {
            tearDownAccounts();
            setupAccounts();
        }
        mServiceState = newState;
    }
};
```

当前服务状态保存在 STATE_IN_SERVICE 中，当 ServiceState 已发生改变时，调用 tearDownAccounts 方法清空已注册的 PhoneAccount，接着调用 setupAccounts 重新设置并注册新的 PhoneAccount，代码逻辑详情如下。

```
private void setupAccounts() {
    Phone[] phones = PhoneFactory.getPhones();//获取 Phone 对象
    final boolean phoneAccountsEnabled = mContext.getResources().getBoolean(
            R.bool.config_pstn_phone_accounts_enabled);

    synchronized (mAccountsLock) {//同步锁
        if (phoneAccountsEnabled) {
            for (Phone phone : phones) {
                int subscriptionId = phone.getSubId();
                if (subscriptionId >= 0 && phone.getFullIccSerialNumber() != null) {
                    mAccounts.add(new AccountEntry(phone, false /* emergency */,
                            false /* isDummy */));
                }
            }
            ......
        }
        ......
    }
}

private void tearDownAccounts() {//关闭 PhoneAccount
    synchronized (mAccountsLock) {
```

```
        for (AccountEntry entry : mAccounts) {
            entry.teardown();
        }
        mAccounts.clear();
    }
}
```

通过 PhoneFactory 获取 Phone 对象数组后，再通过 Phone 对象创建 TelecomAccountRegistry 类的内部类 AccountEntry 对象。

```
AccountEntry(Phone phone, boolean isEmergency, boolean isDummy) {
    mPhone = phone;//GsmCdmaPhone 对象
    mIsEmergency = isEmergency;//是否紧急呼救
    mIsDummy = isDummy;//是否虚拟
    mAccount = registerPstnPhoneAccount(isEmergency, isDummy);//注册 PhoneAccount
    mIncomingCallNotifier = new PstnIncomingCallNotifier((Phone) mPhone);
    mPhoneCapabilitiesNotifier = new PstnPhoneCapabilitiesNotifier((Phone) mPhone,
            this);//回调 Listener 接口，视频电话能力变化后的消息回调
}
```

AccountEntry 类实现了 PstnPhoneCapabilitiesNotifier.Listener 接口，它只有一个方法定义 onVideoCapabilitiesChanged，在视频电话能力变化后将进行消息回调；

在 AccountEntry 的构造方法中，registerPstnPhoneAccount 将创建 PhoneAccount 对象，其主要逻辑简化后如下。

```
private PhoneAccount registerPstnPhoneAccount(boolean isEmergency,
    boolean isDummyAccount) {
    //创建 PhoneAccountHandle 对象
    PhoneAccountHandle phoneAccountHandle =
            PhoneUtils.makePstnPhoneAccountHandleWithPrefix(
                    mPhone, dummyPrefix, isEmergency);

    //通过 Phone 对象获取一些基础数据
    int subId = mPhone.getSubId();
    String subscriberId = mPhone.getSubscriberId();

    //默认所有 SIM 电话账户都能拨打紧急电话
    int capabilities = PhoneAccount.CAPABILITY_SIM_SUBSCRIPTION |
            PhoneAccount.CAPABILITY_CALL_PROVIDER |
            PhoneAccount.CAPABILITY_MULTI_USER;
    //当前 Phone 对象具备的通信能力处理逻辑
    if (mContext.getResources().getBoolean(R.bool.config_pstnCanPlaceEmergencyCalls)) {
        capabilities |= PhoneAccount.CAPABILITY_PLACE_EMERGENCY_CALLS;
    }
    ......
    PhoneAccount account = PhoneAccount.builder(phoneAccountHandle, label)
            .setAddress(Uri.fromParts(PhoneAccount.SCHEME_TEL, line1Number, null))
            .setSubscriptionAddress(
                    Uri.fromParts(PhoneAccount.SCHEME_TEL, subNumber, null))
            .setCapabilities(capabilities)
            .setIcon(icon)
            .setHighlightColor(color)
            .setShortDescription(description)
            .setSupportedUriSchemes(Arrays.asList(//默认有两个 URI 支持列表
                    PhoneAccount.SCHEME_TEL,
                    PhoneAccount.SCHEME_VOICEMAIL))
            .setExtras(extras)
            .setGroupId(groupId)
            .build();
```

```
    // 向Telecom应用注册Account
    mTelecomManager.registerPhoneAccount(account);

    return account;
}
```

上面代码的核心逻辑可总结为以下几点。

（1）同步创建PhoneAccountHandle对象。

PhoneAccountHandle 对象的 id 通过 phone.getFullIccSerialNumber 获取，即当前 SIM 卡的 ICCID；ComponentName 对象的构造方法为：new ComponentName("com.android.phone", "com.android.services.telephony.TelephonyConnectionService")，是不是非常熟悉呢？它就是 TeleService 系统应用中的 IConnectionService 服务；PhoneAccountHandle 对象的构造方法 this(componentName, id, Process.myUserHandle())，用来获取 com.android.phone 进程的 UserHandle，其为 SYSTEM 类型。

（2）通过 GsmCdmaPhone 对象获取一些基础数据，如 SubId、SlotId、SubscriptionInfo 等信息。

（3）通过取值 capabilities 配置信息计算能力。

（4）根据前面获取的信息创建 PhoneAccount 对象。

（5）使用 TelecomManager 调用 registerPhoneAccount 接口注册 PhoneAccount。

3．小结

PhoneAccount 对象在 TeleService 系统应用中创建，它是 Parcelable 类型，支持序列化和反序列化操作，可以跨进程传递此对象。

PhoneAccount 对象的 mAccountHandle 属性为 PhoneAccountHandle 类型，mId 取值是 GsmCdmaPhone 对象读取 SIM 卡的 ICCID，与 GsmCdmaPhone 对象建立了对应关系。

5.3.2　PhoneAccount 注册响应

调用 TelecomManager 的 registerPhoneAccount 方法注册 PhoneAccount，Telecom 系统应用中的调用过程总结如下：TelecomManager.registerPhoneAccount→getTelecomService(). registerPhoneAccount→TelecomServiceImpl.mBinderImpl.registerPhoneAccount→mPhoneAccountRegistrar. registerPhoneAccount→addOrReplacePhoneAccount。

Telecom 系统应用中 PhoneAccountRegistrar 类的 addOrReplacePhoneAccount 方法将响应 TeleService 系统应用发起的注册 PhoneAccount 请求，其关键业务逻辑详情如下。

```
private void addOrReplacePhoneAccount(PhoneAccount account) {
    mState.accounts.add(account);//更新List<PhoneAccount> accounts列表

    write();//写入本地文件，记录PhoneAccount信息
    fireAccountsChanged();//通知PhoneAccount变化
    if (isNewAccount) {//通知PhoneAccount已经注册
        fireAccountRegistered(account.getAccountHandle());
    }
}
```

TeleService 系统应用中创建的 PhoneAccount 对象，在 Telecom 系统应用中有两种表现形式：

- 内存，保存在 PhoneAccountRegistrar 对象的 mState.accounts 属性列表。
- 存储，写入 XML 文件。

>>>>>>>>>> 第 5 章 详解 TeleService

从 Nexus 6P 手机中导出记录 PhoneAccount 的 XML 文件，导出使用的 adb pull 命令及此文件的关键内容信息详情如下。

```
$ adb pull /data/user_de/0/com.android.server.telecom/files/phone-account-registrar-state.xml .
<phone_account_registrar_state version="9">//版本号 9
<default_outgoing />
<accounts>
<phone_account>
<account_handle>
<phone_account_handle>
<component_name>com.android.phone/com.android.services.telephony.TelephonyConnectionService
</component_name>//服务信息
<id>8986011785110158XXXX</id>//SIM 卡的 ICCID
<user_serial_number>0</user_serial_number>//0 UserHandler，即系统用户
</phone_account_handle>
</account_handle>
<handle>tel:%2B86186XXXXXXXX</handle>//手机号码
<subscription_number>tel:%2B86186 XXXXXXXX </subscription_number>
<capabilities>54</capabilities>//GsmCdmaPhone 对象的通话能力取值
<icon>***</icon>
<highlight_color>0</highlight_color>
<label>CHN-UNICOM</label>
<short_description>CHN-UNICOM</short_description>
<supported_uri_schemes length="2">//固定的 URI
<value>tel</value>
<value>voicemail</value>
</supported_uri_schemes>
<extras>......</extras>
<enabled>true</enabled>
<supported_audio_routes>15</supported_audio_routes>
</phone_account>
</accounts>
</phone_account_registrar_state>
```

再次验证了 PhoneAccountHandle 记录的 ICCID、UserHandler 和 component_name 与 TeleService 中创建的对象是一致的。

另外，我们需要重点关注 capabilities 和 supported_uri_schemes，即 GsmCdmaPhone 对象提供的通话能力和支持的 URI。

注意　在不插入 SIM 卡的情况下导出 Nexus 6P 手机中的 phone-account-registrar-state.xml 文件，请读者对比下插卡和不插卡的区别是什么？

5.3.3　PhoneAccount 在拨号流程中的作用分析

前面两个小节仅从 TeleService 系统应用中 PhoneAccount 对象的创建和 Telecom 系统应用中响应 PhoneAccount 对象的注册两个方面进行了分析。那么 PhoneAccount 究竟有什么作用呢？本节继续从拨号流程和来电流程进行梳理和分析。

Telecom 系统应用中响应的拨号请求处理逻辑，最关键的是 CallsManager 对象的 startOutgoingCall 和 placeOutgoingCall 方法调用，它们分别创建 Call 对象和发起 Connection 请求，这两个方法都涉及 PhoneAccount 对象的使用和传递。

- 109 -

1. startOutgoingCall

CallIntentProcessor.processOutgoingCallIntent 作为 Telecom 系统应用响应拨号请求的处理入口,通过 Intent 获取 PhoneAccountHandle 对象,并调用 callsManager.startOutgoingCall 方法创建 Call 对象,因传入的 PhoneAccountHandle 对象为 NULL,连续两次的 mPhoneAccountRegistrar.getPhoneAccount 方法调用均返回了 NULL,从而调用 constructPossiblePhoneAccounts 方法获取到由 PhoneAccountRegistrar 保存的已注册 PhoneAccount 对象,调用过程如下:constructPossiblePhoneAccounts→mPhoneAccountRegistrar.getCallCapablePhoneAccounts→getPhoneAccountHandles→getPhoneAccountHandles→getPhoneAccounts。

其中,最关键的是 getPhoneAccounts 方法中的处理逻辑,详情如下。

```java
private List<PhoneAccount> getPhoneAccounts(
        int capabilities,
        int excludedCapabilities,
        String uriScheme,
        String packageName,
        boolean includeDisabledAccounts,
        UserHandle userHandle) {
    List<PhoneAccount> accounts = new ArrayList<>(mState.accounts.size());
    for (PhoneAccount m : mState.accounts) {//循环 TeleService 注册的 PhoneAccount
        ......
        if (capabilities != 0 && !m.hasCapabilities(capabilities)) {//匹配能力
            // Account doesn't have the right capabilities; skip this one.
            continue;
        }
        if (uriScheme != null && !m.supportsUriScheme(uriScheme)) {//匹配 URI
            // Account doesn't support this URI scheme; skip this one.
            continue;
        }
        ......
        accounts.add(m);
    }
    return accounts;
}
```

startOutgoingCall 中的逻辑根据拨号请求匹配到 TeleService 注册的 PhoneAccount 对象,然后通过 call.setTargetPhoneAccount 调用,将 PhoneAccount 与 Call 产生了关联。

2. placeOutgoingCall

placeOutgoingCall 发起的拨号请求的主要调用过程如下:placeOutgoingCall→call.startCreateConnection→CreateConnectionProcessor.process→attemptNextPhoneAccount。

CreateConnectionProcessor.attemptNextPhoneAccount 方法中有关 PhoneAccount 的处理逻辑,是 mPhoneAccountRegistrar.phoneAccountRequiresBindPermission 调用最终会通过 PackageManager 判断 Telephony 提供的 IConnectionService 是否可用,代码逻辑详情如下。

```java
private List<ResolveInfo> resolveComponent(ComponentName componentName,
        UserHandle userHandle) {
    PackageManager pm = mContext.getPackageManager();
    Intent intent = new Intent(ConnectionService.SERVICE_INTERFACE);
    intent.setComponent(componentName);
    try {
        if (userHandle != null) {
            return pm.queryIntentServicesAsUser(intent, 0, userHandle.getIdentifier());
        } else {
            return pm.queryIntentServices(intent, 0);
```

```
            }
        } ......
}
```

3. createConnection

Telecom 首先匹配到已注册的 PhoneAccount，然后通过 PhoneAccount 判断 TeleService 提供的 IConnectionService 服务是否可用。在这两个条件都满足的请求下，createConnection 继续发起拨号请求；否则调用 notifyCallConnectionFailure 发出拨号失败的通知。

继续查看 ConnectionServiceWrapper.createConnection 方法，主要代码逻辑如下。

```
ConnectionRequest connectionRequest = new ConnectionRequest.Builder()
                .setAccountHandle(call.getTargetPhoneAccount())
......
mServiceInterface.createConnection(
                call.getConnectionManagerPhoneAccount(),
                callId,
                connectionRequest,
                call.shouldAttachToExistingConnection(),
                call.isUnknown(),
                Log.getExternalSession());
```

创建的 ConnectionRequest 对象包含了 PhoneAccount 对象，调用 IConnectionService 服务接口 createConnection 传递 PhoneAccount 对象和 ConnectionRequest 对象。

4. IConnectionService

PhoneAccount 对象转了一圈之后又回到了 TeleService 系统应用，在 IConnectionService 的 createConnection→onCreateOutgoingConnection 方法中，经过 getPhoneForAccount 调用最终获取了 GsmCdmaPhone 对象。

GsmCdmaPhone 与 PhoneAccountHandle 对象是怎么对应起来的呢？原来是通过 PhoneAccountHandle 的 mId 即 ICCID 与 GsmCdmaPhone 对象进行的关联，其核心逻辑详情如下。

```
private static Phone getPhoneFromIccId(String iccId) {
    if (!TextUtils.isEmpty(iccId)) {
        for (Phone phone : PhoneFactory.getPhones()) {//循环 Phone 对象数组
            String phoneIccId = phone.getFullIccSerialNumber();//获取 Phone 的 ICCID
            if (iccId.equals(phoneIccId)) {//匹配 ICCID
                return phone;//返回 GsmCdmaPhone 对象
            }
        }
    }
    return null;
}
```

5. 扩展来电流程

回顾一下 TeleService 系统处理来电请求，它是使用 TelecomManager 发起对 addNewIncomingCall 的调用过程：

PstnIncomingCallNotifier.handleNewRingingConnection→sendIncomingCallIntent，其关键的处理逻辑详情如下。

```
//创建本地 PhoneAccountHandle 对象
PhoneAccountHandle handle = findCorrectPhoneAccountHandle();
TelecomManager.from(mPhone.getContext()).addNewIncomingCall(handle, extras);
```

findCorrectPhoneAccountHandle 方法首先通过 GsmCdmaPhone 对象获取 ICCID，然后创建

"com.android.phone" 和 "com.android.services.telephony.TelephonyConnectionService" 类型的 ComponentName 对象，最终构造 PhoneAccountHandle 对象。

Telecom 系统应用中的 TelecomServiceImpl 响应 TeleService 应用的跨进程服务接口调用，将检查调用方是否合法，关键的代码逻辑详情如下。

```
mAppOpsManager.checkPackage(
        Binder.getCallingUid(),
        phoneAccountHandle.getComponentName().getPackageName());
```

AppOpsManager.checkPackage 是 Android 系统提供的接口，用来判断 UID 与 package name 是否一致，不一致将抛出 SecurityException 异常。Binder.getCallingUid()将获取到 TeleService 系统应用即 com.android.phone 进程的 UID，传入的 package name 则是"com.android.phone"，加上对 TelephonyUtil.isPstnComponentName 的判断。所以非 TeleService 系统应用调用 TelecomManager 的 addNewIncomingCall 方法时将抛出 SecurityException 异常，增强了服务接口的安全性。

Telecom 系统应用接收到来电消息，创建 Call 对象后向 TeleService 系统应用发起 Create Connection 请求，与拨号流程中 PhoneAcount 的作用一致。

5.3.4 小结

PhoneAccount 是在 TeleService 系统应用中创建的，通过 PhoneAccountHandle 的 Id 来标识或关联 GsmCdmaPhone 对象，ComponentName 对象则用来标识 IConnectionService 服务。Telecom 系统应用接收到 PhoneAccount 注册后，将保存在 PhoneAccountRegistrar 对象的 List<PhoneAccount>列表中。

产生通话业务时，Telecom 系统应用根据 List<PhoneAccount>列表判断是否匹配对应的通话能力，最终保存 PhoneAccountHandle 信息到 ConnectionRequest 对象，发起创建 Connection 请求。TeleService 系统应用接收到请求后，解析出 ConnectionRequest 对象中的 PhoneAccountHandle 信息，通过 Id 即 SIM 卡的 ICCID 获取对应的 GsmCdmaPhone 对象，最后使用 GsmCdmaPhone 对象发起对应的通话管理或控制请求。

PhoneAccount 核心类图如图 5-6 所示。

图 5-6 所示的 PhoneAccount 核心类图，我们需要重点掌握以下几点。

- PhoneAccount 是在 TeleService 系统应用中创建的，注册到 Telecom 系统应用。
- PhoneAcount 的唯一标识为 PhoneAccountHandle。
- PhoneAccountHandle 通过 mId 即 ICCID 与 GsmCdmaPhone 对象产生了唯一关联。
- PhoneAccount 的主要功能是描述 GsmCdmaPhone 对象的通话能力，其中最关键的是 capabilities 和 supported_uri。
- Telecom 系统应用通过拨号请求信息匹配对应的 PhoneAcount（能力匹配和 IConnectionService 服务匹配），过滤非法或不支持的拨号请求。
- TeleService 系统应用接收 createConnection 请求，通过 PhoneAcount 找出关联的 GsmCdmaPhone 对象，支持对应的通话管理和控制请求。
- PhoneAcount 在 Telecom 和 TeleService 两个系统应用间流转，保障了通话相关请求的正常响应，过滤掉非法或不支持的通话请求消息。

第 5 章 详解 TeleService

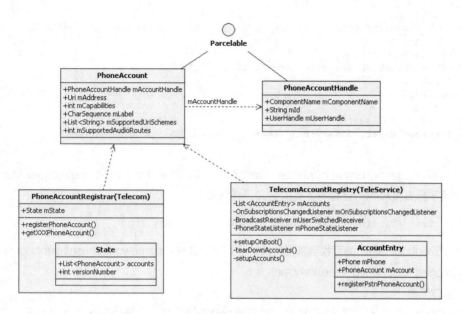

图 5-6 PhoneAccount 核心类图

5.4 TeleService 服务

TeleService 系统应用中提供了两种类型的服务：系统级服务和应用级服务。

● 系统级服务

TeleService 系统应用加载过程中将创建和发布的系统服务，如 phone、isub 等系统服务。Android 系统平台中的所有应用，均可通过 ServiceManager.getService 的方式获取服务的 Binder 对象，从而访问系统服务接口。

● 应用级服务

即 IConnectionService 应用服务，通话业务中 Telecom 系统应用将通过绑定服务的方式访问 TeleService 系统应用提供的服务。

5.4.1 phone 系统服务

phone 系统服务是在 TeleService 系统应用加载过程中创建和发布的，其初始化调用过程如下：PhoneApp.onCreate→PhoneGlobals.onCreate→PhoneInterfaceManager.init，关键处理逻辑详情如下：

```
/* package */ static PhoneInterfaceManager init(PhoneGlobals app, Phone phone) {
    synchronized (PhoneInterfaceManager.class) {//同步锁
        if (sInstance == null) {//单例的实现方式
            sInstance = new PhoneInterfaceManager(app, phone);
        } else {
            Log.wtf(LOG_TAG, "init() called multiple times!  sInstance = " + sInstance);
        }
        return sInstance;
    }
```

```
}
private PhoneInterfaceManager(PhoneGlobals app, Phone phone) {//私有构造方法
    mApp = app;
    mPhone = phone;
    mCM = PhoneGlobals.getInstance().mCM;
    ......
    publish();
}
private void publish() {
    ServiceManager.addService("phone", this);//加入phone名称的系统服务
}
```

ServiceManager.addService("phone", this)增加了系统服务this，即PhoneInterfaceManager对象，它实现了什么接口呢？查看一下类的定义，详情如下。

```
public class PhoneInterfaceManager extends ITelephony.Stub
```

通过ITelephony.Stub类可找到其接口的AIDL定义文件：frameworks/base/telephony/java/com/android/internal/telephony/ITelephony.aidl。

1. 接口汇总

ITelephony.aidl共定义了41个接口，对其进行分类和汇总，主要接口如表5-4所示。

表5-4 ITelephony接口汇总

分类	接口名称	说明
通话	dial\call\endCall\answerRingingCall	通话控制相关接口，如拨号、接听、挂断等
	isOffhook\isRinging\isIdle	获取当前通话状态
状态服务	toggleRadioOnOff\setRadio\setRadioPower	开关飞行模式
	getNetworkType\getSignalStrength\getCellLocation\isRadioOn	获取网络类型、信号量、小区信息等
移动数据	enableDataConnectivity\disableDataConnectivity setDataEnabled\getDataEnabled	开关移动数据
	getDataState\getDataActivity	获取移动数据状态
SIM卡操作	supplyPin\supplyPuk	PIN/PUK解锁

2. PhoneInterfaceManager实现机制分析

PhoneInterfaceManager实现了ITelephony.aidl文件中定义的41个接口，大部分的接口逻辑符合以下两个规律。

- 获取GsmCdmaPhone对象进行对应的操作。

```
public void XXX() {//获取默认的subId进行XXXForSubscriber调用
    XXXForSubscriber(getDefaultSubscription());
}
public void XXXForSubscriber(int subId) {
    enforceModifyPermission();//检查权限
    final Phone phone = getPhone(subId);//通过subId获取GsmCdmaPhone对象
    if (phone != null) {
        phone.XXX();//执行GsmCdmaPhone对象对应的操作
    }
}
```

- GsmCdmaPhone 对象的操作主要有两类：发出控制请求和查询相关属性，其主要的调用方式：phone.setXXX（phone.XXX）或 phone.getXXX。

名字为"phone"的系统服务，是由 PhoneInterfaceManager 实现的 ITelephony.aidl 接口定义，运行在 com.android.phone 进程中。

请读者思考该系统服务为什么取名为"phone"？将 Facade Pattern 应用到 phone 名称的系统服务中。

5.4.2 isub 系统服务

isub 系统服务同样是在 TeleService 系统应用加载过程中创建和发布的，其初始化调用过程如下：

PhoneApp.onCreate → PhoneGlobals.onCreate → PhoneFactory.makeDefaultPhones → makeDefaultPhone→SubscriptionController.init。SubscriptionController 类的静态方法 init 的关键业务逻辑详情如下。

```
public static SubscriptionController init(Context c, CommandsInterface[] ci) {
    synchronized (SubscriptionController.class) {//同步锁
        if (sInstance == null) {//单例的实现方式
            sInstance = new SubscriptionController(c);
        } else {
            Log.wtf(LOG_TAG, "init() called multiple times!  sInstance = " + sInstance);
        }
        return sInstance;
    }
}
protected SubscriptionController(Context c) {
    init(c);
}
protected void init(Context c) {
    mContext = c;
    mCM = CallManager.getInstance();
    ......
    if(ServiceManager.getService("isub") == null) {
            ServiceManager.addService("isub", this);//加入 isub 名称的系统服务
    }
}
```

ServiceManager.addService("isub", this)增加了系统服务 this，即 SubscriptionController 对象，它实现了什么接口呢？查看该类的定义，详情如下。

```
public class SubscriptionController extends ISub.Stub
```

通过 ISub.Stub 类可找到其接口的 AIDL 定义文件：frameworks/base/telephony/java/com/android/internal/telephony/ISub.aidl。

1. 接口汇总

ISub.aidl 共定义了 34 个接口，对其进行分类和汇总，主要接口如表 5-5 所示。

2. SubscriptionController 实现机制分析

SubscriptionController 实现了 ISub.aidl 定义的 34 个接口，主要是对 SubscriptionInfo 的管理和查询接口，以及在多 SIM 卡模式下的默认卡管理接口。

表 5-5　ISub 接口汇总

分类	接口名称	说明
获取 SubscriptionInfo	getAllSubInfoList\getActiveSubscriptionInfoList\getAvailableSubscriptionInfoList\getAccessibleSubscriptionInfoList	获取不同状态的 SubscriptionInfo 列表
	getActiveSubscriptionInfo\getActiveSubscriptionInfoForIccId\getActiveSubscriptionInfoForSimSlotIndex	获取不同状态的 SubscriptionInfo 对象
管理 SubscriptionInfo	addSubInfoRecord\setDisplayName\setDisplayNumber\setDataRoaming	增加、设置显示名称，设置显示号码，设置数据漫游等
默认卡管理接口	setDefaultDataSubId\setDefaultVoiceSubId\setDefaultSmsSubId	默认数据卡、默认拨号、默认短信卡的设置接口（双卡或多卡生效）

SubscriptionInfo 的主要属性有：mId、mIccId、mSimSlotIndex、mDisplayName、mDataRoaming、mMcc、mMnc。

SubscriptionController 类有两个非常重要的私有方法：getSubInfo 和 getSubInfoRecord。getSubInfo 方法通过 SubscriptionManager.CONTENT_URI 即 Uri.parse("content://telephony/siminfo")获取查询 SQLite 数据库的 cursor，再由 getSubInfoRecord 方法使用 cursor 查询到的数据创建 SubscriptionInfo。而 getSubscriptionInfoListForEmbeddedSubscriptionUpdate、getAccessibleSubscriptionInfoList、getAvailableSubscriptionInfoList、getAllSubInfoList、refreshCachedActiveSubscriptionInfoList 这五个方法使用不同的 SQL where 条件语句作为参数调用 getSubInfo 来查询 telephony 数据库的 siminfo 表中的数据，最终返回 SubscriptionInfo 的 List 列表。

导出 Nexus 6P 手机中的 Telephony SQLite 数据库，导出使用的 adb 命令和 siminfo 数据库表的关键信息如下。

```
adb pull /data/user_de/0/com.android.providers.telephony/databases/telephony.db .

----------------------------------------------------------------
Table [siminfo]//siminfo 数据库表
    Fields: 28 //28 个字段，这里做了一些省略
        [_id]: INTEGER
        [icc_id]: TEXT NOT NULL
        [sim_id]: INTEGER DEFAULT '-1'
        [display_name]: TEXT
        [carrier_name]: TEXT
        [number]: TEXT
        ......
    Foreign Keys: 0
    Indexes: 1 //1 个数据库索引
        [] PRIMARY //Primary 类型的数据库索引
            [_id] AUTOINCREMENT //索引创建在字段"_id"上，并且具有自动递增属性
----------------------------------------------------------------
//Nexus 6P 手机是单卡的，表中只有一条记录，这里仅展示五个字段的内容，也请读者导出 telephony.db 文件查看该表的相关信息。
_id         icc_id                  sim_id  display_name    carrier_name
1           8986011785110158XXXX    0       CARD 1          CHN-UNICOM
```

isub 系统服务都是围绕 siminfo 数据库表来运行的，对比 siminfo 数据库表的字段信息与 Java 实体类属

性 SubscriptionInfo，SubscriptionInfo 对象的数据全部可以从 siminfo 数据库表中获取。那么，siminfo 数据库表中的数据是从哪里获取并插入到数据库中的呢？总结其关键流程如图 5-7 所示。

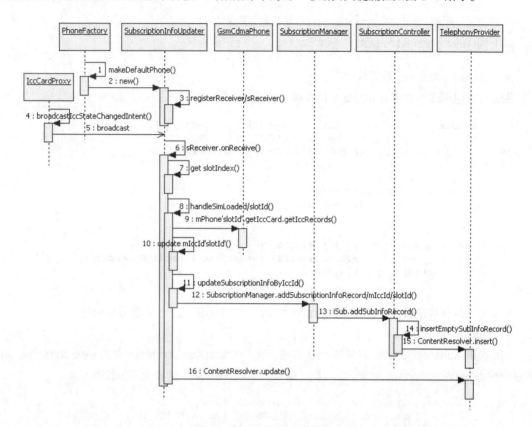

图 5-7　siminfo 数据库管理流程

图 5-7 所示的 siminfo 数据库管理流程，我们需要重点关注以下几点。

- SubscriptionInfoUpdater

在 PhoneFactory 加载 Telephony 业务模型时，将初始化 SubscriptionInfoUpdater，主要是创建广播接收器，接收 TelephonyIntents.ACTION_SIM_STATE_CHANGED 和 IccCardProxy.ACTION_INTERNAL_SIM_STATE_CHANGED 两个广播。

- 响应 SIM 卡状态变化广播

对应 Nexus 6P 手机的插卡和拔卡操作，Telephony 运行的业务模型将通过 IccCardProxy 发送 SIM 卡状态变化广播。图 5-7 中步骤 5 的关键处理逻辑详情如下。

```
Intent intent = new Intent(TelephonyIntents.ACTION_SIM_STATE_CHANGED);
SubscriptionManager.putPhoneIdAndSubIdExtra(intent, mPhoneId);
public static void putPhoneIdAndSubIdExtra(Intent intent, int phoneId) {
    int[] subIds = SubscriptionManager.getSubId(phoneId);//通过 phoneId 获取 subId
    if (subIds != null && subIds.length > 0) {
        putPhoneIdAndSubIdExtra(intent, phoneId, subIds[0]);
    }......
}
public static void putPhoneIdAndSubIdExtra(Intent intent, int phoneId, int subId) {
    intent.putExtra(PhoneConstants.SUBSCRIPTION_KEY, subId);
    intent.putExtra(EXTRA_SUBSCRIPTION_INDEX, subId);
    intent.putExtra(PhoneConstants.PHONE_KEY, phoneId);
```

```
        intent.putExtra(PhoneConstants.SLOT_KEY, phoneId);
}
```

请读者思考，subId、phoneId 有什么关联和区别呢？
同样的，phoneId 与 slotId（SimSlotIndex）是否相同呢？

- phoneId 与 slotId 的关系

图 5-7 中步骤 7、步骤 8 和步骤 9 的关键业务逻辑详情如下。

```
int slotIndex = intent.getIntExtra(PhoneConstants.PHONE_KEY,
            SubscriptionManager.INVALID_SIM_SLOT_INDEX);
sendMessage(obtainMessage(EVENT_SIM_LOADED, slotIndex, -1))
public void handleMessage(Message msg) {
        switch (msg.what) {
            case EVENT_SIM_LOADED:
                handleSimLoaded(msg.arg1);
                break;
}
private void handleSimLoaded(int slotId) {
    IccRecords records = mPhone[slotId].getIccCard().getIccRecords();
    mIccId[slotId] = records.getIccId();
}
```

根据上面的代码逻辑，可以确认 phoneId 与 slotId 的名称不同，但取值是相同的。

3. subId

subId 是 SubscriptionInfo 对象的 mId 取值，在 SubscriptionController 类的 getSubInfoRecord 方法中构造 SubscriptionInfo 对象时，subId 取 siminfo 数据库表中的哪个字段值呢？

```
int id = cursor.getInt(cursor.getColumnIndexOrThrow(
            SubscriptionManager.UNIQUE_KEY_SUBSCRIPTION_ID));
public static final String UNIQUE_KEY_SUBSCRIPTION_ID = "_id";
```

原来，subId 是 siminfo 数据库表中的"_id"字段取值，它唯一标识着该表中保存的 SIM 卡信息记录。

4. 区分 slotId、phoneId、subId

slotId：代码中以 simId、SimSlotIndex、SlotIndex 等方式出现，保存在 siminfo 数据库的 sim_id 字段中，与 phoneId 一一对应；

phoneId：GsmCdmaPhone 对象数组的下标；

subId：siminfo 数据库中的"_id"字段，即 siminfo 数据唯一标识。

因此，phoneId 与 subId 的对应关系可体现在 SubscriptionInfo 对象的 mSimSlotIndex 与 mId 属性，对应 siminfo 数据库的 sim_id 与_id 字段。

5.4.3 IConnectionService 应用服务

回顾一下 Telecom 交互模型绑定 IConnectionService 服务的处理机制，IConnectionService 服务接收到 createConnection 请求后，通过 ConnectionRequest 对象获取相关信息创建 TelephonyConnection 对象，在完成通话相关操作后，通过 TelephonyConnection 对象的相关信息创建 ParcelableConnection 对象并返回给 Telecom 进程。

通话业务流程涉及与 Connection 相关的类名或对象名非常多，很容易混淆，总结其核心类图如

图 5-8 所示。

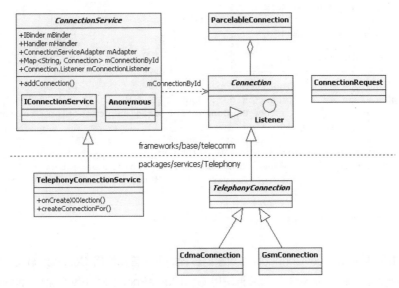

图 5-8　IConnectionService 核心类图

IConnectionService 核心类图需要重点关注以下两点：
- 代码归属

IConnectionService 核心类图的代码归属于两个库：frameworks/base/telecomm（Android Framework）和 packages/services/Telephony（TeleService 系统应用）。
- 区分运行空间

TelephonyConnectionService 继承抽象类 ConnectionService，运行在 TeleService 系统应用空间，承载 IConnectionService 服务；TelephonyConnection 继承抽象类 Connection，同样运行在 TeleService 系统应用空间。

1．区分 Connection

ConnectionRequest 和 ParcelableConnection 对象均实现了 Parcelable 接口，可跨进程在 Telecom 和 TeleService 两个系统应用进程间传递。

ConnectionRequest 对象在 Telecom 系统应用中创建，传递到 TeleService 系统应用；ParcelableConnection 对象在 TeleService 系统应用中创建，传递到 Telecom 系统应用。

TelephonyConnection 对象在 TeleService 系统应用中创建，作为普通 Java 对象，仅在 com.android.phone 进程空间运行。

2．TelephonyConnection 对象的创建过程

TelephonyConnection 对象的创建过程可分为两个逻辑：拨号和来电。

在拨号流程中，TelephonyConnection 对象的创建过程如下：TelephonyConnectionService.createConnection→onCreateOutgoingConnection→getTelephonyConnection，其核心逻辑详情如下。

```
final Phone phone = getPhoneForAccount(request.getAccountHandle(), isEmergencyNumber);
final TelephonyConnection connection =
        createConnectionFor(phone, null, true /* isOutgoing */, request.getAccountHandle(),
            request.getTelecomCallId(), request.getAddress(), request.getVideoState());
//创建 TelephonyConnection 对象
private TelephonyConnection createConnectionFor(
```

```
            Phone phone,
            com.android.internal.telephony.Connection originalConnection,
            boolean isOutgoing,
            PhoneAccountHandle phoneAccountHandle,
            String telecomCallId,
            Uri address,
            int videoState) {
    TelephonyConnection returnConnection = null;
    int phoneType = phone.getPhoneType();
    if (phoneType == TelephonyManager.PHONE_TYPE_GSM) {
        returnConnection = new GsmConnection(originalConnection, telecomCallId,
                isOutgoing);
    } else if (phoneType == TelephonyManager.PHONE_TYPE_CDMA) {
        boolean allowsMute = allowsMute(phone);
        returnConnection = new CdmaConnection(originalConnection,
                mEmergencyTonePlayer, allowsMute, isOutgoing, telecomCallId);
    }
    ......
    return returnConnection;
}
```

拨号流程中创建 TelephonyConnection 对象的过程如下：首先使用 ConnectionRequest 对象选择将要使用的 GsmCdmaPhone 对象，然后再使用 GsmCdmaPhone 对象和 ConnectionRequest 对象的 PhoneAccountHandle、TelecomCallId 和 Address(URI)等属性创建 TelephonyConnection 对象。

来电流程中创建 TelephonyConnection 对象的过程如下：TelephonyConnectionService.createConnection→onCreateIncomingConnection，其核心逻辑详情如下。

```
Phone phone = getPhoneForAccount(accountHandle, isEmergency);
Connection connection =
        createConnectionFor(phone, originalConnection, false /* isOutgoing */,
                request.getAccountHandle(), request.getTelecomCallId(),
                request.getAddress(), videoState);
```

与拨号流程中创建 TelephonyConnection 对象的过程非常相似。

因此，不管是拨号流程还是来电流程，对象 TelephonyConnection 的创建都依赖于 ConnectionRequest 对象的属性（语义上的关系）。

3. setOriginalConnection

TelephonyConnectionService 在成功创建 TelephonyConnection 对象后，还有一个非常关键的操作，就是 setOriginalConnection，即通过该调用将 Telephony Voice Call 业务模型关联在一起。

拨号流程中由 TelephonyConnectionService 类的 onCreateOutgoingConnection 方法创建 TelephonyConnection 对象，再调用 placeOutgoingConnection 方法，使用 GsmCdmaPhone 对象继续发起拨号请求调用，代码逻辑总结如下。

```
originalConnection = phone.dial(number, null, videoState, extras);
connection.setOriginalConnection(originalConnection);
```

GsmCdmaPhone 对象发起拨号请求调用后，返回 com.android.internal.telephony.Connection 对象，即 Telephony Voice Call 语音通话模型的 Connection 对象与 TelephonyConnection 对象产生了关联。

来电流程中由 TelephonyConnectionService 类的 onCreateIncomingConnection 方法首先获取 com.android.internal.telephony.Connection 对象，然后再创建 TelephonyConnection 对象，代码逻辑总结如下。

```
com.android.internal.telephony.Connection originalConnection =
        call.getState() == Call.State.WAITING ?
            call.getLatestConnection() : call.getEarliestConnection();
Connection connection =
        createConnectionFor(phone, originalConnection, false /* isOutgoing */,
                request.getAccountHandle(), request.getTelecomCallId(),
                request.getAddress(), videoState);
```

因此，在通话流程中创建 TelephonyConnection 对象时都会初始化 originalConnection 属性，这样便与 Telephony Voice Call 语音通话模型建立起了关联关系。

4．TelephonyConnection 消息处理机制

TelephonyConnectionService.createConnection 方法响应 Telecom 系统应用发起的接口调用，首先创建 TelephonyConnection 对象，然后调用 addConnection 方法保存 Connection 对象并设置相关的 Listener，其代码逻辑详情如下。

```
private void addConnection(String callId, Connection connection) {
    connection.setTelecomCallId(callId);
    mConnectionById.put(callId, connection);//两个维度保存 Connection 对象
    mIdByConnection.put(connection, callId);
    connection.addConnectionListener(mConnectionListener);
    connection.setConnectionService(this);
}
```

注意

callId 是 Telecom 应用中创建 Call 对象时，为了区分不同 Call 对象生成的唯一编号。既然是唯一编号，自然可以唯一标识 TeleService 系统应用中的 TelephonyConnection 对象。Telecom 系统应用中的 Call 对象与 TeleService 系统应用中的 TelephonyConnection 对象通过 callId 产生了一对一的关联关系。

mConnectionById 和 mIdByConnection 两个列表分别从两个维度保存了 TelephonyConnection 对象，也体现了图 5-8 中 TelephonyConnectionService 与 Connection 的依赖关系。

Telecom 系统应用中保存着 Call 对象，如果当前手机正好有一路通话，不论是进行通话保持或挂断电话，在 Telecom 系统应用中都是通过 callId 找到对应的 Call 对象，然后跨进程调用 TeleService 系统应用提供的 IConnectionService 服务接口：hold 或 disconnect，其接口定义如下。

```
void disconnect(String callId, in Session.Info sessionInfo);
void hold(String callId, in Session.Info sessionInfo);
```

TelephonyConnectionService 响应 Telecom 系统应用发起的通话控制请求，首先是通过 callId 获取 Connection 对象，然后调用 Connection 对象的 onXXX 方法进行通话控制的调用，整理的代码框架详情如下。

```
if (mConnectionById.containsKey(callId)) {
    findConnectionForAction(callId, "disconnect").onXXX();
} else {
    findConferenceForAction(callId, "disconnect").onXXX();//会议电话的处理逻辑
}
private Connection findConnectionForAction(String callId, String action) {
    if (callId != null && mConnectionById.containsKey(callId)) {
        return mConnectionById.get(callId);//获取 TelephonyConnection 对象
    }
    return getNullConnection();
}
```

以接听电话的处理流程为例，通话控制消息的下发关键流程总结如图 5-9 所示。

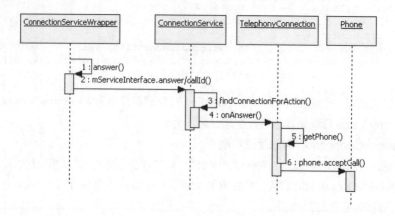

图 5-9　通话控制消息的下发流程总结

图 5-9 所示的通话控制下发流程中，需要重点关注以下两点：
- Telecom 系统应用发起通话控制请求的参数是 callId。
- TeleService 系统应用 ConnectionService 通过 callId 找到 TelephonyConnection 对象，并且通过此对象继续传递 Telecom 系统应用发起的通话控制请求。

那么当通话状态变化后 TelephonyConnectionService 如何通知 Telecom 系统应用呢？答案就在 connection.addConnectionListener(mConnectionListener)，即 TelephonyConnection 与 TelephonyConnectionService 的 mConnectionListener 属性产生的 Listener 消息关联中，图 5-8 中的 TelephonyConnectionService 内部匿名类也体现了此关系，其代码框架详情如下。

```
private final Connection.Listener mConnectionListener = new Connection.Listener() {
    @Override
    public void onStateChanged(Connection c, int state) {
        String id = mIdByConnection.get(c);
        switch (state) {
            case Connection.STATE_ACTIVE:
                mAdapter.setActive(id);
                break;
            case Connection.STATE_DIALING:
                mAdapter.setDialing(id);
                break;
            ......
        }
    }

    @Override
    public void onDisconnected(Connection c, DisconnectCause disconnectCause) {
        String id = mIdByConnection.get(c);
        mAdapter.setDisconnected(id, disconnectCause);
    }
    ......
};
```

使用 TelephonyConnection 对象获取对应的 callId，再通过 mAdapter 调用 Telecom 系统应用中的服务来设置当前通话的最新状态。

总结通话状态变化消息的上报关键流程，如图 5-10 所示。

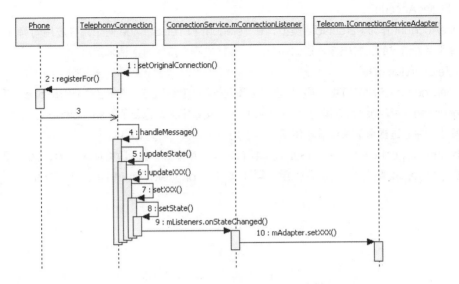

图 5-10　通话状态变化消息的上报关键流程

图 5-10 所示的通话状态变化消息的上报关键流程，我们需要重点关注以下几点。
- 步骤 1：拨号流程或是来电流程都会发起 setOriginalConnection 的操作，来电流程的处理方式与拨号流程稍微有些不同，就是 TelephonyConnection 构造方法中的调用。
- 步骤 3：TelephonyConnection 接收到 GsmCdmaPhone 发起的通话状态变化消息（两个对象在同一个进程：com.android.phone）。
- 步骤 9：ConnectionService 通过 mIdByConnection 列表，使用 TelephonyConnection 对象获取 callId。因此，步骤 10 传递的是 callId。
- Telecom 系统应用收到通话状态变化的信息后，通过 callId 匹配出对应的 Call 对象进行处理。

本 章 小 结

本章从 TeleService 系统应用的加载过程、Telephony Phone 业务模型、PhoneAcount 以及提供的 TeleService 服务四个方面，详细解析了 TeleService 系统应用的核心业务逻辑。
- TeleService 加载过程

TeleService 系统应用常驻内存，PhoneApp.onCreate 是该应用的加载入口，com.android.phone 进程中则运行着以 GsmCdmaPhone 对象为中心的 Telephony 业务模型；

TeleService 加载过程概括如下：加载 Telephony 业务模型（创建以 GsmCdmaPhone 对象为中心的核心业务对象），完成 PhoneAccount 初始化注册操作，发布名字为 phone、isub 的两个系统服务。
- Telephony Phone 业务模型

GsmCdmaPhone 与三大 Tracker：GsmCdmaCallTracker、ServiceStateTracker、DcTracker 的 Composition（组合）关系以及 Façade Pattern（门面设计模式）的应用。GsmCdmaPhone 作为门面提供了 Voice Call 语音通话、ServiceState 网络服务和 Data Call 移动数据业务三大 Telephony 能力。

- PhoneAccount

PhoneAccount 作为 Parcelable 类型，在 Telecom 和 TeleService 两个系统应用间跨进程传递，保障了通话相关请求的正常响应，过滤掉非法、异常或不支持的通话请求消息。

- TeleService 服务

TeleService 系统应用在加载过程中同步加载和发布两个系统服务，服务名称分别是 phone 和 isub。

PhoneInterfaceManager 实现了 ITelephony.aidl 接口定义，包装了 GsmCdmaPhone 对象的部分能力和属性，并通过提供的系统服务供第三方应用使用。

SubscriptionController 实现了 ISub.aidl 接口定义，读取 siminfo 数据库表中的 SIM 卡相关信息，并通过提供的系统服务供第三方应用使用，掌握 slotId、phoneId、subId 的关系与区别。

第 6 章

Voice Call 语音通话模型

学习目标

- 掌握 GsmCdmaCallTracker 的 Handler 消息处理方式。
- 理解 GsmCdmaCallTracker 与 RILJ 的交互机制。
- 掌握 handlePollCalls 业务逻辑。
- 理解各种 Connection 对象的关联与区别。
- 掌握 DriverCall、Call、Connection 语音通话管理模型。
- 掌握 InCallUi 通话界面。
- 验证 Voice Call 语音通话模型并加强理解。

上一章对 TeleService 系统应用进行了非常详细的解析，在 Telephony Phone 业务模型中，介绍了 GsmCdmaPhone 与三大 Tracker：GsmCdmaCallTracker、ServiceStateTracker、DcTracker 的 Composition（组合）关系以及 Facade Paffern（门面设计模式）的应用，GsmCdmaPhone 作为 Facade（门面）提供了三大 Telephony 业务能力：

- Voice Call 语音通话
- ServiceState 网络服务
- Data Call 移动数据业务

接下来我们将围绕这三大 Telephony 业务展开学习，本章主要解析 Voice Call 语音通话业务模型，Voice Call 业务以 GsmCdmaCallTracker 为中心。

6.1 详解 GsmCdmaCallTracker

GsmCdmaCallTracker 从字面意思上可理解为语音通话业务的追踪者（Tracker），GsmCdmaPhone 对象将语音通话业务交给 GsmCdmaCallTracker 对象来管理和维护，主要完成以下两方面的业务逻辑：

- 查询语音通话状态
- 提供语音通话控制能力

GsmCdmaCallTracker 的类层次及继承关系如图 6-1 所示。

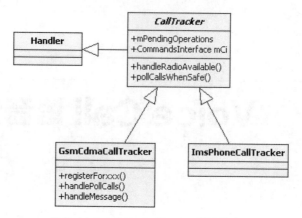

图 6-1　GsmCdmaCallTracker 继承关系图

CallTracker 抽象类是一个自定义的 Handler 消息处理类,它实现了 pollCallsWhenSafe、handleRadioAvailable 等重要方法。

CallTracker 抽象类有两个子类:GsmCdmaCallTracker 和 ImsPhoneCallTracker 类,分别在 CS 域(Circuit Switch,电路交换)域和 PS(Packet Switch,分组交换)域完成通话能力管理和控制的处理逻辑。本章主要解析 CS 域 GsmCdmaCallTracker 对应的语音通话业务。

本节将从代码结构、Handler 消息处理方式和 RILJ 对象交互机制三个方面入手,全面解析 GsmCdmaCallTracker 的运行机制。

6.1.1　代码结构解析

图 6-1 清楚地描述了 GsmCdmaCallTracker 类的继承关系,下面从类的关键属性和方法来总结 GsmCdmaCallTracker 类的代码结构。

1. 关键属性

GsmCdmaCallTracker 类的关键属性如表 6-1 所示。

表 6-1　GsmCdmaCallTracker 关键属性表

属性	类型	说明
mPendingOperations	int	挂起的请求操作计数器
mCi	CommandsInterface	RILJ 对象
mConnections[]	GsmCdmaConnection	GsmCdmaConnection 对象列表
mPendingMO	GsmCdmaConnection	主动拨号创建的 GsmCdmaConnection 对象
mRingingCall	GsmCdmaCall	接收到新来电请求的 Call 对象
mForegroundCall	GsmCdmaCall	第一路通话 Call 对象
mBackgroundCall	GsmCdmaCall	第二路通话 Call 对象
mPhone	GsmCdmaPhone	GsmCdmaPhone 对象
mState	PhoneConstants.State	手机状态

GsmCdmaCallTracker 关键属性表中需要注意以下几点：
- mCi：为 RILJ 对象，GsmCdmaCallTracker 对象与 GsmCdmaPhone 对象一样，通过 mCi 对象具备与 RIL 的交互能力。
- mState 属性：体现手机的通话状态，mRingingCall、mForegroundCall、mBackgroundCall 对象体现了通话状态及基本的通话信息。
- mConnections：最多能够保存 MAX_CONNECTIONS_GSM 或 MAX_CONNECTIONS_CDMA 个 GsmCdmaConnection 通话连接对象。
- mPhone：为 GsmCdmaPhone 对象，GsmCdmaPhone 与 GsmCdmaCallTracker 对象可相互引用对方。

2. 关键方法

GsmCdmaCallTracker 类的关键方法如表 6-2 所示。

表 6-2　GsmCdmaCallTracker 关键方法表

分类	方法	说明
通话控制接口	dial	请求拨号
	acceptCall	接听电话
	rejectCall	拒接电话
	switchWaitingOrHoldingAndActive	交换通话
	conference	多方通话
通话状态判断	canConference	判断是否能发起多方通话
	canDial	判断是否能请求拨号
Handler 消息处理	handleMessage	接收并处理 RILJ 发出的 Handler 消息
	operationComplete	与 RILJ 交互的回调处理
通话状态变化响应	pollCallsWhenSafe	向 RIL 发起 getCurrentCalls 查询 Call List 调用
	handlePollCalls	处理查询 Call List 结果
	updatePhoneState	更新手机状态并发出消息通知

汇总 GsmCdmaCallTracker 关键方法表，有两大类处理能力：
- 通话控制能力

GsmCdmaCallTracker 类提供语音通话控制方法来完成语音通话控制请求，如拨号请求、接听来电、拒接来电等通话能力控制请求。

- Handler 消息处理

GsmCdmaCallTracker 对象接收到 RILJ 对象发出的 Handler 消息后，能及时更新和记录当前时间点通话状态以及通话的基本信息，或是继续向 RILJ 对象发出请求。

6.1.2　Handler 消息处理方式

GsmCdmaCallTracker 本质上是自定义的 Handler 消息处理类，所以 Handler 消息的处理机制非常重要，接下来从它的 Handler 消息处理机制展开分析和学习。

通过前面的学习已经知道 Handler 有三种处理方式，在 GsmCdmaCallTracker 类中使用其中的两种运行和处理机制：

- 基本的 Handler 消息注册和响应处理机制
- Handler 消息 Callback（回调）处理方式

下面逐步分析 GsmCdmaCallTracker 类中的这两种 Handler 消息处理方式。

1. 消息注册

在 GsmCdmaCallTracker 类的构造方法中找到其 Handler 消息注册的代码逻辑，详情如下：

```
mCi.registerForCallStateChanged(this, EVENT_CALL_STATE_CHANGE, null);
mCi.registerForOn(this, EVENT_RADIO_AVAILABLE, null);
mCi.registerForNotAvailable(this, EVENT_RADIO_NOT_AVAILABLE, null);
```

mCi 是 RIL 类型的 Java 对象 RILJ，从上面的代码中可以看出，GsmCdmaCallTracker 对象会被动接收并响应 RILJ 对象发出的三种类型的 Handler 消息：

- EVENT_CALL_STATE_CHANGE 通话状态变化
- EVENT_RADIO_AVAILABLE 无线通信模块可用状态
- EVENT_RADIO_NOT_AVAILABLE 无线通信模块不可用状态

通过 Android 源码查证后可知，EVENT_CALL_STATE_CHANGE 通话状态变化的 Handler 消息，仅在 GsmCdmaCallTracker 类中响应；也就是说，仅有 GsmCdmaCallTracker 对象会接收和响应 RILJ 对象发出的通话状态变化通知。

2. 消息响应

handleMessage 方法接收并响应 RILJ 对象发出的 Handler 消息。

针对在构造方法中注册的这三种 Handler 消息，提取的代码处理逻辑详情如下：

```
case EVENT_CALL_STATE_CHANGE:
    pollCallsWhenSafe();
break;
case EVENT_RADIO_AVAILABLE:
    handleRadioAvailable();
break;
case EVENT_RADIO_NOT_AVAILABLE:
    handleRadioNotAvailable();
```

对上面的代码做进一步分析不难发现，handleRadioAvailable 和 handleRadioNotAvailable 这两个方法中的处理逻辑都会调用其父类 CallTracker 抽象类中实现的 pollCallsWhenSafe 方法，此方法究竟实现了什么逻辑呢？接下来，进入该方法，其处理逻辑详情如下：

```
protected void pollCallsWhenSafe() {//类型是 protected，保证子类能访问
    mNeedsPoll = true;
    if (checkNoOperationsPending()) {
        //创建 Message 消息对象
        mLastRelevantPoll = obtainMessage(EVENT_POLL_CALLS_RESULT);
        mCi.getCurrentCalls(mLastRelevantPoll);
    }
}
```

通过上面的代码可以看出，当 Voice Call 状态和 Radio 状态发生改变时，RILJ 对象会向 GsmCdmaCallTracker 对象发出这三个对应的 Handler 消息；GsmCdmaCallTracker 对象接收到这三个类型的 Handler 消息后，最终调用 mCi.getCurrentCalls 方法，向 RILJ 对象查询当前 Call List（通话列表）。

> 注意

在 GsmCdmaCallTracker 对象中调用 mCi.getCurrentCalls 方法，使用 Handler 创建的 Message 对象作为参数，RIL 对象使用它来完成回调，而 RILJ 对象的消息回调则由 GsmCdmaCallTracker 对象处理 EVENT_POLL_CALLS_RESULT 消息。

3．回调处理机制

GsmCdmaPhone 对象发起的通话管理和控制操作，实际上都是通过调用 mCT 相关的方法来完成的。mCT 即 GsmCdmaCallTracker 对象，如表 6-2 所示的 GsmCdmaCallTracker 关键方法中列举了一些基本的通话控制方法，如 dial、switchWaitingOrHoldingAndActive、acceptCall、rejectCall、hangup 等，这些方法实现的逻辑框架基本一致，全部采用 Handler 消息的回调处理机制，直接向 RILJ 对象发起通话管理和控制请求，主要分成两步完成。

- 调用 obtainCompleteMessage 方法创建 Message 对象。
- 使用 mCi 对象向 RILJ 对象发起通话管理和控制相关方法调用。

GsmCdmaCallTracker、RILJ 提供的通话管理和控制方法的对应关系如表 6-3 所示。

表 6-3　通话管理和控制方法对应关系表

CallTracker 方法	RILJ 方法	Handler 消息
dial	mCi.dial	EVENT_OPERATION_COMPLETE
acceptCall	mCi.acceptCall	EVENT_OPERATION_COMPLETE
rejectCall	mCi.rejectCall	EVENT_OPERATION_COMPLETE
hangup	mCi.hangup	EVENT_OPERATION_COMPLETE
switchWaitingOrHoldingAndActive	mCi.switchWaitingOrHoldingAndActive	EVENT_SWITCH_RESULT
conference	mCi.conference	EVENT_CONFERENCE_RESULT
explicitCallTransfer	mCi.explicitCallTransfer	EVENT_ECT_RESULT
separate	mCi. separateConnection	EVENT_SEPARATE_RESULT

表 6-3 所示的通话管理和控制方法对应关系表需要重点关注以下两点：

- GsmCdmaCallTracker 和 RILJ 提供的通话管理和控制方法的名称保持一致。
- RILJ 消息回调后的处理逻辑。

dial、acceptCall、rejectCall 和 hangup 这四个通话管理和控制方法创建的回调 Message 消息对象的类型都是 EVENT_OPERATION_COMPLETE，其他方法使用的 Message 消息对象类型也都不一样，它们之间有什么关系吗？那就是 RIL 对象发起消息回调后，在 CallTracker 对象中的响应和处理方式不同，其代码逻辑详情如下。

```
case EVENT_OPERATION_COMPLETE:
    operationComplete();
break;

case EVENT_CONFERENCE_RESULT:
    if (isPhoneTypeGsm()) {
        Connection connection = mForegroundCall.getLatestConnection();
        if (connection != null) {
            connection.onConferenceMergeFailed();
        }
    }
case EVENT_SEPARATE_RESULT:
case EVENT_ECT_RESULT:
case EVENT_SWITCH_RESULT:
```

```
        if (isPhoneTypeGsm()) {
            ar = (AsyncResult) msg.obj;
            if (ar.exception != null) {
                mPhone.notifySuppServiceFailed(getFailedService(msg.what));
            }
            operationComplete();
        } else {//CDMA 网络制式
            if (msg.what != EVENT_SWITCH_RESULT) {......
            }
        }
break;
```

上面的代码逻辑需要重点关注以下两点:
- GsmCdmaCallTracker 接收到通话管理和控制 RILJ 发出的回调的任何类型的 Handler 消息，最终都会调用 operationComplete 方法处理，从字面意思理解就是完成了通话管理和控制操作（GSM 网络制式）。
- 接收到非 EVENT_OPERATION_COMPLETE 类型 Handler 消息的情况下，增加了 phone.notify-SuppServiceFailed 方法调用；这个方法发出通话服务失败的消息通知，如无法切换通话、无法进行多方通话等消息通知。

接下来重点解析 operationComplete 方法中的处理逻辑，此方法究竟执行了什么操作？有什么样的处理逻辑呢？其代码详情如下。

```
private void operationComplete() {
    mPendingOperations--;
    if (mPendingOperations == 0 && mNeedsPoll) {
        mLastRelevantPoll = obtainMessage(EVENT_POLL_CALLS_RESULT);
        mCi.getCurrentCalls(mLastRelevantPoll);
    } else if (mPendingOperations < 0) {
        mPendingOperations = 0;
    }
}
```

看出有什么特殊之处了吗？原来在 CallTracker 抽象类中的 pollCallsWhenSafe 方法也有相同的处理逻辑，即向 RILJ 对象查询当前 Call List 列表。

6.1.3 与 RILJ 对象的交互机制

GsmCdmaCallTracker 与 RILJ 对象的交互完成了通话的控制，以及通话状态和通话基本信息的保存、更新，Telephony Call 通话模型中非常重要和关键。交互方式可分为两大类。
- CallTracker 对象主动发起
- CallTracker 对象被动接收

1. 主动发起请求

GsmCdmaCallTracker 对象发起 dial、acceptCall、rejectCall 等通话管理和控制请求方法调用时，会调用 RILJ 对象提供的对应通话管理和控制方法。下面以 rejectCall 拒接来电为例，关键代码详情如下。

```
public void rejectCall() throws CallStateException {//拒接来电请求
    // CHILD指令，通话保持或挂断电话
    // 非响铃状态时，将挂断通话保持的电话
    if (mRingingCall.getState().isRinging()) {
        mCi.rejectCall(obtainCompleteMessage());//调用 RIL 对象的拒接来电请求方法
```

```
        } else {
            throw new CallStateException("phone not ringing");
        }
    }
```

RIL 处理完 GsmCdmaCallTracker 对象发出的通话管理和控制请求后，由 RILJ 对象使用 Message 消息对象发出 Handler 消息通知，GsmCdmaCallTracker 对象中的 handleMessage 方法接收和响应此消息，至此，完成了第一次的 Handler 消息回调处理。

接下来 GsmCdmaCallTracker 对象会进行第二次的 Handler 消息回调处理流程，调用 operationComplete 方法来查询当前最新的 Call List 列表，其代码详情总结如下。

```
lastRelevantPoll = obtainMessage(EVENT_POLL_CALLS_RESULT);
cm.getCurrentCalls(lastRelevantPoll);
```

RIL 处理完 GsmCdmaCallTracker 对象发出的查询当前最新的 Call List 列表的请求后，由 RILJ 对象使用 Message 消息对象发出 Handler 消息通知，GsmCdmaCallTracker 对象中的 handleMessage 方法接收和响应此消息，至此，完成了第二次的 Handler 消息回调处理。

由 GsmCdmaCallTracker 对象主动向 RIL 对象发起通话管理和控制的交互方式，包括两次 Handler 消息回调的处理过程，总结如图 6-2 所示。

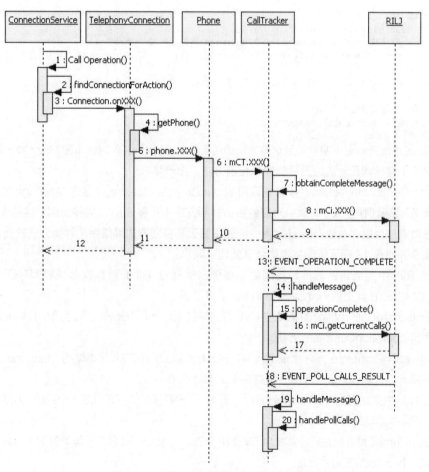

图 6-2　主动发起通话管理和控制的请求流程

图 6-2 中所示的主动发起通话管理和控制的请求流程详细描述了 Telephony Framework 业务模型与 RIL 之间进行通话管理和控制的交互过程，其中最关键的是 GsmCdmaCallTracker 对象与 RILJ 对象间的交互过程。在 Telephony Voice Call 语音通话模型中，GsmCdmaPhone 对象作为 Facade（门面），提供了基本的通话管理和控制方法。

图 6-2 中的几个关键点总结如下。

- ConnectionService 和 TelephonyConnection 是 TeleService 系统应用中的代码，运行在 com.android.phone 进程空间，用来接收 Telecom 系统应用发起的通话控制请求，并通过 Telecom callId 匹配到 TelephonyConnection 对象。
- 步骤 4 中，TelephonyConnection 通过 getPhone 方法，首先获取 GsmCdmaConnection 对象，然后获取到 GsmCdmaPhone 对象；步骤 5 继续调用 GsmCdmaPhone 对象的 dial、acceptCall、rejectCall 等通话管理和控制方法。步骤 4 中 getPhone 方法的简化逻辑如下：

```
Phone getPhone() {
    Call call = getCall();
    if (call != null) {
        //调用 mOwner.getPhone 方法，即 GsmCdmaCallTracker.getPhone 方法
        return call.getPhone();
    }
    return null;
}
protected Call getCall() {
    if (mOriginalConnection != null) {
        //获取 GsmCdmaConnection 的 mParent 属性，即 GsmCdmaCall
        return mOriginalConnection.getCall();
    }
    return null;
}
//getPhone 方法简化后的逻辑
Phone phone = mOriginalConnection.getCall().getPhone();
```

- GsmCdmaPhone 对象接收到这些方法调用请求后，使用 mCT 对象（即 GsmCdmaCallTracker 对象）继续调用对应的通话管理和控制方法，即步骤 6。
- GsmCdmaCallTracker 对象首先调用 obtainCompleteMessage 创建 Message 消息对象（步骤 7）接着调用 RILJ 对象中对应的通话控制方法（步骤 8），同时传递刚创建的 Message 消息对象作为回调参数。RILJ 对象中的通话管理和控制方法都采用了异步方式的处理机制，步骤 9 和步骤 17 都可体现异步的消息处理机制。
- 在 RIL 中完成通话管理和控制后，步骤 13 通过 RILJ 对象发起 EVENT_OPERATION_COMPLETE 类型的 Handler Callback 消息回调。
- GsmCdmaCallTracker 对象响应 EVENT_OPERATION_COMPLETE 类型的 Handler 消息，将调用 operationComplete 方法。
- operationComplete 方法生成 EVENT_POLL_CALLS_RESULT 类型的 Message 消息对象，调用 mCi. getCurrentCalls 开始查询 Call List 列表。
- 在 RIL 中完成查询 Call List 列表后，发起 EVENT_POLL_CALLS_RESULT 类型的 Handler Callback 消息回调。
- GsmCdmaCallTracker 对象响应 EVENT_POLL_CALLS_RESULT 类型的 Handler 消息，最终调用 handlePollCalls 方法。
- GsmCdmaCallTracker 对象向 RILJ 对象发起两次请求，第一次发起管理或控制通话请求，第

二次发起查询当前通话列表请求。RILJ 对象处理这两次请求并发起两次处理结果的 Callback 消息回调，两次消息回调的 Handler 类型分别为 EVENT_OPERATION_COMPLETE 和 EVENT_POLL_CALLS_RESULT。

2．被动接收 Handler 消息

前面分析的 GsmCdmaCallTracker 对象的 Handler 消息处理机制会被动接收和处理 RILJ 对象上报的三种类型的 Handler 消息，这三个消息与通话状态的关系非常密切，其消息处理流程如图 6-3 所示。

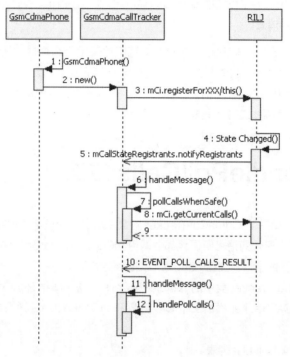

图 6-3　被动接收通话状态变化的消息处理流程

图 6-3 所示的被动接收通话状态变化的消息处理流程，仅表现了 RILJ 与 GsmCdmaCallTracker 两个对象之间的交互流程，共有两次交互过程：

- 第一次交互过程

RILJ 对象向 GsmCdmaCallTracker 对象发起 EVENT_CALL_STATE_CHANGE 类型的 Handler 消息通知。

- 第二次交互过程

GsmCdmaCallTracker 对象调用 mCi.getCurrentCalls 方法，采用 Handler 消息回调的处理方式查询当前最新的 Call List 列表。

被动接收通话状态变化的消息处理流程的说明和描述详情如下。

- 步骤 1 至步骤 3，TeleService 系统应用加载 Telephony 业务模型创建 GsmCdmaCallTracker 对象的同时，向 RILJ 对象注册 EVENT_CALL_STATE_CHANGE 等三个类型的 Handler 消息。
- RILJ 层接收到 Modem 发出的通话状态变化消息，向 GsmCdmaCallTracker 对象发出 EVENT_CALL_STATE_CHANGE 类型的 Handler 消息通知（步骤 4 和步骤 5）。

- GsmCdmaCallTracker 对象调用 pollCallsWhenSafe 方法响应 EVENT_CALL_STATE_CHANGE 类型的 Message 消息。
- pollCallsWhenSafe 方法首先会生成 EVENT_POLL_CALLS_RESULT 类型的 Handler 消息对象,接着使用此对象作为参数发起 mCi.getCurrentCalls 调用,向 RILJ 对象发出查询最新 Call List 列表的消息(步骤 8)。
- 在 RIL 中完成查询 Call List 列表后,发起 EVENT_POLL_CALLS_RESULT 类型的 HandlerCallback 消息回调。
- GsmCdmaCallTracker 对象最终调用 handlePollCalls 方法响应 EVENT_POLL_CALLS_RESULT 类型的 Handler 消息。

CallTracker 对象与 RILJ 对象之间的交互不论采用什么形式,所有处理逻辑都会汇总到 handlePollCalls 方法,这个方法究竟有什么作用呢?原来,它会根据 RILJ 对象返回的 Call List 列表,更新通话状态以及发出相关的消息通知。可以说 CallTracker 类中的核心处理逻辑都集中到了此方法中,可见此方法的重要。

6.2 handlePollCalls 方法

通过前面的学习可知,GsmCdmaCallTracker 对象主动发起的通话管理或控制请求和被动接收到 RILJ 对象发起的通话状态变化的消息回调,都会调用 mCi.getCurrentCalls 方法查询 Call List 获取当前所有的通话连接,查询的结果将交给 handlePollCalls 方法处理。

因此,GsmCdmaCallTracker 类中的 handlePollCalls 方法将根据 RILJ 对象上报的 Call List 列表对象,更新 GsmCdmaCallTracker 类的三个 Call 对象和 mConnections 通话连接列表,主要有以下几个处理逻辑:

- 准备阶段,获取 Call List 列表。
- 更新通话状态及对应信息。更新 GsmCdmaCallTracker 对象中的 mState、mConnections、mForegroundCall、mBackground- Call 和 mRingingCall 等属性。
- 继续传递消息。根据当前最新的通话状态,发出通话状态变化的消息通知。

本节重点解析 handlePollCalls 方法,主要是获取 conn 对象和 dc 对象,根据它们之间的状态组合关系,分为四个处理流程来完成通话相关基本信息的更新,最后根据新的通话基本信息,更新 Phone.State 状态及发出相关通话状态变化的消息通知。

6.2.1 准备阶段

在 handlePollCalls 方法的准备工作中,最重要的是接收到 Message 消息后,获取 Call List 对象列表,然后声明一些重要的变量,其详情如下。

```
List polledCalls; //声明 Call List 对象,它将保存 RILJ 查询出的当前 Call 对象列表
if (ar.exception == null) {//mCi.getCurrentCalls 执行成功,RILJ 无异常
    polledCalls = (List)ar.result;//强制类型转换
} else if (isCommandExceptionRadioNotAvailable(ar.exception)) {
    polledCalls = new ArrayList();//Radio 不可用状态,创建一个空的列表
} else {
```

```
            pollCallsAfterDelay();//调用父类方法, 延迟处理, 将继续调用pollCallsWhenSafe
            return;
    }
    Connection newRinging = null;  //来电请求创建的通话连接对象
    boolean hasAnyCallDisconnected = false;//通话连接断开标志
    boolean needsPollDelay = false;//延迟查询标志
    boolean unknownConnectionAppeared = false;//出现未知通话连接的标志
    ......
```

上述代码中最关键的是对 polledCalls 变量的处理,RILJ 对象发给 GsmCdmaCallTracker 对象 Handler 消息中的 result 属性,通过强制类型转换成 List 列表对象。

mCi.getCurrentCalls 执行成功后,polledCalls 作为 List 对象,它究竟保存了什么类型的对象列表?DriverCall 或 GsmCdmaCall 对象?原来,它保存的是 DriverCall 对象列表,是在 RILJ 对象中创建的。

6.2.2 更新通话相关信息

GsmCdmaCallTracker 对象通过 Handler 消息获取最新的 DriverCall List 对象列表,紧接着遍历 mConnections 保存通话连接对象列表,再通过 DriverCall List 中对应的 DriverCall 对象,更新 GsmCdmaCallTracker 对象中的通话相关信息。

整理遍历 GsmCdmaCallTracker 对象的 mConnections 列表的处理逻辑,其代码框架详情如下:

```
//完成 MAX_CONNECTIONS_GSM 或 MAX_CONNECTIONS_CDMA 次循环
for (int i = 0, curDC = 0, dcSize = polledCalls.size()
            ; i < mConnections.length; i++) {//关注i、curDC下标的对应关系
    GsmCdmaConnection conn = mConnections[i];
    DriverCall dc = null;

    if (curDC < dcSize) {
            //再次确认polledCalls 列表保存的是DriverCall 对象
            dc = (DriverCall) polledCalls.get(curDC);
            //DriverCall 对象匹配成功,curDC++,下一次循环时获取下一个DriverCall对象
            if (dc.index == i+1) {
                    curDC++;
            } else {//DriverCall对象未匹配成功,则不取出dc对象,等待下一次循环获取
                    dc = null;
            }
    }
    if (conn == null && dc != null) {......//通话状态变化的处理}
}//mConnections.length 循环结束
```

for 循环遍历 mConnections 数组的处理逻辑需重点关注以下两点:

1. 获取 conn 对象和 dc 对象

根据 i 数组 mConnections 的下标取值获取 GsmCdmaConnection 对象,然后通过 curDC 索引值获取 polledCalls 列表中的 DriverCall 对象 dc,匹配 DriverCall 对象是通过 dc.index == i+1 关系来完成的,如果无法匹配到 DriverCall 对象,则进入下一次循环完成 DriverCall 对象的匹配,需要注意以下两点:

● 循环次数

通过 MAX_CONNECTIONS_GSM 和 MAX_CONNECTIONS_CDMA 两个常量的定义,mConnections 数组最多可保存 19 个或 8 个 GsmCdmaConnection 对象。在循环 connections 数组的过程中,不论

获取的 conn 对象如何取值，都会完成 19 次或 8 次循环。
- 下标

这里涉及两个非常重要的下标：i 和 curDC，其初始值都是 0。conn 对象通过 i 可获取 mConnections 数组中的 GsmCdmaConnection 对象；dc 对象通过 curDC 获取 polledCalls 列表中的 DriverCall 对象。i 的取值为 MAX_CONNECTIONS_GSM 或 MAX_CONNECTIONS_CDMA，curDC 的取值为 0 到 dcSize。

conn 对象与 dc 对象的关系是：

```
dc.index == i+1
```

为什么会相差 1 呢？因为数组的下标从 0 开始。难道 polledCalls 中 DriverCall 对象的下标从 1 开始，才能满足这个关系？

打开 Nexus 6P 手机，建立两路通话，一路为通话中，一路为通话保持，查看 radio 日志，摘录的关键信息如下：

```
D/RILJ ( 7778): [4136]> GET_CURRENT_CALLS [SUB0]
D/RILJ ( 7778): [4136]< GET_CURRENT_CALLS{
[id=1, HOLDING,toa=129,norm,mo,0,voc,noevp,,cli=1,,1]
[id=2, ACTIVE,toa=129,norm,mt,0,voc,noevp,,cli=1,,1] } [SUB0]
```

通过上面的日志信息可以推断出 dc 对象的 index 是从 1 开始递增的。

因此，在遍历 mConnections 列表的过程中，通过 i 变量可以不断获取 mConnections 列表中的 GsmCdmaConnection 对象 conn，同时获取 polledCalls 列表中的 DriverCall 对象 dc。

一旦满足条件 dc.index == i+1 时，因为当前的 conn 对象由 dc 对象创建，只要 dc 对象发生了变化，conn 对象也将进行相应的调整和更新。

注意　从 mConnections 数组中取出的 conn 对象可代表老的通话连接，而 dc 对象则代表新的通话基本信息；比如 conn 为 null，dc 不为 null，则表示查询到新的通话连接或查询到新的来电。具体的通话状态变化，且看后面的分析。

2. 通话状态的变化

根据 conn 和 dc 这两个对象基本信息的组合关系，可得出四种通话状态的变化，分别是：
- 出现新的通话
- 通话连接断开
- 通话连接断开并且有新的来电
- 通话状态发生变化

这四个处理逻辑完全覆盖了通话状态变化的所有应用场景，作为本章最为关键的内容。接下来，我们将展开介绍这四种通话状态变化的处理逻辑。

（1）出现新的通话

条件判断：

```
conn == null && dc != null
```

老的通话连接不存在为 null，dc 不为 null 代表查询出新的通话 DriverCall，说明接收到新的通话连接，这会是什么样的通话应用场景呢？最容易理解的就是接收到新的来电请求。具体处理逻辑详情如下。

```
if (conn == null && dc != null) {
    //主动拨号查询到新的 DriverCall
    if (mPendingMO != null && mPendingMO.compareTo(dc)) {
        mConnections[i] = mPendingMO;
        mPendingMO.mIndex = i;
        mPendingMO.update(dc);
        mPendingMO = null;
        if (mHangupPendingMO) {
            ......//hangup(mConnections[i])挂断拨号
        }
    } else {
        //通过dc创建新的GsmCdmaConnection对象,并保存到mConnections数组
        mConnections[i] = new GsmCdmaConnection(mPhone, dc, this, i);

        Connection hoConnection = getHoConnection(dc);//获取Handover
        if (hoConnection != null) {
            ......//SRVCC 处理逻辑
        } else {
            newRinging = checkMtFindNewRinging(dc,i);
            ......
        }
    }
    hasNonHangupStateChanged = true; //只要出现新的通话连接,均取true值
}
private Connection checkMtFindNewRinging(DriverCall dc, int i) {
    Connection newRinging = null;
    if (mConnections[i].getCall() == mRingingCall) {//验证来电Connection信息
        newRinging = mConnections[i];
    } else {......}
    return newRinging;
}
```

通过上面的代码,非常容易看出它有两个分支处理逻辑。

● 主动发起拨号请求后,第一次查询到 DriverCall 后,会进入此分支处理。

首先,在拨号的时候,会创建一个 GsmCdmaConnection 对象,并将此对象保存在 GsmCdmaCallTracker 对象的 mPendingMO 属性中,详情如下。

```
mPendingMO = new GsmCdmaConnection(mPhone, checkForTestEmergencyNumber(dialString),
        this, mForegroundCall, isEmergencyCall);
```

需要重点关注 GsmCdmaConnection 对象的 parent 属性为 mForegroundCall 对象,也就是主动拨号默认在第一路通话中;并且,其 mIndex 默认为-1。GsmCdmaConnection 对象的 index 取值范围通常都是大于等于 0 的,只有在拨号时创建的对象的 index 取值为-1。最后在此分支中通过 RIL 返回的 DriverCall 对象 dc 更新它的基本信息,最关键的是设置 mConnections[i]为此对象,并且将对象的 index 设置为 i。

● 接收到新来电请求。

通过 dc 对象创建 GsmCdmaConnection 对象,保存到 mConnections[i]数组中,其构造方法的关键处理逻辑详情如下:

```
/** This is probably an MT call that we first saw in a CLCC response or a hand over. */
public GsmCdmaConnection (GsmCdmaPhone phone, DriverCall dc, GsmCdmaCallTracker ct,
int index) {
    mOwner = ct;//此对象属于CallTracker

    mAddress = dc.number;
    mIsIncoming = dc.isMT;
```

```
        ......//通过DriverCall更新Connection基本属性

        mIndex = index;// mConnections数组下标

        mParent = parentFromDCState(dc.state);//通过DriverCall状态获取CallTracker的Call对象
        mParent.attach(this, dc);//Call对象增加Connection和更新状态
}
private GsmCdmaCall parentFromDCState (DriverCall.State state) {
    switch (state) {
        case ACTIVE:
        case DIALING:
        case ALERTING:
            return mOwner.mForegroundCall;//CallTracker的第一路通话
        case HOLDING:
            return mOwner.mBackgroundCall; //CallTracker的第二路通话
        case INCOMING:
        case WAITING:
            return mOwner.mRingingCall; //CallTracker来电通话
        default:
            throw new RuntimeException("illegal call state: " + state);
    }
}
//GsmCdmaCall中的attach方法
public void attach(Connection conn, DriverCall dc) {
    mConnections.add(conn);
    mState = stateFromDCState (dc.state);//转换DriverCall状态为Call.State状态
}
```

上面的代码逻辑中，使用 dc 对象创建 GsmCdmaConnection 对象。首先会获取 dc 对象的基本信息构建 GsmCdmaConnection 对象的基本信息，然后根据 dc 对象的状态获取 GsmCdma-CallTracker 的三个 Call 对象之一，并设置为 GsmCdmaConnection 对象的 mParent 属性。当前逻辑中，mParent 属性为 CallTracker 对象的 mRingingCall 对象。

最后，只要出现新的通话连接，hasNonHangupStateChanged 的状态标志都会设置为 true。

（2）通话连接断开

判断条件：

```
conn != null && dc == null
```

老的通话连接对象已经存在，匹配到的 DriverCall 对象为空，也就是说这个通话连接已经清空，即通话连接已断开，其处理逻辑详情如下。

```
else if (conn != null && dc == null) {
    if (isPhoneTypeGsm()) {//GSM网络制式
        mDroppedDuringPoll.add(conn); //增加老的conn对象到删除通话连接列表中
    } else {//CDMA网络制式
        int count = mForegroundCall.mConnections.size();
        for (int n = 0; n < count; n++) {
            GsmCdmaConnection cn =
                (GsmCdmaConnection)mForegroundCall.mConnections.get(n);
            mDroppedDuringPoll.add(cn);
        }
        ......
    }
    mConnections[i] = null; //设置老的通话连接对象为null，与dc匹配
}
```

通话连接已断开的处理逻辑相对简单，GSM 和 CDMA 两种不同网络制式的处理稍微有些差异，

最终将 conn 对象加入 mDroppedDuringPoll 列表，取消 GsmCdmaCallTracker 对象 mConnections 属性的引用。

注意

这里的处理逻辑相对简单，真正的通话连接断开的处理逻辑将在后面的小节中展开讲解。

（3）通话连接断开并且有新的来电（GSM 网络）

在通话应用中，这样的情景出现的概率比较少。conn 和 dc 都不为 null，但是它们的信息并不匹配，主要体现在电话号码发生了改变，其处理逻辑详情如下。

```
else if (conn != null && dc != null && !conn.compareTo(dc) && isPhoneTypeGsm()) {
    mDroppedDuringPoll.add(conn); //增加老的 conn 对象到删除通话连接列表中
    mConnections[i] = new GsmCdmaConnection (mPhone, dc, this, i);

    if (mConnections[i].getCall() == mRingingCall) {
        newRinging = mConnections[i]; //非常关键，设置 newRinging
    }
    hasNonHangupStateChanged = true;
}
```

通过上面的代码不难发现，它的处理逻辑结合了新来通话请求和通话连接断开这两种通话连接状态变化的处理逻辑。

（4）通话状态发生变化

这种情况非常多，如拨号状态变为通话中状态、通话中状态变为通话保持状态等，其处理逻辑详情如下。

```
else if (conn != null && dc != null) { /* implicit conn.compareTo(dc) */
    boolean changed;
    changed = conn.update(dc);
    hasNonHangupStateChanged = hasNonHangupStateChanged || changed;
}

else if (conn != null && dc != null) { /* implicit conn.compareTo(dc) */
    //Call 状态冲突处理分支
    if (!isPhoneTypeGsm() && conn.isIncoming() != dc.isMT) {
        if (dc.isMT == true) {
            // MT 通话优先级高于从 0 拨号，删除从 0 拨号对应的连接
            mDroppedDuringPoll.add(conn);
            // 使用 dc 在 mConnections 数组中匹配 MT 来电或未知类型通话连接
            newRinging = checkMtFindNewRinging(dc,i);
            if (newRinging == null) {
                unknownConnectionAppeared = true;
                newUnknownConnectionCdma = conn;
            }
            checkAndEnableDataCallAfterEmergencyCallDropped();
        } else {......}
    } else {//正常的通话更新处理逻辑
        boolean changed;
        //通过 dc 对象更新 conn 对象，返回通话状态是否已经更新的标志
        changed = conn.update(dc);
        //只要通话状态已经更新，hasNonHangupStateChanged 取值为 true
        hasNonHangupStateChanged = hasNonHangupStateChanged || changed;
    }
}
```

回顾这四种通话状态发生变化的判断条件及处理逻辑，conn 与 dc 之间的关系如表 6-4 所示。

表 6-4 conn 与 dc 的状态组合

条件	说明
conn == null && dc != null	出现新的通话，主要有两种情况：主动拨号后第一次查询到此通话连接和接收到来电请求
conn != null && dc == null	通话连接已经断开
conn != null && dc != null && !conn.compareTo(dc) && isPhoneTypeGsm()	GSM 网络下，通话连接断开的同时，接收到新的来电请求
conn != null && dc != null	通话状态发生了变化

6.2.3 发出通知

收尾处理工作将根据最新的通话状态发出通话状态变化的消息通知。

```
if (newRinging != null) {//接收到来电请求
    //通过 GsmCdmaPhone 对象发出接收到新来电请求的消息通知
    mPhone.notifyNewRingingConnection(newRinging);//参数为 GsmCdmaConnection 对象
}
......//省略通话连接断开、mHandoverConnections 的逻辑处理
if (needsPollDelay) {//再次查询 Call 列表
    pollCallsAfterDelay();
}
//清理通话连接已断开的 GsmCdmaConnection 对象
if (newRinging != null || hasNonHangupStateChanged || hasAnyCallDisconnected) {
    internalClearDisconnected();
}

updatePhoneState();//非常关键，无任何条件限制，同步更新 mState 属性

if (unknownConnectionAppeared) {//出现未知的通话连接
    //通过 GsmCdmaPhone 对象发出消息通知
    if (isPhoneTypeGsm()) {
        for (Connection c : newUnknownConnectionsGsm) {
            mPhone.notifyUnknownConnection(c);
        }
    } else {
        mPhone.notifyUnknownConnection(newUnknownConnectionCdma);
    }
}
//非通话连接断开、接收到新的来电请求、通话连接断开的情况
if (hasNonHangupStateChanged || newRinging != null || hasAnyCallDisconnected) {
    mPhone.notifyPreciseCallStateChanged();//发出通话状态变化的消息通知
    updateMetrics(mConnections);
}
```

其实 handlePollCalls 方法中的收尾处理逻辑比较简单：完成循环 mConnections 数组更新 GsmCdmaCallTracker 对象通话相关信息后，根据最新的通话基本信息发出通话状态变化的相关消息通知。

其中最关键的内容如下。

● conn 与 dc 对象的匹配关系，dc.index == i + 1 说明 conn 对象是通过 dc 对象创建的。

- conn 与 dc 对象的状态组合关系，这些组合关系构成了通话状态变化的所有应用场景。
- mConnections 数组的下标与其保存的 GsmCdmaConnection 对象的 index 的对应关系，通过代码分析，可知它们是一一对应的。即 mConnections[0]引用的 Connection 对象的 mIndex 取值也一定是 0。

6.2.4 更新 mState

通过前面对 handlePollCalls 方法的分析，可知调用 updatePhoneState 方法是没有任何条件限制的，此方法将更新 GsmCdmaCallTracker 对象的 mState 状态，其代码详情如下。

```
private void updatePhoneState() {
    PhoneConstants.State oldState = mState; //记录当前的状态
    //根据三个 GsmCdmaCall 对象的状态，更新 mState 状态
    if (mRingingCall.isRinging()) {
        mState = PhoneConstants.State.RINGING; //来电状态
    } else if (mPendingMO != null ||
            !(mForegroundCall.isIdle() && mBackgroundCall.isIdle())) {
        mState = PhoneConstants.State.OFFHOOK; //摘机状态
    } else {
        ......
        mState = PhoneConstants.State.IDLE; //待机状态
    }
    //通话状态改变，并且当前状态为待机状态，说明通话结束
    if (mState == PhoneConstants.State.IDLE && oldState != mState) {
        // mVoiceCallEndedRegistrants 发出通知，移动数据进入可用状态
        mVoiceCallEndedRegistrants.notifyRegistrants(
            new AsyncResult(null, null, null));
    } else if (oldState == PhoneConstants.State.IDLE && oldState != mState) {
        //与上一个条件相反，如果通话状态转变为非待机状态，开始通话
        // mVoiceCallStartedRegistrants 发出通知，移动数据进入不可用状态
        mVoiceCallStartedRegistrants.notifyRegistrants (
                new AsyncResult(null, null, null));
    }
    if (mState != oldState) {//mState 状态已经更新
        mPhone.notifyPhoneStateChanged();//通过 mPhone 对象发出状态变化通知
        mMetrics.writePhoneState(mPhone.getPhoneId(), mState);
    }
}
```

上面的代码处理逻辑主要分为两个部分。

- 通过 Call 对象获取其 mState 状态从而更新 GsmCdmaCallTracker 的 mState 属性，它们之间的状态是保持同步的。
- 发出 mState 状态变化的消息通知。

Call.State 和 PhoneConstants.State 之间的状态转换过程如图 6-4 所示。

图 6-4 所示的 Call.State 与 PhoneConstants.State 状态转换，重点关注以下几点：

- Call.State 共有九个状态，可对应 PhoneConstants.State 的三个状态。
- 除去非 IDLE 和 INCOMING 这两个状态，Call.State 剩余的七个状态将对应 PhoneConstants.State 的 OFFHOOK 状态。
- PhoneConstants.State.RINGING 可理解为特殊的 OFFHOOK 状态。
- Call.Sate.DISCONNECTED 可转换为 PhoneConstants.State.IDLE 状态。

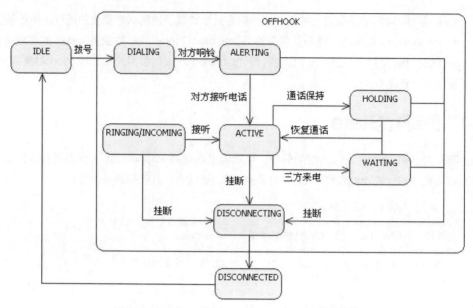

图 6-4 Call.State 与 PhoneConstants.State 状态转换

注意

这些状态之间的转换关系，读者可以自行在 Nexus 6P 手机上进行验证，以帮助理解。GsmCdmaCallTracker 和 GsmCdmaCall 类中都有 mConnections 数组，可保存多个 GsmCdmaConnection 通话连接对象，它们之间的关系是：GsmCdmaCallTracker 对象的 mConnections 数组将保存所有的通话 Connection 对象；保存三个 GsmCdmaCall 对象的 mConnections 数组的集合，也同样保存了当前所有的通话 Connection 对象：GsmCdmaCallTracker.mConnections = mRingingCall.mConnections + mForegroundCall.mConnections + mBackgroundCall. mConnections。

6.3 通话管理模型分析

创建 GsmCdmaCallTracker 对象的同时，将同步创建三个 GsmCdmaCall 对象，这三个对象分别代表了三路通话。

- mRingingCall　来电
- mForegroundCall　第一路通话
- mBackgroundCall　第二路通话

因此，Telephony Call 通话模型最多支持三路通话，在每一路通话中可以包含多个通话连接。每个 GsmCdmaCall 对象都有独立的 mState 状态和 mConnections 通话连接对象列表。以 GsmCdmaCall 为核心的通话模型主要集中在 GsmCdmaCall、GsmCdmaConnection 和 DriverCall 三个关键类，它们共同构建了 Telephony Voice Call 语音通话模型；GsmCdmaConnection 和 DriverCall 对象作为基石则构建了以 GsmCdmaCall 对象为中心的通话模型框架。

当 DriverCall List 列表中的 DriverCall 对象发生变化时，根据 DriverCall 对象的基本信息创建或更新 GsmCdmaConnection 对象，与此同时，会同步更新 GsmCdmaConnection 对象的 parent 属性，即更新 GsmCdmaCall 对象的 mState 和 mConnections 属性，从而同步更新通话模型中的通话信息。

―――――――――――――――――― >>>>>>>>>> 第 6 章　Voice Call 语音通话模型

因此，GsmCdmaCall 对象的状态更新的驱动入口是在 GsmCdmaConnection 对象的创建或更新的方法中。

6.3.1　GsmCdmaCall

GsmCdmaCall 类继承于 Call 抽象类 frameworks/opt/telephony/src/java/com/android/internal/telephony/Call.java，注意区分 Telecom 系统应用中的 Call 对象。

1．关键属性

GsmCdmaCall 类的关键属性如表 6-5 所示。

表 6-5　GsmCdmaCall 关键属性

属性	类型	说明
mState	State	当前通话的状态
mConnections	ArrayList<Connection>	通话的连接对象列表
mOwner	GsmCdmaCallTracker	所有者是 GsmCdmaCallTracker

表 6-5 中最关键的是 mState 和 mConnections 两个属性，分别记录了通话状态和通话连接对象列表。

2．关键方法

GsmCdmaCall 类的关键方法如表 6-6 所示。

表 6-6　GsmCdmaCall 类关键方法

分类	方法	说明
更新 Call State 通话状态	update	通过 DriverCall 对象状态更新 mState
	attach	增加 mConnections，并通过 DriverCall 对象状态更新 mState
	detach	删除 mConnections，根据 mConnections 列表更新 mState
	clearDisconnected	清除 mConnections 列表，并更新 mState
挂断电话的处理	hangup	挂断电话
	onHangupLocal	处理本地挂断电话 Call 状态更新，更新通话连接 Connection 为 DisconnectCause.LOCAL，并将 mState 修改为 DISCONNECTING
获取通话连接	getEarliestConnection	获取最早创建的 Connection
	getLatestConnection	获取最晚创建的 Connection

注意　GsmCdmaCall 对象提供的方法的主要逻辑是更新 mState 和 mConnections 这两个关键属性。

6.3.2　GsmCdmaConnection

GsmCdmaCall 类中涉及的 GsmCdmaConnection 类继承于 Connection 抽象类，即 frameworks/

opt/telephony/src/java/com/android/internal/telephony/Connection.java，要注意与 TeleService 系统应用中的 Connection 加以区分。

1．关键属性

GsmCdmaConnection 类的关键属性如表 6-7 所示。

表 6-7　GsmCdmaConnection 关键属性

属性	类型	说明
mIndex	int	GsmCdmaCallTracker.connections 数组的下标值
mOwner	GsmCdmaCallTracker	它的所有者是 GsmCdmaCallTracker
mParent	GsmCdmaCall	它的来源是 GsmCdmaCall，即 GsmCdmaCall 对象拥有它
mTelecomCallId	String	通话连接的一些基本信息是在 com.android.internal.telephony.Connection 父类中实现的，包括 mAddress 通话连接对方的电话号码、mIsIncoming 是否来电、mCause 通话断开原因编号、mConnectTime 通话建立连接起始时间等
mAddress	String	
mDialString	String	
mPostDialString	String	
mIsIncoming	boolean	
mCause	int	
mConnectTime	long	
userData	Object	
mPartialWakeLock	PowerManager.WakeLock	电源管理

对比 GsmCdmaCall 类可以发现，GsmCdmaConnection 类的关键属性用来保存通话连接比较全面的一些基本信息，如电话号码、通话计时时间等；GsmCdmaConnection 类和 GsmCdmaCall 类的 mOwner 属性都是 GsmCdmaCallTracker 对象。

2．关键方法

GsmCdmaConnection 类的关键方法如表 6-8 所示。

表 6-8　GsmCdmaConnection 关键方法

分类	方法	说明
更新状态	update	根据 DriverCall 对象更新 GsmCdmaConnection 对象的相关信息
	updateParent	更新 Call 状态
获取 Call 对象	parentFromDCState	根据 DriverCall 对象的状态获取对应的 GsmCdma-CallTracker 的三个 Call 对象
获取通话连接的基本信息	getXXX	获取电话号码、通话开始时间、是否是来电等 GsmCdma-Connection 对象的基本属性
获取通话状态的相关方法	getState	获取通话状态
	getCall	获取所属 Call 对象
通话断开	onDisconnect	本地挂断通话的处理
	onRemoteDisconnect	远端挂断电话的处理

表 6-8 中的 GsmCdmaConnection 关键方法，需要注意以下几点。

- 重点关注 GsmCdmaConnection 类的构造方法和 update 方法。

- parent 属性，即 GsmCdmaCall 对象，说明任意一个 Connection 对象都归属于一个 Call 对象。通俗来讲，任意一个通话连接都可归为 GsmCdmaCallTracker 三路通话中的一路。

> 通过 GsmCdmaConnection 对象也可获取当前通话状态，其实现原理是首先获取它的 parent 对象，parent 对象即 GsmCdmaCall 对象，然后获取 Call.mState 通话状态。

3. Call 与 Connection 的关系

通过前面对 GsmCdmaCall 和 GsmCdmaConnection 两个类代码结构的分析，可以看到它们的功能划分非常清晰。GsmCdmaConnection 作为基石，保存了通话连接的基本信息，多个 GsmCdmaConnection 对象组合成一路通话；GsmCdmaCall 作为框架，主要管理 mState 当前通话状态和 mConnections 列表，一路通话中有多个通话连接。

GsmCdmaCall 与 GsmCdmaConnection 类之间的关系比较复杂，可以相互引用，总结它们之间的关系，详情如图 6-5 所示。

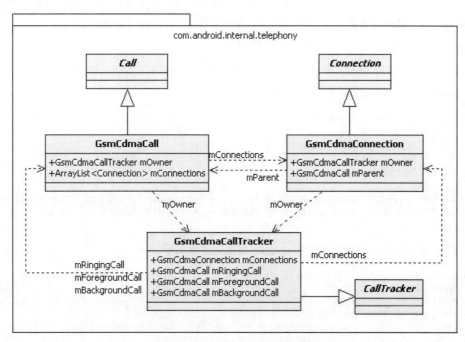

图 6-5　Call 与 Connection 的关系图

- 图 6-5 中涉及的类，全部在同一个包：com.android.internal.telephony。
- GsmCdmaCall 与 GsmCdmaConnection 的对应关系。GsmCdmaCall 类的 mConnections 属性可以保存多个 GsmCdmaConnection 对象；而 GsmCdmaConnection 对象的 mParent 属性，则是它对应的唯一 GsmCdmaCall 对象。这里体现了基本的通话模型，一路通话中可以有多个通话连接，电话会议就是一个典型的实例。任意一个通话连接都属于 GsmCdmaCallTracker 的三路通话中的一路。
- GsmCdmaCall 和 GsmCdmaConnection 两个类的 mOwner 属性，都是 GsmCdmaCallTracker 对象。
- GsmCdmaCallTracker 的应用关系。

GsmCdmaCallTracker 和 GsmCdmaCall 类中都有 mConnections 数组，可保存多个 GsmCdmaConnection 通话连接对象，它们之间有什么关系呢？

6.3.3 DriverCall、Call、Connection

DriverCall 是 RILJ 中创建的对象，Modem 查询完通话列表后，返回给 RIL 固定格式的字符串，RILJ 对象将这些字符串信息拼装成 DriverCall 对象列表，一行字符串信息可创建一个 DriverCall 对象。DriverCall 对象列表能够真实反映出 Modem 无线通信模块中所有通话连接的真实信息。

1. DriverCall 与 GsmCdmaConnection

通过前面的学习知道，一个 GsmCdmaConnection 对象保存着一个通话连接的基本信息，这些基本信息在 Telephony Voice Call 语音通话模型中是如何获取的呢？

再次回顾 GsmCdmaConnection 的构造方法，原来 GsmCdmaConnection 对象是根据 DriverCall 对象的一些基本信息创建的，并且在 GsmCdmaConnection 对象创建成功后，还能通过新的 DriverCall 对象更新自身的一些基本信息，它们之间的匹配关系是通过 mIndex 下标来标识的。

GsmCdmaConnection 类的构造方法详情如下。

```
public GsmCdmaConnection (GsmCdmaPhone phone, DriverCall dc, GsmCdmaCallTracker ct,
int index) {
    mOwner = ct;//此对象属于 CallTracker

    mAddress = dc.number;
    mIsIncoming = dc.isMT;
    ......//通过 DriverCall 更新 Connection 基本属性

    mIndex = index;// mConnections 数组下标

    mParent = parentFromDCState(dc.state);//通过 DriverCall 状态获取 CallTracker 的 Call 对象
    mParent.attach(this, dc);//Call 对象增加 Connection 和更新状态
}
```

GsmCdmaConnection 对象提供的 update 方法的详情如下。

```
public boolean update (DriverCall dc) {
    GsmCdmaCall newParent;
    boolean changed = false;
    boolean wasConnectingInOrOut = isConnectingInOrOut();
    boolean wasHolding = (getState() == GsmCdmaCall.State.HOLDING);
    //通过 DriverCall 状态获取新对应的 GsmCdmaCall 对象
    newParent = parentFromDCState(dc.state);
    ......//更新 Connection 属性
    if (newParent != mParent) {//Call 发生了变化
        if (mParent != null) {//更新 GsmCdmaCall
            mParent.detach(this);
        }
        newParent.attach(this, dc);
        mParent = newParent;
        changed = true;
    } else {
        boolean parentStateChange;
        parentStateChange = mParent.update (this, dc);
        changed = changed || parentStateChange;
    }
    return changed;
}
```

有关 GsmCdmaConnection 类的构造方法和 update 方法的主要处理逻辑，重点关注以下三点内容。

- GsmCdmaConnection 对象的创建或更新，其数据来源和依据都是 DriverCall 对象。
- GsmCdmaConnection 对象在创建或更新的同时，同步调用 mParent 的更新方法来更新它所属的 GsmCdmaCall 对象，主要是 mParent.attach、mParent.detach 和 mParent.update 这三个方法调用。
- mParent 的切换。

在 parentFromDCState 方法中可找到通过 DriverCall.State 获取对应 mParent 的逻辑代码。DriverCall 对象的基本信息发生变化以后，由它创建的或通过 mIndex 下标对应的 GsmCdmaConnection 对象也会跟着调整。其中，最重要的就是更新了 GsmCdmaConnection 对象的 mParent。需要注意的是，在这个过程中不会创建新的 GsmCdmaCall 对象，只会在 GsmCdmaCallTracker 对象的三个 GsmCdmaCall 对象之间进行切换，其处理逻辑在 parentFromDCState 方法中可体现出来，总结如表 6-9 所示。

表 6-9 GsmCdmaConnection 对象的 mParent 取值与 DriverCall 状态的对应关系

GsmCdmaConnection 对象的 mParent 取值	DriverCall.State	说明
mForegroundCall	ACTIVE DIALING ALERTING	第一路通话包含 ACTIVE、DIALING 和 ALERTING 三种状态的通话连接
mBackgroundCall	HOLDING	第二路通话仅包含 HOLDING 状态的通话连接，即第二路通话的状态只能是 HOLDING
mRingingCall	INCOMING WAITING	来电仅包含 INCOMING 和 WAITING 这两种状态的通话连接，在有一路或两路通话的情况下接收到来电，此时 DriverCall.State 的状态为 WAITING

这里不太容易理解，详细解释一下。GsmCdmaCallTracker 对象接收到新的来电请求消息后，首先会创建一个 GsmCdmaConnection 对象，通过上面的关系可知，此对象的 mParent 属性为 mRingingCall 对象。然后，我们接听此来电请求，进入正在通话中状态，此时 GsmCdmaCallTracker 对象会更新之前创建的 GsmCdmaConnection 对象。在更新过程中，mParent 的引用会更改为 mForegroundCall 对象，同时将 mRingingCall 对象中的 mConnections 中的 GsmCdmaConnection 对象删除，完成 GsmCdmaConnection 对象的 mParent 的切换和更新。

2. DriverCall 与 GsmCdmaCall

GsmCdmaCall 对象可以保存多个 GsmCdmaConnection 通话连接基本信息对象，并且可以在 GsmCdmaConnection 对象中调用 attach、detach 和 update 这三个方法来更新 GsmCdmaCall 对象，其代码详情如下。

```
public void attach(Connection conn, DriverCall dc) {
    mConnections.add(conn); //增加 GsmCdmaConnection 对象
    mState = stateFromDCState (dc.state); //通过 DriverCall.State 更新 GsmCdmaCall.State
}
public void detach(GsmCdmaConnection conn) {
    mConnections.remove(conn); //移除 GsmCdmaConnection 对象
    if (mConnections.size() == 0) {//没有一个 GsmCdmaConnection 对象
```

```
            mState = State.IDLE; //更新状态为 IDLE 待机状态
        }
    }
    boolean update (GsmCdmaConnection conn, DriverCall dc) {
        State newState;
        boolean changed = false; //GsmCdmaCall 状态默认未更新

        newState = stateFromDCState(dc.state); //通过 DriverCall.State 获取新状态
        if (newState != mState) {//Call 状态发生变化
            mState = newState;//更新状态
            changed = true; //状态已更新
        }

        return changed; //返回 GsmCdmaCall 状态是否已经更新
    }
```

GsmCdmaCall 类中的三个更新方法 attach、detach 和 update，主要完成两方面的更新操作。
- 更新 mState
- 更新 mConnections

attach、detach 和 update 方法更新 GsmCdmaCall 对象时，仅在 GsmCdmaConnection 类中有相关方法的调用，意味着 RIL 上报的 DriverCall 对象发生变化后，首先通过创建或更新 GsmCdmaConnection 对象来同步调用 GsmCdmaCall 对象的更新方法，从而更新 GsmCdmaCall 对象的 mState 和 mConnections。

stateFromDCState 方法非常重要，它的处理逻辑能够体现出 Call.State 与 DriverCall.State 的对应关系，其代码详情如下。

```
    public static State stateFromDCState (DriverCall.State dcState) {
        switch (dcState) {
            case ACTIVE:           return State.ACTIVE;
            case HOLDING:          return State.HOLDING;
            case DIALING:          return State.DIALING;
            case ALERTING:         return State.ALERTING;
            case INCOMING:         return State.INCOMING;
            case WAITING:          return State.WAITING;
            default:               throw new RuntimeException ("illegal call state:" + dcState);
        }
    }
```

原来，Call.State 与 DriverCall.State 的枚举类型关系是一一对应的。需要我们注意的是，在 Call.State 和 DriverCall.State 枚举类型的定义中，Call.State 中相对多了有关 IDLE、DISCONNECTED 和 DISCONNECTING 这三种状态的定义，通过字面意思理解，它们分别代表待机、通话连接已经断开和通话连接正在断开三种状态。

Call.State 相对多了 IDLE、DISCONNECTED 和 DISCONNECTING 这三种状态的定义，那么什么情况下会产生呢？请读者先自行思考，将在本章给出答案。

Telephony Voice Call 语音通话模型分析过程中，我们遇见了三种与 Call 相关的 State，分别是：
- Call.State
- DriverCall.State
- PhoneConstants.State

有关这三个 State 的定义，总结如图 6-6 所示。

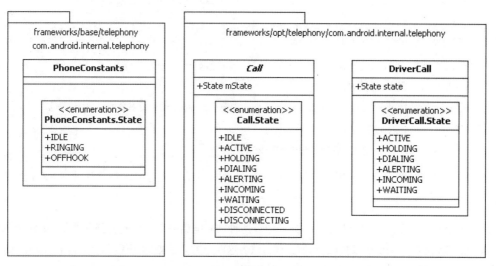

图 6-6　Call.State、DriverCall.State 和 PhoneConstants.State

我们重点关注以下几点：
- 代码基本属性

三个 State 的 package 相同，都是 com.android.internal.telephony，但是 PhoneConstants.State 的代码在 frameworks/base/telephony 代码库中，将编译到 framework.jar 中，而 Call.State 和 DriverCall.State 的代码在 frameworks/opt/telephony 代码库中，将编译出 telephony-common.jar。

- 枚举类型的对应关系

Call.State 和 DriverCall.State 的状态一一对应，但 Call.State 多出 IDLE、DISCONNECTED 和 DISCONNECTING 三种状态；PhoneConstants.State 和 Call.State 的对应关系可参考图 6-6 中 Call.State 与 PhoneConstants.State 的状态转换；PhoneConstants.State 与 DriverCall.State 没有直接的转换关系，需要通过 Call.State 进行中转。

6.4　补充通话连接断开处理机制

通话连接断开有两种应用场景：
- 本地主动挂断通话
- 远端断开通话连接（包括网络断开和对方挂断）

那么在 GsmCdmaCallTracker 类中，是如何实现这两种通话连接断开的呢？

6.4.1　本地主动挂断通话

ConnectionService 接收 Telecom 系统应用发起的挂断通话请求，通过 Telecom callId 匹配 TelephonyConnection 对象并调用其 onDisconnect 方法进行挂断电话请求的响应，关键处理逻辑详情如下：

```
public void onDisconnect() {
    hangup(android.telephony.DisconnectCause.LOCAL);//设置挂断电话原因
}
protected void hangup(int telephonyDisconnectCode) {
    if (mOriginalConnection != null) {
        try {
            if (isValidRingingCall()) {//拒接来电
                Call call = getCall();
                if (call != null) {
                    call.hangup();//调用 GsmCdmaCall 对象的 hangup 方法
                } else {......}
            } else {
                mOriginalConnection.hangup();
            }
        } catch (CallStateException e) {......}
    } else {......}
}
```

TelephonyConnection 类的 hangup 方法中有两种挂断电话的实现机制,分别调用 GsmCdmaCall 对象和 GsmCdmaConnection 对象的 hangup 方法,这两个方法的处理逻辑都是调用 mOwner.hangup(this),mOwner 是对 GsmCdmaCallTracker 的引用。

1. 主动挂断电话的请求过程

查看 GsmCdmaCallTracker 类中的两个 hangup 方法,它们的处理逻辑分为三步:

(1)调用 mCi 对象 hangupXXX 挂断电话的相关方法,向 RIL 请求挂断电话。

(2)调用 GsmCdmaCall.onHangupLocal 或 GsmCdmaConnection.onHangupLocal 方法,完成 Call 对象或 Connection 对象的更新。

(3)调用 mPhone.notifyPreciseCallStateChanged 方法,发出通话状态变化通知。

下面重点讲解和区分 GsmCdmaCall 和 GsmCdmaConnection 类的 onHangupLocal 方法的处理逻辑,代码详情如下。

```
void onHangupLocal() {//GsmCdmaCall
    for (int i = 0, s = mConnections.size(); i < s; i++) {
        //循环获取 GsmCdmaCall 对象的 GsmCdmaConnection 对象
        GsmCdmaConnection cn = (GsmCdmaConnection)mConnections.get(i);
        cn.onHangupLocal();//设置通话连接断开原因为 DisconnectCause.LOCAL
    }
    mState = State.DISCONNECTING;  //修改状态为挂断中
}
void onHangupLocal() {//GsmCdmaConnection
    mCause = DisconnectCause.LOCAL;//设置通话断开原因
    mPreciseCause = 0;
    mVendorCause = null;
}
```

通过上面的代码可知,本地主动挂断电话的处理逻辑可分为两种:

● 挂断来电

更新 GsmCdmaCall 对象的状态为 DISCONNECTING,这个状态在 DriverCall.State 中是没有定义的;同时设置 GsmCdmaCall 对象包含的所有 GsmCdmaConnection 对象的通话连接断开原因,设置为 DisconnectCause.LOCAL。

● 挂断通话

仅设置 GsmCdmaConnection 的 mCause 为 DisconnectCause.LOCAL。

2. 响应通话连接断开的处理逻辑

接着展开 GsmCdmaCallTracker 类中响应通话连接断开的处理方法。在 handlePollCalls 方法中，通话连接断开的 GsmCdmaConnection 对象会保存到 mDroppedDuringPoll 列表中，然后将围绕此列表对通话断开的 GsmCdmaConnection 对象进行处理，其详情如下。

```
ArrayList<GsmCdmaConnection> locallyDisconnectedConnections = new ArrayList<>();
for (int i = mDroppedDuringPoll.size() - 1; i >= 0 ; i--) {
    GsmCdmaConnection conn = mDroppedDuringPoll.get(i);//取出通话断开连接
    //CDMA
    boolean wasDisconnected = false;
    //拒接来电或漏接来电，关注判断条件
    if (conn.isIncoming() && conn.getConnectTime() == 0) {
        int cause;
        if (conn.mCause == DisconnectCause.LOCAL) {
            cause = DisconnectCause.INCOMING_REJECTED;//拒接
        } else {
            cause = DisconnectCause.INCOMING_MISSED;//漏接
        }
        mDroppedDuringPoll.remove(i); // 从 mDroppedDuringPoll 列表中移除
        hasAnyCallDisconnected |= conn.onDisconnect(cause);//再次更新 Connection 对象
        wasDisconnected = true;
        locallyDisconnectedConnections.add(conn);//记录拒接来电或漏接来电
    } else if (conn.mCause == DisconnectCause.LOCAL
            || conn.mCause == DisconnectCause.INVALID_NUMBER) {//挂断通话
        mDroppedDuringPoll.remove(i); // 从 mDroppedDuringPoll 列表中移除
        hasAnyCallDisconnected |= conn.onDisconnect(conn.mCause);
        wasDisconnected = true;
        locallyDisconnectedConnections.add(conn);
    }
    ......
}
if (newRinging != null || hasNonHangupStateChanged || hasAnyCallDisconnected) {
    internalClearDisconnected();//清理已经断开的通话连接
}
```

在 mDroppedDuringPoll 列表的循环逻辑中，有两个主要的流程处理分支，分析其代码，发现处理逻辑基本相同。其中，最关键的逻辑是获取通话连接断开的原因，最终调用 conn.onDisconnect 处理通话连接断开。

3. onDisconnect 方法

下面介绍 GsmCdmaConnection 类中的 onDisconnect 方法，它的主要处理逻辑如下。

```
public boolean onDisconnect(int cause) {
    boolean changed = false;
    mCause = cause;

    if (!mDisconnected) {
        doDisconnect();
        mOwner.getPhone().notifyDisconnect(this);
        if (mParent != null) {
            //GsmCdmaCall 对象的更新，关键是设置其状态为 DISCONNECTED
            changed = mParent.connectionDisconnected(this);
        }
        mOrigConnection = null;
    }
    return changed;
}
private void doDisconnect() {
    mIndex = -1;
```

```
            mDisconnectTime = System.currentTimeMillis();//记录通话连接断开时间
            mDuration = SystemClock.elapsedRealtime() - mConnectTimeReal; //计算通话时长
            mDisconnected = true; //设置通话连接已经断开标志
            clearPostDialListeners();
    }
```

4．小结

主动挂断电话的处理机制可分为三个处理过程，详情如下。

- 发起挂断通话连接的请求

```
CallTracker.hangup->mCi.hangupXXX
              ->GsmCdmaCall.onHangupLocal
              ->GsmCdmaConnection.onHangupLocal
```

- 接收到 RIL 发出的通话状态变化

GsmCdmaCallTracker.handlePollCalls→conn.onDisconnect→mParent.onDisconnect，mParent 为 GsmCdmaCall 对象。

- 清理通话连接状态为 DISCONNECTED 的 Connection 对象

```
CallTracker.internalClearDisconnected->Call.clearDisconnected
public void clearDisconnected() {
    for (int i = mConnections.size() - 1 ; i >= 0 ; i--) {
        Connection c = mConnections.get(i);
        if (c.getState() == State.DISCONNECTED) {
            mConnections.remove(i);
        }
    }
    if (mConnections.size() == 0) {
        setState(State.IDLE);
    }
}
```

6.4.2　远端断开通话连接

远端断开通话连接的处理机制与本地主动挂断通话的处理机制是否相似呢？本地主动挂断通话连接时，会将对应的 conn 对象的通话断开原因设置为 Local，而从远端断开通话连接时，GsmCdmaCallTracker 对象会向 RILJ 对象查询通话连接断开的原因。

继续分析 GsmCdmaCallTracker 类中的 handlePollCalls 方法。

1．handlePollCalls 方法

在 handlePollCalls 方法中，处理完本地主动挂断通话连接的请求之后，接着会处理是否从远端挂断电话的逻辑，详情如下。

```
//确定非本地断开通话连接的原因，需要查询 Modem
if (mDroppedDuringPoll.size() > 0) {
    mCi.getLastCallFailCause(
        obtainNoPollCompleteMessage(EVENT_GET_LAST_CALL_FAIL_CAUSE));
}
```

GsmCdmaCallTracker 对象会向 RILJ 对象查询最后一路通话连接断开的原因，RIL 处理完成后，回调的 Handler 消息类型为 EVENT_GET_LAST_CALL_FAIL_CAUSE。

2．EVENT_GET_LAST_CALL_FAIL_CAUSE 消息响应

GsmCdmaCallTracker 对象响应 EVENT_GET_LAST_CALL_FAIL_CAUSE 类型的 Handler 消息

回调，其处理逻辑详情如下。

```
case EVENT_GET_LAST_CALL_FAIL_CAUSE:
    int causeCode; //声明通话连接断开原因编号
    String vendorCause = null;
    ar = (AsyncResult)msg.obj;
    operationComplete();//完成查询操作
    if (ar.exception != null) {
    ......异常处理 causeCode = CallFailCause.NORMAL_CLEARING
    } else {
        LastCallFailCause failCause = (LastCallFailCause)ar.result;
        causeCode = failCause.causeCode;//通话断开查询结果
        vendorCause = failCause.vendorCause;
    }
    ......异常原因日志记录
    for (int i = 0, s = mDroppedDuringPoll.size(); i < s ; i++) {
        GsmCdmaConnection conn = mDroppedDuringPoll.get(i);
        conn.onRemoteDisconnect(causeCode, vendorCause);
    }
    updatePhoneState();//更新 mState 状态并发出消息通知
    mPhone.notifyPreciseCallStateChanged();
    mDroppedDuringPoll.clear();//清理 mDroppedDuringPoll 列表
break;
```

上面的代码处理逻辑可分为两个重要部分：

- 获取 causeCode，作为参数调用 conn.onRemoteDisconnect 方法完成与通话连接从远端断开相关的处理。
- 更新 mState 并发出通话状态变化的消息通知。

3. conn.onRemoteDisconnect 方法

conn.onRemoteDisconnect 方法的处理逻辑如下。

```
/*package*/ void onRemoteDisconnect(int causeCode, String vendorCause) {
    this.mPreciseCause = causeCode;
    this.mVendorCause = vendorCause;
    onDisconnect(disconnectCauseFromCode(causeCode));
}
```

首先，将 causeCode 通话连接断开原因的编号转变成 DisconnectCause 的枚举类型，其处理逻辑采用了常用的 switch case 方式，请读者自行阅读和总结其对应的关系。最后，调用 onDisconnect 方法处理。

不论是本地主动挂断通话还是远端断开通话连接，其差异仅在于获取通话连接断开的原因，调用 conn. onDisconnect 来更新 conn 及 mParent（GsmCdmaCall）等通话相关信息，最后调用 GsmCdmaCallTracker.internalClearDisconnected 方法清理所有与通话连接断开相关的信息。本地主动挂断通话中，首先将对应某一路通话对象 GsmCdmaCall 的状态修改为 DISCONNECTING，同时更新对应的 GsmCdmaConnection 对象断开通话连接的原因是 LOCAL；而远端断开通话连接中，GsmCdmaCall 对象并不会进入 DISCONNECTING 状态而是直接变为 DISCONNECTED 状态，对应的 GsmCdmaConnection 对象断开通话连接的原因可通过 RIL 查询 Modem 获取。

现在可以回答 DriverCall.State 与 Call.State 的对应关系中为什么缺少 DISCONNECTING 和 DISCONNECTED 状态的对应。

IDLE 待机状态是创建 GsmCdmaCallTracker 时，构造 mRingingCall、mForegroundCall、mBackgroundCall 三个 Call 对象的默认状态。

6.5　区分 Connection

TeleService 系统应用中的 IConnectionService 接收 Telecom 系统应用发起的创建连接请求，其中涉及 ConnectionRequest、ParcelableConnection、TelephonyConnection 等 Connection 对象，它们之间的关系和区别，请读者回顾图 5-8 所示的 IConnectionService 核心类图。

而 TelephonyConnection 与 Telephony Call 通话模型中的 Connection 对象产生了交互和关系，总结它们之间的差异和关联，如图 6-7 所示。

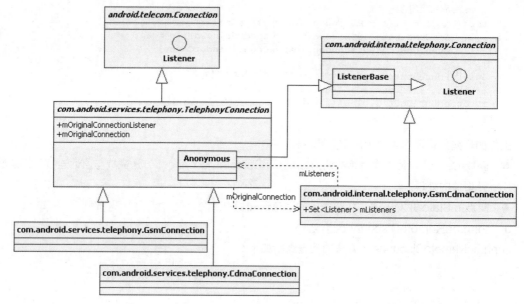

图 6-7　区分 Connection 类

重点关注以下几点：

- 两个 Connection 抽象类

两个 Connection 抽象类对应的代码分别在 frameworks/base/telecomm 和 frameworks/opt/telephony 代码库中，重点区分它们的包名。

- android.telecom.Connection

android.telecom.Connection 的子类代码在 TeleService 代码库中，有 TelephonyConnection 抽象类，GsmConnection 和 CdmaConnection 作为 TelephonyConnection 抽象类的子类。

- com.android.internal.telephony.Connection

com.android.internal.telephony.Connection 的子类代码在 frameworks/opt/telephony 代码库中，仅有一个子类 com.android.internal.telephony.GsmCdmaConnection。

- 运行空间

GsmConnection、CdmaConnection 和 GsmCdmaConnection 对象全部运行在 com.android.phone 进程空间。

- 两个 Connection 的关系

TelephonyConnection 对象的 mOriginalConnection 属性是 GsmCdmaConnection 对象的引用，

在创建此对象时便建立起了依赖关系；GsmCdmaConnection 对象的 mListeners 列表中，保存着 TelephonyConnection 内部匿名类对象 mOriginalConnectionListener，它重写了 com.android.internal.telephony.Connection.ListenerBase 类的方法。

- 消息流转

TelephonyConnection 通过 mOriginalConnection 属性访问 Telephony Voice Call 语音通话模型中的属性和方法。

GsmCdmaConnection 通过 mListeners 列表，调用其 Listener 接口，从而访问 TelephonyConnection 主类的方法，传递通话变化相关信息。

TelephonyConnection 对象接收通话变化消息有两个通道：
GsmCdmaPhone 对象的 RegistrantList 消息回调（setOriginalConnection 方法调用时注册 Handler 消息）。
GsmCdmaConnection 对象的 mListeners 接口调用。

两个通道接收到的通话变化消息，最后都是通过 TelephonyConnection 的 mListeners 消息调用进行传递的。TelephonyConnection mListeners 保存着 android.telecom.Connection.Listener 对象，它有 8 个子类，其中最关键的是 ConnectionService 和 TelephonyConferenceController 类的 mConnectionListener 内部匿名类对象。

6.6 扩展 InCallUi

Dialer 应用中的 InCallUi 通话界面，用来展示通话信息并提供通话控制交互界面；代码主要集中在 packages/apps/Dialer/java/com/android/incallui 路径下，运行在 Dialer 应用空间。

根据界面加载顺序，InCallUi 通话界面的关键处理逻辑可划分为三个阶段：
- 初始化过程
- addCall
- updateCall

6.6.1 初始化过程

InCallUi 通话界面的初始化过程可理解为 Telecom 系统应用绑定 IInCallService 的过程，其关键流程可参考图 3-5 所示的 InCallService onBind&setInCallAdapter 流程图，分为两个重要逻辑：
- onBind
- setInCallAdapter

InCallService 的 onBind 响应逻辑，首先创建 InCallPresenter、CallList 和 ExternalCallList 等关键对象，并调用 InCallPresenter 对象的 setUp 方法建立与 CallList 的消息传递框架结构，总结其中的关键类如图 6-8 所示。

图 6-8 所示的 InCallPresenter 与 CallList 的关系，我们重点关注以下几点：
- InCallPresenter 的本质

InCallPresenter 类实现了 CallList.Listener 接口，在 Dialer 应用进程中，可通过 getInstance 方法获取单例的 InCallPresenter 对象。

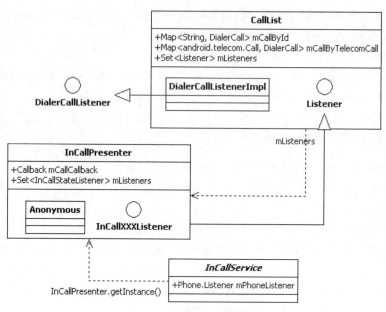

图 6-8　InCallPresenter 与 CallList

- InCallPresenter 的内部接口定义

　　InCallPresenter 类一共定义了七个内部接口：InCallStateListener、IncomingCallListener、CanAddCallListener、InCallDetailsListener、InCallOrientationListener、InCallEventListener 和 InCallUiListener，这些接口仅有一个方法定义，并且该类有七个列表属性，分别用来保存这七个内部 Listener 监听回调接口对象，如图 6-9 所示。

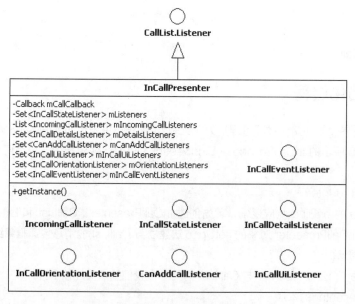

图 6-9　InCallPresenter 与 Listener

>>>>>>>>>> 第6章 Voice Call 语音通话模型

注意

结合 InCallPresenter 的命名，在这里做一个推测，InCallPresenter 是 InCallUi 通话界面的消息中转和处理中心。

- CallList 的 mListeners

在 Dialer 应用进程中，可通过 CallList.getInstance 方法获取单例的 CallList 对象，其 mListeners 列表中有一个 CallList.Listener 对象是 InCallPresenter，在 InCallPresenter 的 setUp 方法中的调用逻辑为 mCallList.addListener(this)。

接下来，InCallService 的 onBind 响应逻辑将启动 InCallActivity 界面，其调用入口是：InCallServiceImpl.onBind→InCallPresenter.getInstance().maybeStartRevealAnimation（为拨号流程的调用过程，来电流程则是通过 com.android.incallui.StatusBarNotifier 发送状态栏通知消息来加载 InCallActivity），InCallActivity 的配置信息如下：

```
<activity
    android:directBootAware="true"
    android:excludeFromRecents="true"
    android:exported="false"//对外不可用
    android:label="@string/phoneAppLabel"
    android:launchMode="singleInstance"//单例
    android:name="com.android.incallui.InCallActivity"
    android:resizeableActivity="true"
    android:screenOrientation="nosensor"
    android:taskAffinity="com.android.incallui"
    android:theme="@style/Theme.InCallScreen">
</activity>
```

注意

InCallActivity 并未配置 intent-filter，再加上 exported="false" 的定义，要启动此 Activity 必须在 Dialer 应用内部进行，因此保障了通话界面的安全加载。

InCallActivity 有一个辅助工具类 InCallActivityCommon，在 Activity 的生命周期管理方法 onXXX 中通过对 common.onXXX 的调用，交给 InCallActivityCommon 对象处理。

packages/apps/Dialer/java/com/android/incallui/res/layout/incall_screen.xml 是 InCallActivity 的布局文件，其中并没有定义通话界面的相关控件，那么通话界面展示的通话信息和提供的通话控制是在什么地方加载的呢？请读者先思考一下。

InCallUi 初始化过程中最后的逻辑是响应 Telecom 系统应用发起的 setInCallAdapter 接口调用，创建 Phone 对象并初始化其 Listener 消息处理框架。

总结 InCallUi 初始化过程中关键对象的创建和加载过程，其中涉及的关键类如图 6-10 所示。

对于图 6-10 所示的 InCallUi 初始化过程中的关键类图，重点关注以下几点：

- android.telecom.Phone 的本质

对应的 Java 代码是 frameworks/base/telecomm/java/android/telecom/Phone.java，是一个普通的 Java 类，注意与 Telephony 应用模型中的 GsmCdmaPhone 加以区分。

- android.telecom.Phone.Listener

InCallService 的内部匿名类对象 mPhoneListener 实现了 android.telecom.Phone.Listener 接口。

- android.telecom.Phone 与 InCallService 的关系

InCallService 在响应 setInCallAdapter 接口调用的逻辑中，首先创建 android.telecom.Phone 对象，通过 mPhone 持有此对象的引用，然后调用 mPhone.addListener(mPhoneListener)，android.telecom.Phone 对象的 mListeners 列表保存 InCallService 的内部匿名类对象 mPhoneListener。

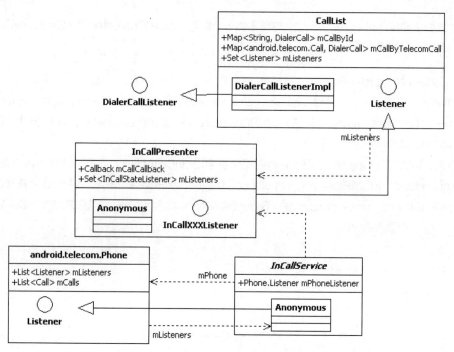

图 6-10 InCallUi 初始化过程中的关键类图

InCallUi 通话界面的初始化过程已经完成，建立了 InCallPresenter、CallList 和 Phone 等对象之间的 Listener 消息处理框架。

InCallService 接收到 Telecom 系统应用发出的 addCall、updateCall 等接口调用后，将通过 mPhone 即 android.telecom.Phone 对象发出对应的 mPhone.internalXXX 调用。

请读者思考 CallList 与 InCallPresenter 的消息来源。

6.6.2 addCall

Telecom 系统应用处理拨号请求或是接收到来电消息时，都会创建对象 com.android.server.telecom.Call，并将其转换为 ParcelableCall 对象，再通过调用 IInCallService 服务的 addCall 方法，传递 ParcelableCall 对象给 InCallUi 通话界面以进行通话信息的展示；关键流程可参考图 3-6 所示的 InCallService addCall 流程。

继续查看 Phone 对象的 internalAddCall 方法，关键处理逻辑详情如下：

```
final void internalAddCall(ParcelableCall parcelableCall) {
    //创建 android.telecom.Call 对象，注意 this 和 telecom callId
    Call call = new Call(this, parcelableCall.getId(), mInCallAdapter,
            parcelableCall.getState(), mCallingPackage, mTargetSdkVersion);
    mCallByTelecomCallId.put(parcelableCall.getId(), call);
    mCalls.add(call);//保存 android.telecom.Call 对象
    checkCallTree(parcelableCall);
    //根据 ParcelableCall 更新 android.telecom.Call 对象
    call.internalUpdate(parcelableCall, mCallByTelecomCallId);
    fireCallAdded(call);//发出已经增加了 Call 对象的消息通知
}
```

Phone 对象 fireCallAdded 方法的处理逻辑，主要是创建以 android.telecom.Call.Callback 为中心的消息处理框架，总结如图 6-11 所示。

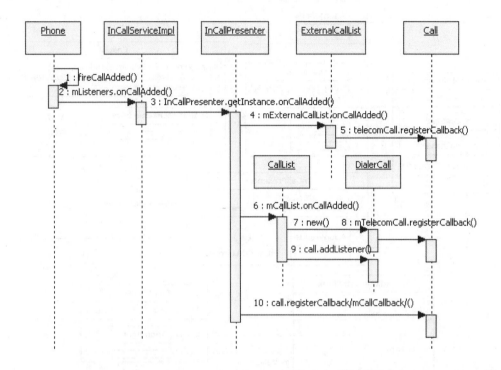

图 6-11 android.telecom.Call.Callback 消息处理框架

对于图 6-11 所示的 android.telecom.Call.Callback 消息处理框架，重点关注以下几点：
- onCallAdded 消息处理中心 InCallPresenter

步骤 3： InCallPresenter 中的 onCallAdded 方法将 onCallAdded 消息通过步骤 4、步骤 6 和步骤 10 分发出去，最终的处理全部进入 android.telecom.Call 对象的 registerCallback 方法调用。
- android.telecom.Call.mCallbackRecords

mCallbackRecords 属性将保存 android.telecom.Call.Callback 列表的详情：
步骤 5：注册 ExternalCallList 的匿名内部类对象 mTelecomCallCallback；
步骤 8：注册 DialerCall 的匿名内部类对象 mTelecomCallCallback；
步骤 10：注册 InCallPresenter 的匿名内部类对象 mCallCallback。
- DialerCall

DialerCall 作为 android.telecom.Call 与 CallList 的消息中转站，可将 Call 对象发出的 Listener 消息调用通过以下消息路径，最终发送到 InCallPresenter 对象，详情如下：

android.telecom.Call.mCallbackRecords → DialerCall.mListeners → CallList.mListeners → InCallPresenter

汇总 InCallUi 通话界面消息处理的关键类如图 6-12 所示。

InCallUi 通话界面的 addCall 响应逻辑的核心是创建了以 android.telecom.Call 对象为中心的 Call 消息监听回调框架，Phone 和 CallList 对象均保存了 android.telecom.Call 对象列表。

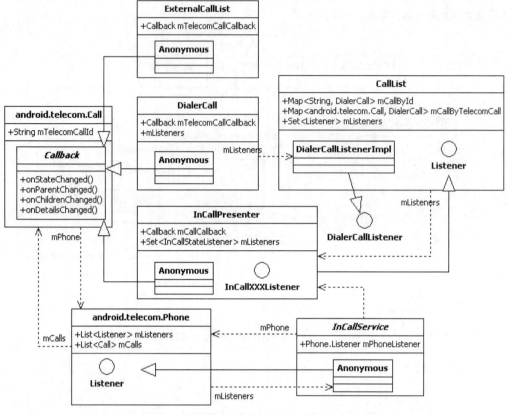

图 6-12 InCallUi 通话界面消息处理的关键类

6.6.3　InCallUi 通话界面

InCallActivity 通话界面的加载和显示可根据通话类型和状态分为两种情况：

- 通话中的 InCallUi 界面
- 来电响铃中的 InCallUi 界面

上面的两种应用场景都将加载 InCallActivity 通话界面，因为其特殊定义，只能在 Dialer 应用中通过 Class 类的方式进行加载，而在 InCallActivity 的 getIntent 方法中将获取加载此 Activity 的 Intent 对象，详情如下：

```
public static Intent getIntent(
    Context context, boolean showDialpad, boolean newOutgoingCall, boolean isForFullScreen) {
    Intent intent = new Intent(Intent.ACTION_MAIN, null);
    intent.setFlags(Intent.FLAG_ACTIVITY_NO_USER_ACTION |
            Intent.FLAG_ACTIVITY_NEW_TASK);
    intent.setClass(context, InCallActivity.class);
    InCallActivityCommon.setIntentExtras(intent, showDialpad,
            newOutgoingCall, isForFullScreen);//更新 InCallActivity 辅助工具类信息
    return intent;
}
```

在 InCallActivity 的 onStart 方法中，将调用 showMainInCallFragment 方法来展示通话界面相关 View 控件。

1. 加载入口

在 showMainInCallFragment 方法中可以通过 CallList 获取当前通话信息，选择加载的第一级 Fragment 共有三个选项：

- 来电界面
- 通话界面
- 视频通话界面

这三个 InCallActivity 的第一级 Fragment 分别是 AnswerFragment、InCallFragment 和 VideoCallFragment，可以使用 FragmentTransaction 进行 Fragment 的显示和删除操作。以加载通话界面 InCallFragment 为例，其主要的处理逻辑详情如下：

```
private boolean showInCallScreenFragment(FragmentTransaction transaction) {
    //创建 InCallFragment 对象
    InCallScreen inCallScreen = InCallBindings.createInCallScreen();
    transaction.add(R.id.main, inCallScreen.getInCallScreenFragment(),
            TAG_IN_CALL_SCREEN);//加入 Fragment
    didShowInCallScreen = true;
    return true;
}
```

2. InCallFragment

总结 InCallFragment 的加载流程如图 6-13 所示。

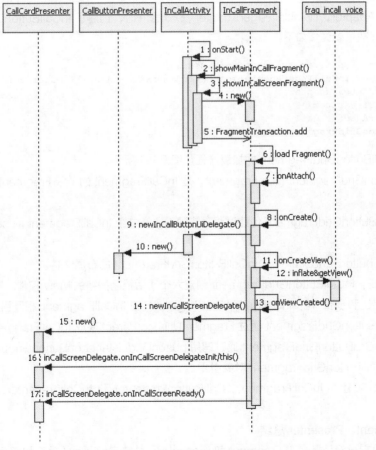

图 6-13　InCallFragment 加载流程

InCallFragment 作为 Fragment，其加载过程将依次调用并重写父类的 onAttach、onCreate、onCreateView、onViewCreated 和 onResume 方法，重点关注以下几点：

- inCallButtonUiDelegate 和 inCallScreenDelegate 对象的创建。步骤 9 和步骤 14：在 InCallActivity 中分别创建两个 Delegate 对象的关键代码，详情如下。

```
inCallButtonUiDelegate =
        FragmentUtils.getParent(this, InCallButtonUiDelegateFactory.class)
            .newInCallButtonUiDelegate();
inCallScreenDelegate =
        FragmentUtils.getParent(this, InCallScreenDelegateFactory.class)
            .newInCallScreenDelegate();
```

通过分析 FragmentUtils.getParent 方法中的处理逻辑，其返回 InCallActivity 对象，对 newInCallButtonUiDelegate 和 newInCallScreenDelegate 两个方法的调用将分别创建 CallButtonPresenter 和 CallCardPresenter 对象。

- frag_incall_voice.xml 作为 InCallFragment 的布局文件。
- 步骤 12：由 InCallFragment 创建 contactGridManager、pager 和 endCallButton 等 View 控件。
- 步骤 16 和步骤 17：建立 InCallFragment、CallCardPresenter 和 InCallPresenter 之间的消息处理框架。

3. InCallButtonGridFragment

InCallButtonGridFragment 的加载过程相对要复杂一些，是在 InCallFragment 的 pager 控件中设置 InCallPagerAdapter，而 InCallPagerAdapter 获取 Item 时，创建了 InCallButtonGridFragment 对象，简化后的代码逻辑详情如下：

```
InCallFragment.setPrimary->setAdapterMedia
adapter = new InCallPagerAdapter(getChildFragmentManager(), multimediaData);
pager.setAdapter(adapter);
pager.setCurrentItem
InCallPagerAdapter.getItem
InCallButtonGridFragment.newInstance();
```

总结 InCallButtonGridFragment 的加载流程如图 6-14 所示。

InCallButtonGridFragment 作为 Fragment，是 InCallFragment 的子级 Fragment，重点关注以下几点：

- InCallButtonGridFragment 创建过程。步骤 5：集中在 InCallFragment 的 setAdapterMedia 方法中。
- incall_button_grid.xml 作为 InCallButtonGridFragment 的布局文件。
- 步骤 12：InCallButtonGridFragment 创建六个通话控制按钮的 View 控件。
- 步骤 10 和步骤 14：InCallButtonGridFragment 与 InCallFragment 相互引用。步骤 10 中 InCallButtonGridFragment 通过 FragmentUtils.getParent 获取 InCallFragment 的引用；步骤 14：InCallButtonGridFragment 通过调用 buttonGridListener.onButtonGridCreated(this)，保存 InCallButtonGridFragment 对象引用。
- 步骤 16：建立 InCallFragment、CallButtonPresenter 和 InCallPresenter 之间的消息处理框架。

4. Fragment、Presenter 总结

通话界面目前已经涉及多个 Fragment 和 Presenter，总结它们之间的关系如图 6-15 所示。

>>>>>>>>>> 第 6 章 Voice Call 语音通话模型

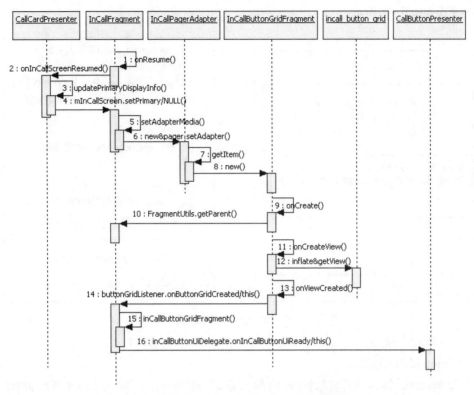

图 6-14 InCallButtonGridFragment 的加载流程

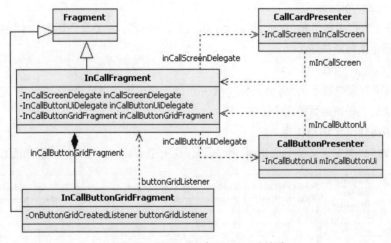

图 6-15 Fragment 与 Presenter 关系

图 6-15 所示的 Fragment 与 Presenter 关系，重点关注以下几点：
- InCallFragment 与 CallCardPresenter 相互引用
- InCallFragment 与 CallButtonPresenter 相互引用
- InCallFragment 与 InCallButtonGridFragment 相互引用

5．通话界面缩略图

packages/apps/Dialer/java/com/android/incallui/incall/impl/res/layout/frag_incall_voice.xml　作为 InCallActivity 通话中 Fragment 的布局文件，再结合 InCallFragment 的代码逻辑，可得 View 控件

- 163 -

对应的示意图如图 6-16 所示。

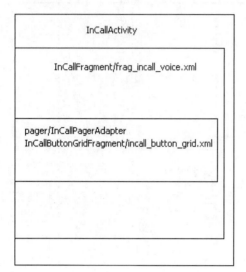

图 6-16　InCallFragment View 控件示意图

图 6-16 所示的 InCallFragment View 控件示意图，重点关注以下几点：

- Fragment 的层级关系

InCallFragment 作为 InCallActivity 的第一级 Fragment，InCallButtonGridFragment 作为 InCallFragment 的下一级 Fragment，将是 InCallActivity 的第二级 Fragment。

- frag_incall_voice.xml

图 6-16 右上方框中显示 frag_incall_voice.xml 的布局内容，主要是通话信息（包括两路通话）、数字键盘和挂断电话控制按钮。

- incall_button_grid.xml

图 6-16 右下方框中显示 incall_button_grid.xml 的布局内容，主要是通话控制按钮，包括静音、扬声器、增加通话、保持通话等六个按钮。

6．InCallPresenter 消息传递

InCallPresenter 类中定义了七个 Listener 接口，这七个 Listener 接口的实现如图 6-17 所示。

	Conference Manager Presenter	VideoCall Presenter	CallButton Presenter	CallCard Presenter	Proximity Sensor	Dialpad Presenter	VideoPause Controller	StatusBar Notifier	VideoPause Controller	ReturnToCall Controller
InCallStateListener	√	√	√	√	√	√	√	√		
IncomingCallListener		√	√	√					√	
CanAddCallListener			√							
InCallDetailsListener	√	√		√						
InCallOrientationListener		√								
InCallEventListener		√		√						
ReturnToCallController										√

图 6-17　InCallPresenter Listener 汇总表

表 6-10 所示的 InCallPresenter Listener 汇总信息，重点关注以下几点：

- 七个 InCallPresenter Listener 接口有十个实现类。
- InCallStateListener 接口的实现最多，共有八个类实现了此接口。
- CallCardPresenter 和 CallButtonPresenter 等类实现了多个 InCallPresenter.Listener 接口。

InCallPresenter 对象共提供了七对有关 mXXXListener 的管理方法：addXXXListener 和 removeXXXListener，那么这些方法在什么时候调用并向 InCallPresenter 注册 Listener 呢？

图 6-13 中的步骤 16 和步骤 17 和图 6-14 中的步骤 16，它们对应的关键代码详情如下：

```
public void onInCallScreenReady() {//CallCardPresenter
    // 注册最新通话状态更改的监听器
    InCallPresenter.getInstance().addListener(this);
    InCallPresenter.getInstance().addIncomingCallListener(this);
    InCallPresenter.getInstance().addDetailsListener(this);
    InCallPresenter.getInstance().addInCallEventListener(this);
}
public void onInCallButtonUiReady(InCallButtonUi ui) {//CallButtonPresenter
    //注册最新通话状态更改的监听器
    final InCallPresenter inCallPresenter = InCallPresenter.getInstance();
    inCallPresenter.addListener(this);
    inCallPresenter.addIncomingCallListener(this);
    inCallPresenter.addDetailsListener(this);
    inCallPresenter.addCanAddCallListener(this);
    inCallPresenter.getInCallCameraManager().addCameraSelectionListener(this);
}
```

通过上面的代码逻辑可知，在加载 InCallFragment 和 InCallButtonGridFragment 两个 Fragment 的过程中，CallButtonPresenter 和 CallCardPresenter 向 InCallPresenter 注册了 Listener 监听。

7．小结

InCallService 的 addCall 响应逻辑可归纳为两点：

- 创建以 android.telecom.Phone 和 android.telecom.Call 对象为中心的 Call.Callback 消息处理框架。
- 加载 InCallFragment 和 InCallButtonGridFragment 界面，同时创建 CallCardPresenter、CallButtonPresenter 与 InCallPresenter 的消息处理框架。

CallList 对象接收到 onCallAdded 方法调用后的主要处理逻辑详情如下：

```
if (call.getState() == DialerCall.State.INCOMING
        || call.getState() == DialerCall.State.CALL_WAITING) {
    onIncoming(call);//处理来电请求
} else {
    dialerCallListener.onDialerCallUpdate();//继续调用 notifyGenericListeners
}
private void notifyGenericListeners() {
    for (Listener listener : mListeners) {//将调用 InCallPresenter 的 onCallListChange 方法
        listener.onCallListChange(this);
    }
}
```

InCallPresenter 将信息转发给 CallButtonPresenter 和 CallCardPresenter 等与通话界面相关的对象，从而变更 InCallFragment 和 InCallButtonGridFragment 的通话界面信息显示。

6.6.4 updateCall

InCallUi 无论展示的是来电界面还是通话中界面，接收到通话信息改变的消息通知时，比如：来电界面用户接听后进入到通话中状态、通话中被对方 Hold 或本地主动 Hold 当前通话等，都是 Telecom 系统应用向 IInCallService 服务发起 updateCall 接口调用，总结其响应流程如图 6-18 所示。

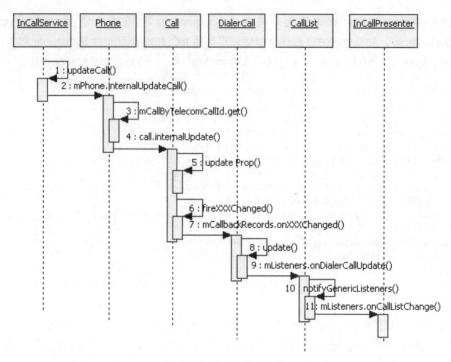

图 6-18 updateCall 流程

对于图 6-17 所示的 updateCall 流程，我们重点关注以下几点：
- 步骤 3：使用接收到的 ParcelableCall 对象的 Telecom CallId 匹配 Phone 对象的 mCallByTelecomCallId 列表，找到对应的 android.telecom.Call 对象。
- 步骤 4：调用 call.internalUpdate(parcelableCall, mCallByTelecomCallId)，步骤 5 则通过传入的 parcelableCall 对象属性判断通话是否有更新以及更新类型，如有变化，则更新 android.telecom.Call 对象对应的属性。
- 步骤 6：根据通话信息变更的类型，调用不同的 fireXXXChanged，发出对应的通知。回顾图 6-12，Call.Callback 有三个类接口实现和注册监听，重点关注 DialerCall 的通知响应逻辑。
- 步骤 11：InCallPresenter 的 onCallListChange 方法响应 Telecom 系统应用发出的 updateCall 接口调用，从而更新当前 InCallUi 的显示。

6.7 验证 Call 运行模型

Android Voice Call 语音通话业务涉及多个 Call 对象，根据它们不同的用途和运行空间，可分为以下三个：
- 以 GsmCdmaCall 对象为核心的 Telephony Voice Call 语音通话业务模型
- 以 com.android.server.telecom.Call 为中心的 Telecom 交互模型
- 围绕 android.telecom.Call 对象的通话界面 InCallUi

我们将修改 Telephony 业务模型、Telecom 系统应用和 Dialer 应用中的相关代码，在 Nexus 6P 手机上建立通话的相关应用场景，运行和验证 Voice Call 语音通话业务模型。

6.7.1 Telephony Voice Call

GsmCdmaCallTracker 类的三个 Call 对象 mRingingCall、mForegroundCall、mBackgroundCall，作为 Telephony Voice Call 业务模型的核心；参考 GsmCdmaCallTracker 类中有关 dump 和 dumpState 方法的逻辑，打印出我们关注的信息，从而完成 Call 通话业务模型验证。

frameworks/opt/telephony/src/java/com/android/internal/telephony/GsmCdmaCall.java 中的 toString 方法打印的内容较少，可增加打印输出，代码详情如下：

```java
public String getMainInfos() {
    StringBuffer infos = new StringBuffer("GsmCdmaCall info mState:" + mState.toString());
    infos.append(", mConnections[\n");
    for(Connection conn:mConnections) {//增加 mConnections 的信息输出
        if (conn != null) {
            infos.append(conn.toString());
            infos.append("\n");
        }
    }
    infos.append("]");
    return infos.toString();
}
```

通话信息发生改变后，可打印当前通话模型的 mRingingCall、mForegroundCall、mBackgroundCall 等对象信息，修改 GsmCdmaCallTracker.java 文件中的 handlePollCalls 方法，在此方法最后加入 dumpCallMode 方法调用，代码详情如下：

```java
public void dumpCallMode() {
    StringBuffer callModeInfos = new StringBuffer("CallTracker mState{");
    callModeInfos.append(mState.toString());
    callModeInfos.append("\nmRingingCall:" + mRingingCall.getMainInfos());
    callModeInfos.append("\nmForegroundCall:" + mForegroundCall.getMainInfos());
    callModeInfos.append("\nmBackgroundCall:" + mBackgroundCall.getMainInfos());
    callModeInfos.append("mConnections[\n");
    for(Connection conn:mConnections) {//遍历 CallTracker 的 mConnections 列表
        if (conn != null) {
            callModeInfos.append(conn.toString());
            callModeInfos.append("\n");
        }
    }
    callModeInfos.append("]\n}");
    android.util.Log.d("XXX", callModeInfos.toString());
}
```

编译并导入 Nexus 6P 手机，建立常用的三种类型的通话场景：拨号、来电和两路通话。

1. 拨号

使用 Nexus 6P 手机拨打电话，对方接听电话后进入通话中应用场景的验证日志，详情如下：

```
CallTracker mState:OFFHOOK
mRingingCall:GsmCdmaCall info mState:IDLE, mConnections[
]
mForegroundCall:GsmCdmaCall info mState:DIALING, mConnections[
  callId: TC@2_1 isExternal: N incoming: false state: DIALING post dial state: NOT_STARTED
]
mBackgroundCall:GsmCdmaCall info mState:IDLE, mConnections[
]
mConnections[
```

```
    callId: TC@2_1 isExternal: N incoming: false state: DIALING post dial state: NOT_STARTED
]

CallTracker mState:OFFHOOK
mRingingCall:GsmCdmaCall info mState:IDLE, mConnections[
]
mForegroundCall:GsmCdmaCall info mState:ACTIVE, mConnections[
    callId: TC@2_1 isExternal: N incoming: false state: ACTIVE post dial state: COMPLETE
]
mBackgroundCall:GsmCdmaCall info mState:IDLE, mConnections[
]
mConnections[
    callId: TC@2_1 isExternal: N incoming: false state: ACTIVE post dial state: COMPLETE
]
```

对于验证的输出日志，我们重点关注以下几点：

- mForegroundCall 状态的变化过程：DIALING→ALERTING→ACTIVE
- GsmCdmaCallTracker 的 mState 为 OFFHOOK
- 对比 GsmCdmaCallTracker.mConnections 与 Call.mConnections 的对应关系

2. 来电

使用其他手机向 Nexus 6P 手机拨打电话，本地接听来电后进入通话中应用场景的验证日志，详情如下：

```
CallTracker mState:RINGING
mRingingCall:GsmCdmaCall info mState:INCOMING, mConnections[
    callId: null isExternal: N incoming: true state: INCOMING post dial state: NOT_STARTED
]
mForegroundCall:GsmCdmaCall info mState:IDLE, mConnections[
]
mBackgroundCall:GsmCdmaCall info mState:IDLE, mConnections[
]
mConnections[
    callId: null isExternal: N incoming: true state: INCOMING post dial state: NOT_STARTED
]
//接听电话后的信息变化
CallTracker mState:OFFHOOK
mRingingCall:GsmCdmaCall info mState:IDLE, mConnections[
]
mForegroundCall:GsmCdmaCall info mState:ACTIVE, mConnections[
    callId: TC@3_1 isExternal: N incoming: true state: ACTIVE post dial state: NOT_STARTED
]
mBackgroundCall:GsmCdmaCall info mState:IDLE, mConnections[
]
mConnections[
    callId: TC@3_1 isExternal: N incoming: true state: ACTIVE post dial state: NOT_STARTED
]
```

对于验证的输出日志，我们重点关注以下几点：

- mRingingCall 保存来电信息。
- 本地接听电话后，通话状态变化为 ACTIVE，Connection 从 mRingingCall 迁移到了 mForegroundCall。
- GsmCdmaCallTracker 的 mState 状态转换：RINGING→OFFHOOK。
- 对比 GsmCdmaCallTracker.mConnections 与 Call.mConnections 的对应关系。

3. 两路通话

使用 Nexus 6P 手机先建立一路通话（拨号或是来电均可），再增加一路通话的拨号请求或是接听一路来电的通话请求，即可建立起两路通话。当然，这两路通话一路是 ACTIVE 通话中，一路是

HOLDING 通话保持。

通话中，接收到来电的验证日志，详情如下：

```
CallTracker mState:RINGING
mRingingCall:GsmCdmaCall info mState:WAITING, mConnections[
  callId: null isExternal: N incoming: true state: WAITING post dial state: NOT_STARTED
]
mForegroundCall:GsmCdmaCall info mState:ACTIVE, mConnections[
  callId: TC@4_1 isExternal: N incoming: false state: ACTIVE post dial state: COMPLETE
]
mBackgroundCall:GsmCdmaCall info mState:IDLE, mConnections[
]
mConnections[
  callId: TC@4_1 isExternal: N incoming: false state: ACTIVE post dial state: COMPLETE
  callId: null isExternal: N incoming: true state: WAITING post dial state: NOT_STARTED
]
```

对于验证的输出日志，我们重点关注以下几点：

- GsmCdmaCallTracker 的 mState 状态为 RINGING，它的级别高于 OFFHOOK。
- mRingingCall 保存来电信息，其状态是 WAITING 呼叫等待，等待接听本地电话。
- mForegroundCall 保存 ACTIVE 通话中信息。
- GsmCdmaCallTracker.mConnections 列表中包含了两个 Connection 对象，分别对应 mRingingCall.mConnections 和 mForegroundCall.mConnections 中的对象。

接着接听本地来电请求，验证日志的详情如下：

```
CallTracker mState:OFFHOOK
mRingingCall:GsmCdmaCall info mState:IDLE, mConnections[
]
mForegroundCall:GsmCdmaCall info mState:ACTIVE, mConnections[
  callId: TC@5_1 isExternal: N incoming: true state: ACTIVE post dial state: NOT_STARTED
]
mBackgroundCall:GsmCdmaCall info mState:HOLDING, mConnections[
  callId: TC@4_1 isExternal: N incoming: false state: HOLDING post dial state: COMPLETE
]
mConnections[
  callId: TC@4_1 isExternal: N incoming: false state: HOLDING post dial state: COMPLETE
  callId: TC@5_1 isExternal: N incoming: true state: ACTIVE post dial state: NOT_STARTED
]
```

对于验证的输出日志，我们重点关注以下几点：

- GsmCdmaCallTracker 的 mState 状态转换为 OFFHOOK。
- mRingingCall 保存来电信息，转换到 mForegroundCall 后其状态为 ACTIVE，而之前 mForegroundCall 的通话转移到 mBackgroundCall，状态为 HOLDING 通话保持。
- GsmCdmaCallTracker.mConnections 列表中包含了两个 Connection 对象，分别对应 mForegroundCall.mConnections 和 mBackgroundCall.mConnections 中的对象。

在通话界面中触发交换通话时，mForegroundCall 和 mBackgroundCall 的 mConnections 属性会进行交换，验证日志的详情如下：

```
CallTracker mState:OFFHOOK
mRingingCall:GsmCdmaCall info mState:IDLE, mConnections[
]
mForegroundCall:GsmCdmaCall info mState:ACTIVE, mConnections[
  callId: TC@4_1 isExternal: N incoming: false state: ACTIVE post dial state: COMPLETE
]
```

```
mBackgroundCall:GsmCdmaCall info mState:HOLDING, mConnections[
  callId: TC@5_1 isExternal: N incoming: true state: HOLDING post dial state: NOT_STARTED
]
mConnections[
  callId: TC@4_1 isExternal: N incoming: false state: ACTIVE post dial state: COMPLETE
  callId: TC@5_1 isExternal: N incoming: true state: HOLDING post dial state: NOT_STARTED
]
```

对于验证的输出日志，我们重点关注以下几点：
- 通过 callId 验证通话状态的切换。
- mForegroundCall 状态为 ACTIVE，mBackgroundCall 状态为 HOLDING，为了保持该关系状态，mConnections 对象将进行切换。

6.7.2 Telecom Call

Telecom 系统应用以 com.android.server.telecom.Call 为中心，建立了与 TeleService 和 InCallUi 两个应用的通话交互模型。

修改 packages/services/Telecomm/src/com/android/server/telecom/Call.java 代码，详情如下：

```java
public String getMainInfos() {
    StringBuffer callInfo = new StringBuffer("com.android.server.telecom.Call info:");
    callInfo.append("\nmId:" + mId);
    callInfo.append("\nmConnectionId:" + mConnectionId);
    callInfo.append("\nmState:" + CallState.toString(mState));
    callInfo.append("\nmCallDirection:" + mCallDirection);
    callInfo.append("\nmHandle:" + mHandle);
    callInfo.append("\nmIsConference:" + mIsConference);
    callInfo.append(", mChildCalls[\n");
    for(Call call:mChildCalls) {
        if (call != null) {
            callInfo.append(call.toString());
            callInfo.append("\n");
        }
    }
    callInfo.append("]");
    return callInfo.toString();
}
```

修改 packages/services/Telecomm/src/com/android/server/telecom/CallIdMapper.java 代码，详情如下：

```java
public void dumpCalls() {
    //打印调用堆栈信息，可验证两个 CallIdMapper 的更新过程
    android.util.Log.e("XXX", "CallIdMapper dumpCalls Stack:", new RuntimeException());
    StringBuffer callInfo = new StringBuffer("CallIdMapper mCalls{");
    Iterator iterator = mCalls.mPrimaryMap.keySet().iterator();
    while (iterator.hasNext()) {
        String key = (String)iterator.next();
        callInfo.append("\n" + mCalls.getValue(key).getMainInfos());
    }
    callInfo.append("\n}");
    android.util.Log.d("XXX", callInfo.toString());
}
```

需要在 CallIdMapper 的 addCall 和 removeCall 方法中增加对 dumpCalls 的调用，在 Call 管理的过程中打印关键信息；并且在 ConnectionServiceWrapper 的内部类 Adapter 的 setActive、 setRinging 和 setOnHold 等方法中加入 CallIdMapper 对象的 dumpCalls 方法调用。

使用 Nexus 6P 手机建立两路通话，验证 Telecom Call 模型，验证日志的详情如下：

```
CallIdMapper mCalls{
com.android.server.telecom.Call info:
mId:TC@6
mConnectionId:TC@6_1
mState:ACTIVE
mCallDirection:1
mHandle:tel:00000001
mIsConference:false, mChildCalls[
]
com.android.server.telecom.Call info:
mId:TC@7
mConnectionId:TC@7_1
mState:RINGING
mCallDirection:2
mHandle:tel:00000002
mIsConference:false, mChildCalls[
]}
```

对于 Telecom 系统应用输出的日志，我们重点关注以下几点：

- dumpCalls 的两次调用

CallIdMapper 有两个对象：InCallController 和 ConnectionServiceWrapper，它们都通过 mCallId-Mapper 属性来管理 Call 对象的列表。因此，dumpCalls 的两次调用是由两个不同的 CallIdMapper 对象发起的。

- 两路通话状态

一路通话状态为 ACTIVE 通话中，另一路通话状态为 RINGING 来电响铃。

6.7.3 InCallUi Call

InCallUi 通话界面是围绕 android.telecom.Call 对象信息来展示通话信息和提供通话控制交互界面的。修改 android.telecom.Phone 类，在 internalAddCall、internalRemoveCall 和 internalUpdateCall 方法中增加 dumpCalls 调用，代码详情如下：

```
void dumpCalls() {
    StringBuffer callInfo = new StringBuffer("InCallUi Phone mCalls{");
    for (Call call:mCalls) {
        if (call != null) {
            callInfo.append("\n" + call.toString() + "\nChild Call[");
            for (Call child:call.getChildren()) {
                callInfo.append("\n" + child.toString());
            }
            callInfo.append("]\n");
        }
    }
    callInfo.append("}");
    android.util.Log.d("XXX", callInfo.toString());
}
```

使用 Nexus 6P 手机建立两路通话，将两路通话合并为一路会议电话（三方通话），再增加一路通话，InCallUi 通话界面的截图如图 6-19 所示。

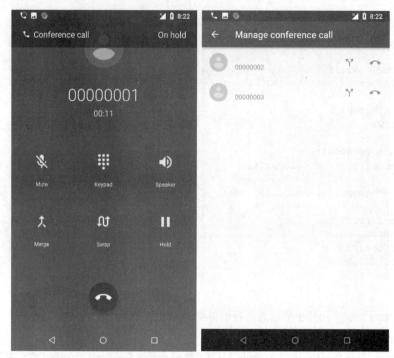

图 6-19 会议电话通话界面

图 6-19 左图显示了两路通话：通话中的是正与 00000001 通话，通话保持的是会议电话，可以通过 Swap 交换通话，来切换两路通话的通话状态。将会议电话切换为通话中后，会出现 Manage 会议电话管理按键，点击后进入图 6-19 右图，可见会议电话有两个通话连接：00000002 和 00000003。

Telephony Voice Call 语言通话业务模型的验证日志，详情如下：

```
CallTracker mState{OFFHOOK
mRingingCall:GsmCdmaCall info mState:IDLE, mConnections[]
mForegroundCall:GsmCdmaCall info mState:ACTIVE, mConnections[
  callId: TC@3_1 isExternal: N incoming: false state: ACTIVE post dial state: COMPLETE
  callId: TC@4_1 isExternal: N incoming: false state: ACTIVE post dial state: COMPLETE
]
mBackgroundCall:GsmCdmaCall info mState:HOLDING, mConnections[
  callId: TC@5_1 isExternal: N incoming: false state: HOLDING post dial state: COMPLETE
]mConnections[
  callId: TC@3_1 isExternal: N incoming: false state: ACTIVE post dial state: COMPLETE
  callId: TC@4_1 isExternal: N incoming: false state: ACTIVE post dial state: COMPLETE
  callId: TC@5_1 isExternal: N incoming: false state: HOLDING post dial state: COMPLETE
]
}
```

总共有三个通话连接：两个通话连接是 ACTIVE 通话中状态，在 mForegroundCall 中；另一个通话连接是 HOLDING 通话保持状态，在 mBackgroundCall 中。

Telecom Call 模型的验证日志，详情如下：

```
CallIdMapper mCalls{
com.android.server.telecom.Call info:
mId:9a26bf64-0741-4927-a463-c2af2e0c2303//Conference Call Id
mConnectionId:9a26bf64-0741-4927-a463-c2af2e0c2303
mState:ACTIVE
mCallDirection:0
mHandle:null
```

```
    mIsConference:true, mChildCalls[//Conference Call 有两个子 Call 对象
    [TC@4, ACTIVE, ......]
    [TC@3, ACTIVE, ......]
    ]
    com.android.server.telecom.Call info:
    mId:TC@3
    mConnectionId:TC@3_1
    mState:ACTIVE
    mCallDirection:1
    mHandle:tel: 00000002
    mIsConference:false, getChildCalls:0, mChildCalls[]
    com.android.server.telecom.Call info:
    mId:TC@4
    mConnectionId:TC@4_1
    mState:ACTIVE
    mCallDirection:1
    mHandle:tel: 00000003
    mIsConference:false, mChildCalls[]
    com.android.server.telecom.Call info:
    mId:TC@5
    mConnectionId:TC@5_1
    mState:ON_HOLD
    mCallDirection:1
    mHandle:tel: 00000001
    mIsConference:false, mChildCalls[]
    }
```

Telecom 系统应用共有四个 Call 对象：一个 Conference Call 对象，三个通话连接创建的 Call 对象，两个状态是 ACTIVE 通话中状态，归属于 Conference Call 对象，一个状态是 ON_HOLD 通话保持状态。

InCallUi 通话界面 Call 模型的验证日志，详情如下：

```
InCallUi Phone mCalls{
Call [id: TC@3, state: ACTIVE, details: [......], Child Call[]
Call [id: TC@4, state: ACTIVE, details: [......], Child Call[]
Call [id: 9a26bf64-0741-4927-a463-c2af2e0c2303, state: ACTIVE, details: [......], Child Call[
Call [id: TC@4, state: ACTIVE, details: [......]
Call [id: TC@3, state: ACTIVE, details: [......]
]
Call [id: TC@5, state: HOLDING, details: [......], Child Call[]
}
```

InCallUi 通话界面共有四个 Call 对象：一个 Conference Call 对象，三个通话连接创建的 Call 对象，两个状态是 ACTIVE 通话中状态，归属于 Conference Call 对象，一个状态是 HOLDING 通话保持状态。

本 章 小 结

本章以 GsmCdmaCallTracker 为中心，从 Handler 消息处理方式、与 RILJ 的交互机制和 handlePollCalls 关键业务逻辑三个方面进行了详细的分析和总结，分析和验证了 Voice Call 语音通话业务模型，并扩展讲解了 InCallUi 通话界面的核心实现机制，需要重点掌握：

- GsmCdmaCallTracker 的 Handler 消息处理方式

GsmCdmaCallTracker 的本质是 Handler 子类，响应 RILJ 对象发出的 Handler Message 消息回调，最终汇总到 handlePollCalls 方法调用。

- GsmCdmaCallTracker 与 RILJ 的交互机制

GsmCdmaCallTracker 与 RILJ 的交互机制可重点关注图 6-2 和图 6-3 总结的关键流程：GsmCdmaCallTracker 主动发起通话管理和控制请求和被动接收 RILJ 对象发出的通话状态变化消息。

- handlePollCalls 方法中 Call、Connection 对象的处理机制

掌握 GsmCdmaCallTracker 对象的 mConnections、mRingingCall、mForegroundCall、mBackgroundCall 和 mState 属性的更新过程。

- DriverCall、Call、Connection 的通话管理模型

掌握图 6-5 总结的 CallTracker、Call 与 Connection 之间的关系。

- 区分 State

掌握图 6-6 总结的 Call.State、DriverCall.State 和 PhoneConstants.State 之间的转换过程和对应关系。

- 区分 Connection

掌握图 6-7 总结的两个 Connection 之间的区别和关联。

- InCallUi 通话界面

InCallUi 通话界面运行在 Dialer 应用空间，其核心是接收 Telecom 系统应用的 ParcelableCall 对象后将其转换为 android.telecom.Call 对象，并围绕此对象进行通话界面的展示，提供对应的通话管理和控制操作界面。

- 验证 Call 模型

在 Telephony Call、Telecom Call 和 InCallUi Call 三个关键的通话模型代码中打印 Call 模型来验证 Android Voice Call 语音通话业务模型。重点掌握图 6-19 所示的会议电话业务，可加强对 Android Voice Call 语音通话业务模型的直观认识和深入理解。

第 7 章

ServiceState 网络服务

学习目标

- 认识网络服务信息 ServiceState 对象。
- 掌握 ServiceStateTracker 运行机制。
- 掌握 handlePollStateResult 和 pollStateDone 核心业务逻辑。
- 参考 4636 工具 Phone information 实现机制。
- 掌握 ITelephonyRegistry 实现原理和消息流转方式。
- 理解飞行模式的实现方式。
- 了解 A3A8 和 Milenage 驻网鉴权算法原理和相关规范。
- 了解 SIM 卡业务扩展 eSIM 和 SoftSim。

通过前面的学习，知道 Android Telephony 三大业务能力分别是：
- Voice Call 语音通话
- ServiceState 网络服务
- Data Call 移动数据业务

上一章我们对 Voice Call 语音通话业务模型进行了详细的分析和总结，本章将围绕 ServiceStateTracker 类展开对 ServiceState 网络服务业务模型的学习，通过解析 ServiceStateTracker 对 ServiceState 管理的机制和流程，认识和理解 Android Telephony 网络服务业务的核心运行机制。

ServiceState 网络服务的管理和更新由 ServiceStateTracker 对象完成，主要维护 ServiceState 类的两个实体对象：mSS 和 mNewSS，从而完成 ServiceState 服务状态的管理和更新。

本章从两方面解析 ServiceState 网络服务业务的关键运行流程和核心处理机制。
- 解析 ServiceState 类。
- ServiceStateTracker 的运行机制。

7.1 ServiceState

Android 手机插入 SIM 卡后,将有一个驻网的过程,来完成 SIM 卡中信息的验证和运营商移动网络的注册,这样我们的手机才能使用运营商提供的网络服务,如接打电话、发送接收短信彩信、移动数据上网等。

ServiceState 从字面意思可以理解为服务状态。在 Android Telephony 业务模型中,使用 ServiceState 实体类来保存 SIM 卡网络注册成功后运营商网络的一些基本服务信息,如服务状态(Voice Call 语音通话和 Data Call 移动数据)、运营商信息、Radio 无线通信模块使用的技术类型和状态、运营商网络是否处于漫游状态等网络服务的基本信息。

ServiceState 类作为保存网络服务基本信息的实体类,其代码源文件为:frameworks/base/telephony/java/android/telephony/ServiceState.java,本节从以下三个方面来展开对 ServiceState 实体类的解析。

- ServiceState 类的本质。
- ServiceState 类的关键常量定义及属性。
- ServiceState 类的关键方法。

7.1.1 ServiceState 类的本质

ServiceState 类实现了 Parcelable 接口,主要由以下两个过程组成。

- writeToParcel 序列化过程,将对象数据写入外部提供的 Parcel 中。
- createFromParcel 反序列化过程,通过外部提供的 Parcel 获取基本数据来创建 ServiceState 对象。

继续分析 ServiceState 类中是如何实现序列化和反序列化的?其代码运行框架如下。

```
// ServiceState 实现了 Parcelable 接口
public class ServiceState implements Parcelable {
    public static final Parcelable.Creator<ServiceState> CREATOR =
            new Parcelable.Creator<ServiceState>() {//静态的 Parcelable.Creator 接口
        public ServiceState createFromParcel(Parcel in) {//反序列化过程
            return new ServiceState(in);//创建 ServiceState 对象,关注其构造方法
        }
        //ServiceState 数组对象序列化方法
        public ServiceState[] newArray(int size) {
            return new ServiceState[size];
        }
    };

    public ServiceState(Parcel in) {//构造方法,参数 Parcel
        mVoiceRegState = in.readInt();
        ......//Parcel 序列化读取过程
    }
    public void writeToParcel(Parcel out, int flags) {//序列化过程
        out.writeInt(mVoiceRegState);
        ......//Parcel 序列化写入过程
    }
}
```

通过上面的代码可知,Android 中使用 Parcelable 接口实现序列化和反序列化的过程非常简单,必须实现 writeToParcel 方法和 Parcelable.Creator 接口,只需要对照上面的代码框架,便能实现自定

义的序列化和反序列化对象。

7.1.2 关键常量信息

ServiceState 类的关键常量，总结如表 7-1 所示。

表 7-1 ServiceState 类的关键常量定义

分类	定义	描述
网络服务状态	STATE_IN_SERVICE	服务状态正常
	STATE_OUT_OF_SERVICE	不在服务中
	STATE_EMERGENCY_ONLY	只能呼叫紧急号码
	STATE_POWER_OFF	无线通信模块已经关闭
网络注册信息	RIL_REG_STATE_NOT_REG	网络未注册
	RIL_REG_STATE_HOME	注册本地网络服务
	RIL_REG_STATE_DENIED	拒绝注册
	RIL_REG_STATE_ROAMING	注册漫游网络服务
	总计 10 个 RIL_REG_STATE_XXX（其余省略）	
无线通信网络类型	RIL_RADIO_TECHNOLOGY_GPRS	2G GSM 网络
	RIL_RADIO_TECHNOLOGY_TD_SCDMA	3G TD_SCMA 网络
	RIL_RADIO_TECHNOLOGY_LTE	4G LTE 网络
	总计 19 个 RIL_RADIO_TECHNOLOGY_XXX（其余省略），参考 rilRadioTechnologyToString 方法中匹配网络类型的对应关系	

ServiceState 类中共定义了 4 种网络服务状态、10 个网络注册信息和 19 种无线通信网络类型。19 种无线通信网络类型覆盖了 GSM、CDMA、UMTS 和 LTE 等全球主流的无线通信技术，从 2G 到 4G，它们的对应关系可在 rilRadioTechnologyToString 方法中获取。

7.1.3 关键属性

ServiceState 类的关键属性定义如表 7-2 所示。

表 7-2 ServiceState 类的关键属性

属性	类型	描述
mVoiceRegState	int	网络服务状态（语音）
mDataRegState	int	移动数据服务状态
mVoiceRoamingType	int	语音漫游类型
mDataRoamingType	int	移动数据漫游类型
mVoiceOperatorAlphaLong mDataOperatorAlphaLong	String	运营商名称
mVoiceOperatorAlphaShort mDataOperatorAlphaShort	String	运营商名称

续表

属性	类型	描述
MVoiceOperatorNumeric mDataOperatorNumeric	String	运营商编号
mIsManualNetworkSelection	boolean	手动选择运营商标志
mIsEmergencyOnly	boolean	仅有紧急呼救状态标志

ServiceState 实体类主要从 Voice Call 语音通话业务和 Data Call 移动数据业务两个方面，保存运营商网络服务状态以及运营商的基本信息。

通过表 7-1 和表 7-2 可以直观地认识 ServiceStateTracker 抽象类中 ServiceState 的两个对象 mSS 和 mNewSS，它们保存服务状态和运营商的基本信息。但是，网络服务信息不只有这些内容，还有如当前注册的运营商网络信号量强度、驻网的小区信息等，那么这些信息保存在什么地方呢？ServiceStateTracker 类中的一些关键属性将保存这些信息。

7.1.4 关键方法

ServiceState 类实现了 Parcelable 接口，使用 Parcel 序列化方式，关键方法主要体现在以下三个方面。
- get/set 方法
- rilRadioTechnologyToString 方法获取手机网络类型对应关系
- toString 方法

ServiceState 类的序列化过程和反序列化过程主要包括 writeToParcel 序列化过程和 createFromParcel 反序列化过程，如表 7-3 所示。

表 7-3 ServiceState 对象的序列化过程和反序列化过程

ServiceState 实体类属性	writeToParcel 序列化过程	createFromParcel 反序列化过程
mVoiceRegState	out.writeInt	in.readInt
mDataRegState	out.writeInt	in.readInt
mVoiceRoamingType	out.writeInt	in.readInt
mDataRoamingType	out.writeInt	in.readInt
mVoiceOperatorAlphaLong	out.writeString	in.readString
mVoiceOperatorAlphaShort	out.writeString	in.readString
mVoiceOperatorNumeric	out.writeString	in.readString
mDataOperatorAlphaLong	out.writeString	in.readString
mDataOperatorAlphaShort	out.writeString	in.readString
mDataOperatorNumeric	out.writeString	in.readString
mIsManualNetworkSelection	out.writeInt	in.readInt
mRilVoiceRadioTechnology	out.writeInt	in.readInt
mRilDataRadioTechnology	out.writeInt	in.readInt
mCssIndicator	out.writeInt	in.readInt
mNetworkId	out.writeInt	in.readInt

表 7-3 所示 ServiceState 对象的序列化过程和反序列化过程，重点关注以下两点。
- ServiceState 共有 24 个属性，表 7-3 仅列出了 15 个属性的序列化过程和反序列化过程。
- 在序列化过程和反序列化过程中，对序列化流的操作将保持一致。

rilRadioTechnologyToString 方法获取手机网络类型对应关系，toString 方法将主要的网络服务信息转换为字符串，其逻辑相对简单，请读者自行阅读和学习源代码中的处理逻辑，可帮助我们理解 ServiceState 网络服务信息。

7.2　ServiceStateTracker 运行机制详解

前面认识了 ServiceState 实体类，本节学习以 ServiceStateTracker 类为核心的 ServiceState 网络服务信息的管理和运行机制。

ServiceStateTracker 从字面意思可理解为网络服务信息业务管理的跟踪者，GsmCdmaPhone 对象将网络服务信息交给它来管理和维护，主要完成两方面的处理逻辑。
- 管理网络服务信息
- 提供网络服务控制管理能力

7.2.1　核心类图

在 Telephony 业务模型中，ServiceStateTracker 对象主要管理网络服务的基本信息，有着非常重要的作用，其核心类如图 7-1 所示。

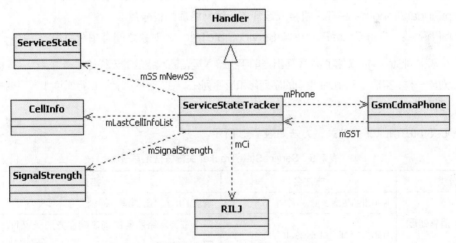

图 7-1　ServiceStateTracker 核心类图

如图 7-1 所示的 ServiceStateTracker 核心类图，以 ServiceStateTracker 类为中心，还有其他五个非常关键的类，分别是 RILJ、GsmCdmaPhone、ServiceState、SignalStrength 和 CellInfo 类，这六个类共同完成网络服务信息的管理，重点关注以下几点：
- ServiceStateTracker 作为核心类，网络服务信息的管理将围绕它来运行，与 GsmCdma-CallTracker 类相同，它继承于 Handler 类，其实质是自定义 Handler 消息处理类。
- ServiceState、CellInfo 和 SignalStrength 三个类均实现了 Parcelable 接口，对象都可以跨进

程传递。

- ServiceStateTracker 的 mPhone 是 GsmCdmaPhone 对象，而 GsmCdmaPhone 对象的 mSST 是 ServiceStateTracker 对象，它们之间相互引用。

7.2.2 代码结构

ServiceStateTracker 类的关键属性如表 7-4 所示。

表 7-4 ServiceStateTracker 关键属性汇总表

属性	类型	描述
mCi	CommandsInterface	RILJ 对象
mSS	ServiceState	当前 ServiceState 对象
mNewSS	ServiceState	最新 ServiceState 对象
mSignalStrength	SignalStrength	手机无线网络信号量
mCellLoc	CellLocation	小区信息
mNewCellLoc	CellLocation	最新小区信息
mPhone	GsmCdmaPhone	GSMPhone 对象

对表 7-4 所示的 ServiceStateTracker 关键属性，重点关注以下几点：

- mCi：RILJ 对象，说明此对象与 GsmCdmaPhone 对象一样，通过 mCi 对象具备与 RIL 的交互能力。
- mSS 和 mNewSS，保存接入的运营商网络服务状态以及运营商的基本信息，主要包括两个方面：Voice Call 语音通话业务和 Data Call 移动数据业务的网络服务信息。
- mSignalStrength：保存手机接入运营商无线网络后的信号量。
- mPhone：GsmCdmaPhone 与 ServiceStateTracker 对象之间有相互引用。

注意

mSignalStrength 对象能够体现出当前手机接入运营商网络后的无线信号量，它和 ServiceState 对象一样，实现了 Parcel 类型的序列化和反序列化。感兴趣的读者请自行阅读和学习相关代码处理逻辑。

ServiceStateTracker 类的关键方法如表 7-5 所示。

表 7-5 ServiceStateTracker 关键方法汇总表

分类	方法名	描述
Handler 消息处理	handleMessage	响应 RILJ 发出的消息回调
	handlePollStateResult	四个查询网络服务的消息响应方法 EVENT_POLL_STATE_XXX
更新网络服务信息	pollState/modemTriggeredPollState	查询基本网络服务信息，包括 getOperator 查询电信运营商信息、getDataRegistrationState 查询移动数据注册状态、getVoiceRegistrationState 查询语音注册状态、getNetworkSelectionMode 查询网络模式
	pollStateDone	根据 pollState 的查询结果，完成 mSS 信息的更新并发出 ServiceState 变化的消息通知

续表

分类	方法名	描述
更新网络服务信息	updateSpnDisplay	更新网络运营商显示名称
	queueNextSignalStrengthPoll	查询当前无线信号
	onSignalStrengthResult	根据 queueNextSignalStrengthPoll 的查询结果，更新信号
网络服务控制	setRadioPower	开关 Radio 无线通信模块
	enableLocationUpdates	开启位置更新消息上报
	disableLocationUpdates	关闭位置更新消息上报

表 7-5 中的 pollStateDone 方法根据查询网络服务的结果，更新 mSS 和 mSignalStrength 对象来完成网络服务信息的更新，主要处理逻辑与 GsmCdmaCallTracker 对象的处理方式非常相似，比如 handlePollCalls 与 handlePollStateResult 方法相对应，都是处理 RILJ 对象返回的查询信息。

另外，网络服务的控制由 GsmCdmaPhone 对外提供统一的方法，通过 mSST 对象调用对应的方法来实现，mSST 即 ServiceStateTracker 对象。

7.2.3　Handler 消息处理机制

ServiceStateTracker 类的本质是自定义的 Handler 消息处理类，Handler 消息的处理逻辑是当前类的核心业务。

通过前面的学习，我们已经知道 Handler 的三种处理方式，在 ServiceStateTracker 类中只使用其中两种运行和处理机制。

- 基本的 Handler 消息注册和响应处理机制
- Handler 消息 Callback 回调处理方式

接下来逐步展开对 ServiceStateTracker 类中这两种 Handler 消息处理方式的分析和学习。

1. 消息注册

在 ServiceStateTracker 类的构造方法中找到了其 Handler 消息注册的代码逻辑，详情如下。

```
mCi.setOnSignalStrengthUpdate(this, EVENT_SIGNAL_STRENGTH_UPDATE, null);
mCi.registerForCellInfoList(this, EVENT_UNSOL_CELL_INFO_LIST, null);
mCi.registerForImsNetworkStateChanged(this, EVENT_IMS_STATE_CHANGED, null);
mCi.registerForRadioStateChanged(this, EVENT_RADIO_STATE_CHANGED, null);
mCi.registerForNetworkStateChanged(this, EVENT_NETWORK_STATE_CHANGED, null);
mCi.setOnNITZTime(this, EVENT_NITZ_TIME, null);
mCi.setOnRestrictedStateChanged(this, EVENT_RESTRICTED_STATE_CHANGED, null);
```

mCi 即 RILJ 对象，原来，ServiceStateTracker 对象会被动接收并响应 RILJ 对象发出的七种类型的 Handler 消息。

- EVENT_RADIO_STATE_CHANGE 无线通信模块状态变化
- EVENT_NETWORK_STATE_CHANGE 无线网络状态变化
- EVENT_SIGNAL_STRENGTH_UPDATE 信号量变化
- EVENT_UNSOL_CELL_INFO_LIST 小区信息上报
- EVENT_NITZ_TIME 获取网络时间
- EVENT_RESTRICTED_STATE_CHANGED 驻网信息变化

- EVENT_IMS_STATE_CHANGED IMS 驻网状态变化

通过源码查证得知,这几种服务状态变化的 Handler 消息都定义在 ServiceStateTracker 抽象类中,并且仅在 ServiceStateTracker 相关的类中产生响应。也就是说,有且仅有 ServiceStateTracker 对象会接收和响应 RILJ 对象发出的这几种与网络服务状态相关的 Handler 消息通知。

2. 消息响应

针对在构造方法中注册的七种 Handler 消息,handleMessage 响应逻辑中提取这七种 Handler 消息的响应逻辑,详情如下。

```
case EVENT_RADIO_STATE_CHANGED:
    modemTriggeredPollState();//查询网络服务基本信息
    break;
case EVENT_NETWORK_STATE_CHANGED:
    modemTriggeredPollState();//查询网络服务基本信息
    break;
case EVENT_SIGNAL_STRENGTH_UPDATE:
    ar = (AsyncResult) msg.obj;
    mDontPollSignalStrength = true;
    onSignalStrengthResult(ar); //响应新的无线网络信号
    break;
case EVENT_UNSOL_CELL_INFO_LIST:
    ar = (AsyncResult) msg.obj;
    List<CellInfo> list = (List<CellInfo>) ar.result;
    mLastCellInfoListTime = SystemClock.elapsedRealtime();
    mLastCellInfoList = list;//更新小区信息列表
    mPhone.notifyCellInfo(list);//发出通知
    break;
case EVENT_RESTRICTED_STATE_CHANGED:
    if (mPhone.isPhoneTypeGsm()) {
        ar = (AsyncResult) msg.obj;
        onRestrictedStateChanged(ar);//响应网络注册状态变化
    }
    break;
```

在五个重要的 Handler 类型消息的响应过程中,会产生两次 modemTriggeredPollState 方法调用来查询网络服务信息,onSignalStrengthResult 方法调用更新网络信号,onRestrictedStateChanged 方法调用更新网络注册信息。

首先查看 modemTriggeredPollState 方法,通过字面理解,是 Modem 触发的查询 PollState 调用,将发起对 pollState(true) 的调用来查询当前最新网络服务信息,其逻辑主要分为两个部分。

- 查询网络服务逻辑
- Radio 无线通信模块状态异常的处理逻辑

在 Radio 无线通信模块状态正常的情况下,pollState 方法会有什么样的处理逻辑呢?原来是使用 mCi 对象连续向 RILJ 对象发出四个关于网络服务信息的查询请求,其处理逻辑的详情如下。

```
mPollingContext[0]++; //查询运营商信息
mCi.getOperator(obtainMessage(EVENT_POLL_STATE_OPERATOR, mPollingContext));

mPollingContext[0]++;//查询注册的移动数据类型
mCi.getDataRegistrationState(obtainMessage(EVENT_POLL_STATE_GPRS, mPollingContext));

mPollingContext[0]++;//查询注册的语音类型
mCi.getVoiceRegistrationState(obtainMessage(EVENT_POLL_STATE_REGISTRATION,
```

```
            mPollingContext));
if (mPhone.isPhoneTypeGsm()) {//GSM 网络制式
    mPollingContext[0]++;//查询网络类型
    mCi.getNetworkSelectionMode(obtainMessage(
            EVENT_POLL_STATE_NETWORK_SELECTION_MODE, mPollingContext));
}
```

上面的代码逻辑中，mCi 对象的调用方式是不是非常熟悉呢？它使用了 Handler Message 消息的 Callback 处理机制。

Radio 无线通信模块状态异常的处理逻辑的代码详情如下。

```
switch (mCi.getRadioState()) {//获取 Radio 状态
    case RADIO_UNAVAILABLE://Radio 不可用
        mNewSS.setStateOutOfService();//mNewSS 设置为 STATE_OUT_OF_SERVICE
        mNewCellLoc.setStateInvalid();//设置小区信息状态异常
        setSignalStrengthDefaultValues();//设置无线信号为默认值
        mGotCountryCode = false;//未获取国家代码
        mNitzUpdatedTime = false;//时间未更新
        pollStateDone();//更新 mSS
        break;
    case RADIO_OFF:
        mNewSS.setStateOff();//mNewSS 设置为 STATE_POWER_OFF
        mNewCellLoc.setStateInvalid();
        setSignalStrengthDefaultValues();
        mGotCountryCode = false;
        mNitzUpdatedTime = false;
        if (mDeviceShuttingDown ||
                (!modemTriggered && ServiceState.RIL_RADIO_TECHNOLOGY_IWLAN
                != mSS.getRilDataRadioTechnology())) {
            pollStateDone();//更新 mSS
            break;
        }
```

上面的代码逻辑，可重点关注以下几点：

● 更新 mNewSS 状态

RADIO_UNAVAILABLE 设置 mNewSS 状态为 STATE_OUT_OF_SERVICE

RADIO_OFF 设置 mNewSS 状态为 STATE_POWER_OFF

● 相同处理

mNewCellLoc 设置为小区信息不可用状态

设置无线信号量为默认值

获取国家代码的标志为 false

mNitzUpdatedTime 时间更新标志为 false

调用 pollStateDone 完成更新网络服务相关信息

3. Callback 处理机制

在 GsmCdmaCallTracker 对象中，直接使用 Message 对象作为参数访问 RILJ 对象的通话管理接口，RILJ 再通过 Message 对象发起 Callback 回调；在 ServiceStateTracker 类中，同样也采用了这种消息处理机制。

Handler 消息处理机制的实现逻辑框架基本一致，主要分两步完成。

● 调用 obtainMessage 方法创建 Message 对象。

● 通过 mCi 对象向 RILJ 对象发起查询网络服务信息的方法调用。

在 RIL 中处理完 ServiceStateTracker 对象发起的 ServiceState 相关的请求后，使用 Message 对象发起 Callback 回调消息通知。ServiceStateTracker 对象中的响应和处理方式全部一致，其代码详情如下。

```
case EVENT_POLL_STATE_REGISTRATION:
case EVENT_POLL_STATE_GPRS:
case EVENT_POLL_STATE_OPERATOR:
    ar = (AsyncResult) msg.obj;
    handlePollStateResult(msg.what, ar);
    break;
case EVENT_POLL_STATE_NETWORK_SELECTION_MODE:
    ar = (AsyncResult) msg.obj;
    if (mPhone.isPhoneTypeGsm()) {
        handlePollStateResult(msg.what, ar);
    } else { ......}//CMDA 网络制式处理逻辑
    break;
```

对四个查询网络服务的请求的响应方式都一样，都是调用 handlePollStateResult 方法。接下来，重点解析 handlePollStateResult 方法中的处理逻辑。

handlePollStateResult 与 GsmCdmaCallTracker 类中的 handlePollCalls 方法的处理机制非常相似，都是处理向 RILJ 对象发起查询请求返回的结果。这里先不讲，留在后面展开讲解。

7.2.4　与 RILJ 对象的交互机制

ServiceStateTracker 与 RILJ 对象的交互完成了服务信息的管理和控制，并将服务信息保存在两个属性对象 mSS 和 mNewSS 中，在 Telephony ServiceState 网络服务业务中非常重要和关键。与 GsmCdmaCallTracker 一样，其交互方式可分为两大类。

- ServiceStateTracker 对象主动发起
- ServiceStateTracker 对象被动接收

1. 被动接收 Handler 消息

前面分析和学习了 ServiceStateTracker 对象的 Handler 消息处理机制，它会被动接收 RILJ 对象上报的七种类型的 Handler 消息，其中 EVENT_RADIO_STATE_CHANGED 和 EVENT_NETWORK_STATE_CHANGED 类型的 Handler 消息非常关键，将触发 modemTriggeredPollState 调用，关键处理流程，如图 7-2 所示。

图 7-2 所示的被动接收网络服务变化消息处理流程，即 ServiceStateTracker 对象接收 RILJ 对象发起的网络服务变化的消息处理流程，重点关注以下几点：

- ServiceStateTracker 与 GsmCdmaPhone 对象具有相同的生命周期。TeleService 系统应用加载 Telephony 业务模型的过程中，

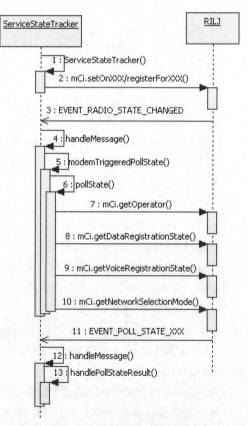

图 7-2　被动接收网络服务变化消息处理流程

同步完成 ServiceStateTracker 对象的创建，调用 mCi 对象的 registerForXXX 和 setOnXXX 方法，共完成七个不同类型的 Handler 消息注册；ServiceStateTracker 与 GsmCdmaPhone 相互持有对方的引用。

- 网络信息和 Radio 状态若发生改变，由 RILJ 发出 EVENT_RADIO_STATE_CHANGED 和 EVENT_NETWORK_STATE_CHANGED 消息回调，ServiceStateTracker 则调用 modemTriggered-PollState 方法响应。
- pollState 方法会连续向 RILJ 对象发起四个查询网络服务信息的请求：
  getOperator 获取接入网络运营商信息
  getDataRegistrationState 获取接入移动数据网络基本信息
  getVoiceRegistrationState 获取接入语音网络信息
  getNetworkSelectionMode 获取接入的网络类型
- 四个查询网络服务的 Message 回调消息类型：
  EVENT_POLL_STATE_OPERATOR
  EVENT_POLL_STATE_GPRS
  EVENT_POLL_STATE_REGISTRATION
  EVENT_POLL_STATE_NETWORK_SELECTION_MODE。
- ServiceStateTracker 对象响应这四种不同类型的查询网络服务信息的 Handler 消息，统一交给 handlePollStateResult 方法处理。
- 在 handlePollStateResult 方法中，对四种不同类型 Handler 消息区分处理，主要是更新 mNewSS 对象对应的属性，最后调用 pollStateDone 方法完成收尾工作。
- pollStateDone 方法主要是通过 mNewSS 更新 mSS 对象，若网络服务信息发生了改变，将发起对应的消息通知。

2. 主动发起请求

ServiceStateTracker 对象主动向 RILJ 对象发起网络服务管理控制请求，其中最重要的是调用 setRadioPower 方法打开或关闭 Radio 无线通信模块，ServiceStateTracker 对象则会调用 RILJ 对象中对应的服务状态控制方法，简化后的代码框架如下。

```
public void setRadioPower(boolean power) {
    mDesiredPowerState = power;
    setPowerStateToDesired();
}
protected void setPowerStateToDesired() {
//根据当前状态打开或关闭 Radio 无线通信模块
if (mDesiredPowerState && !mRadioDisabledByCarrier
        && mCi.getRadioState() == CommandsInterface.RadioState.RADIO_OFF) {
    mCi.setRadioPower(true, null);
} else ......
```

mCi.setRadioPower 中并没有定义该操作的回调 Message 对象，RIL 处理完 ServiceStateTracker 对象发出的打开或关闭 Radio 无线通信模块的请求之后，在 RILJ 对象中发出 EVENT_RADIO_STATE_CHANGED 类型的 Handler 消息。是不是很熟悉呢？原来图 7-2 中的步骤 3 就体现了此消息的处理逻辑。

因此，由 ServiceStateTracker 对象主动向 RILJ 对象发起 ServiceState 管理和控制的交互流程中，可分为两次交互过程：

第一次，mCi.setRadioPower 交互，没有直接的 Message 消息回调处理过程；
第二次，响应 EVENT_RADIO_STATE_CHANGED 类型的 Handler 消息。
总结两次交互过程的关键业务流程，如图 7-3 所示。

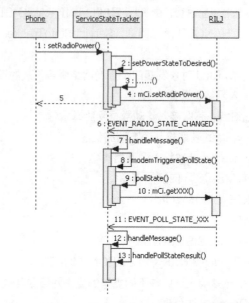

图 7-3　主动发起的网络服务管理和控制的请求流程

图 7-3 所示的主动发起的网络服务管理和控制的请求流程中，重点关注以下两点：

- 步骤 4 的 mCi.setRadioPower

mCi.setRadioPower 的 Message 参数为 NULL，即不支持 Message 消息 Callback 回调处理机制；当然此操作会触发 RILJ 发出 EVENT_RADIO_STATE_CHANGED 消息的通知。

- 简化的 modemTriggeredPollState 响应逻辑（步骤 6 到步骤 13）

可对比或参考图 7-2 中的步骤 3 到步骤 13。

7.3　handlePollStateResult 方法

RIL 完成 ServiceStateTracker 对象发起的查询最新网络服务信息的请求后，RILJ 对象使用 ServiceStateTracker 对象创建的 Message 发起 Callback 消息回调。在 ServiceStateTracker 对象中，由 handlePollStateResult 方法进行网络服务信息处理，将查询出的当前最新网络服务信息更新并保存在 ServiceStateTracker 对象的多个属性中，可分成三个处理逻辑。

- 异常处理。
- 调用 handlePollStateResultMessage 方法，分别处理四个不同网络信息查询的返回结果。
- 收尾工作，四个网络信息查询工作全部处理完成后，再更新 mNewSS 属性和调用 pollStateDone 方法。

7.3.1　异常处理

RILJ 对象返回的异常信息处理逻辑的详情如下：

```
if (ar.userObj != mPollingContext) return; //请求与返回的消息不匹配，直接返回，不做处理
if (ar.exception != null) {//判断 RILJ 是否返回了异常信息
    CommandException.Error err=null;
    //匹配异常类型，具体可查看 CommandException 类中 Error 枚举类型的定义
    if (ar.exception instanceof CommandException) {
        //强制转换类型 CommandException，并获取 Error
        err = ((CommandException)(ar.exception)).getCommandError();
    }
    //无线通信模块的工作状态不可用
    if (err == CommandException.Error.RADIO_NOT_AVAILABLE) {
        // Radio has crashed or turned off
        cancelPollState();//清理查询上下文 mPollingContext
        return;
    }
    ......//记录异常日志信息
}
```

7.3.2 handlePollStateResultMessage

handlePollStateResultMessage 方法将分别处理四个不同网络服务信息查询的返回结果，其处理框架总结如下：

```
switch (what) {
    case EVENT_POLL_STATE_REGISTRATION: { ......break;}
    case EVENT_POLL_STATE_GPRS: { ......break;}
    case EVENT_POLL_STATE_OPERATOR: { ......break;}
    case EVENT_POLL_STATE_NETWORK_SELECTION_MODE: { ......break;}
    default:
        loge("handlePollStateResultMessage: Unexpected RIL response received: " + what);
}
```

1. EVENT_POLL_STATE_REGISTRATION

EVENT_POLL_STATE_REGISTRATION 逻辑分支将处理当前注册的语音网络服务信息的查询结果，其代码详情如下。

```
//获取 VoiceRegStateResult 对象，重点关注此对象的定义
VoiceRegStateResult voiceRegStateResult = (VoiceRegStateResult) ar.result;
//RegState 转换为 ServiceState.RIL_REG_STATE_XXX 状态
int registrationState = getRegStateFromHalRegState(voiceRegStateResult.regState);
//设置 mNewSS 的 mVoiceRegState 状态，注意 regCodeToServiceState 状态的转换
mNewSS.setVoiceRegState(regCodeToServiceState(registrationState));
//设置 mNewSS 网络服务信息的 RadioTechnology
mNewSS.setRilVoiceRadioTechnology(voiceRegStateResult.rat);

//Denial reason if registrationState = 3
int reasonForDenial = voiceRegStateResult.reasonForDenial;//拒绝驻网原因
if (mPhone.isPhoneTypeGsm()) {//GSM
    mGsmRoaming = regCodeIsRoaming(registrationState);//漫游状态
    mNewRejectCode = reasonForDenial;
    //是否只能拨打紧急呼救电话，重点关注 ServiceState.RIL_REG_STATE_XXX 信息
    if (registrationState==ServiceState.RIL_REG_STATE_XXX......) {
        mEmergencyOnly = true;
    } else {
        mEmergencyOnly = false;
    }
} else {......}//CDMA 处理逻辑
```

通过上面的代码，可以非常直观地看出语音网络服务的注册信息 VoiceRegStateResult 对象非常

重要，通过此对象可以得到注册的语音网络服务信息、漫游标志和紧急呼救标志。重点关注 getRegStateFromHalRegState 和 regCodeToServiceState 方法。

getRegStateFromHalRegState 方法中的逻辑可体现 RegState 注册状态的常量定义与 ServiceState.RIL_REG_STATE_XXX 常量定义的关系，其关键处理逻辑如下：

```
private int getRegStateFromHalRegState(int regState) {
    switch (regState) {
        case RegState.NOT_REG_MT_NOT_SEARCHING_OP:
            return ServiceState.RIL_REG_STATE_NOT_REG;
        case RegState.REG_HOME:
            return ServiceState.RIL_REG_STATE_HOME;
        case RegState.NOT_REG_MT_SEARCHING_OP:
            return ServiceState.RIL_REG_STATE_SEARCHING;
        case RegState.REG_DENIED:
            return ServiceState.RIL_REG_STATE_DENIED;
        case RegState.UNKNOWN:
            return ServiceState.RIL_REG_STATE_UNKNOWN;
        case RegState.REG_ROAMING:
            return ServiceState.RIL_REG_STATE_ROAMING;
        ......
        default:
            return ServiceState.REGISTRATION_STATE_NOT_REGISTERED_AND_NOT_SEARCHING;
    }
}
```

对于上面的代码，我们重点关注以下两点：
- getRegStateFromHalRegState 名称与 VoiceRegStateResult 对象类以，来源于 HAL 层。
- 语音网络服务注册状态 RegState 与网络服务状态 ServiceState.RIL_REG_XXX 的对应关系。

regCodeToServiceState 方法中的逻辑可体现 ServiceState.STATE_IN_SERVICE 和 ServiceState.STATE_OUT_OF_SERVICE 与 RegState 的关系，其关键处理逻辑如下：

```
private int regCodeToServiceState(int code) {
    switch (code) {
        case ServiceState.RIL_REG_STATE_HOME:
        case ServiceState.RIL_REG_STATE_ROAMING:
            return ServiceState.STATE_IN_SERVICE;
        default:
            return ServiceState.STATE_OUT_OF_SERVICE;
    }
}
```

通过上面的信息，可重点关注 ServiceState.RIL_REG_STATE_HOME 和 ServiceState.RIL_REG_STATE_ROAMING 的情况，两个状态都成功则代表完成了网络注册，区别是：本地网络和漫游网络。

注意

VoiceRegStateResult 的定义文件是：out/target/common/gen/JAVA_LIBRARIES/android.hardware.radio-V1.0-java-static_intermediates/android/hardware/radio/V1_0/VoiceRegStateResult.java，在 ServiceStateTracker 代码中导入了 android.hardware.radio.V1_0.DataRegStateResult 对象。

我们当前用到了该对象的 regState 和 rat 属性，它们是 HAL 层中 RIL 的处理信息，我们在解析 RIL 时，将分析其数据的来源和对象的创建。

2. EVENT_POLL_STATE_GPRS

EVENT_POLL_STATE_GPRS 逻辑分支将处理当前注册的移动数据网络服务信息的查询结果，其代码详情如下。

```
        //获取 DataRegStateResult 对象,名字与 VoiceRegStateResult 非常相似
        DataRegStateResult dataRegStateResult = (DataRegStateResult) ar.result;
        //获取注册状态值,转换的方法和 REGISTRATION 分支相同
        int regState = getRegStateFromHalRegState(dataRegStateResult.regState);
        int dataRegState = regCodeToServiceState(regState);
        int newDataRat = dataRegStateResult.rat;
        //设置 mNewSS 的 mDataRegState 状态,注意 regCodeToServiceState 状态的转换
        mNewSS.setDataRegState(dataRegState);
        //设置 mNewSS 的 mRilDataRadioTechnology 属性为移动数据网络 RadioTechnology
        mNewSS.setRilDataRadioTechnology(newDataRat);

        if (mPhone.isPhoneTypeGsm()) {//GSM
            mNewReasonDataDenied = dataRegStateResult.reasonDataDenied;
            mNewMaxDataCalls = dataRegStateResult.maxDataCalls;//最大支持的数据连接数
            mDataRoaming = regCodeIsRoaming(regState);
            // Save the data roaming state reported by modem registration before resource
            // overlay or carrier config possibly overrides it.
            mNewSS.setDataRoamingFromRegistration(mDataRoaming);//更新移动数据漫游标志
        } else if (mPhone.isPhoneTypeCdma()) {......} //CDMA
        else {......}

        updateServiceStateLteEarfcnBoost(mNewSS, getLteEarfcn(dataRegStateResult));
        break;
```

通过上面的信息,我们可以发现 EVENT_POLL_STATE_REGISTRATION 和 EVENT_POLL_STATE_GPRS 类型的 Handler 消息的处理机制非常相似,重点关注以下几点:

- 更新 mNewSS 的 mVoiceRegState 和 mDataRegState 的状态一致,通过调用 getRegStateFromHalRegState 和 regCodeToServiceState 方法对 HAL 层网络服务的驻网状态进行转换,即语音通话 Voice Call 和移动数据 Data Call 的驻网状态是一致的。
- ServiceStateTracker 对象的 mGsmRoaming 和 mDataRoaming 保持一致,通过调用 regCodeIsRoaming 方法对 HAL 层网络服务的漫游状态进行转换,即语音通话 Voice Call 和移动数据 Data Call 的漫游标志保持一致。

3. EVENT_POLL_STATE_OPERATOR

EVENT_POLL_STATE_OPERATOR 逻辑分支将处理当前注册的运营商网络服务信息的查询结果,其代码详情如下。

```
        if (mPhone.isPhoneTypeGsm()) {//GSM 网络
            String opNames[] = (String[]) ar.result; //运营商基本信息
            if (opNames != null && opNames.length >= 3) {
                //获取 SIM 卡中的运营商信息
                String brandOverride = mUiccController.getUiccCard(getPhoneId()) != null ?
                    mUiccController.getUiccCard(getPhoneId()).getOperatorBrandOverride():null;
                if (brandOverride != null) {//SIM 卡中运营商信息设置
                    mNewSS.setOperatorName(brandOverride, brandOverride, opNames[2]);
                } else {
                    mNewSS.setOperatorName(opNames[0], opNames[1], opNames[2]);
                }
            }
        } else {......}//CDMA 网络
        break;
        //setOperatorName 的处理逻辑,关注 mVoiceOperatorXXX 与 mDataOperator 的对应关系
        public void setOperatorName(String longName, String shortName, String numeric) {
            mVoiceOperatorAlphaLong = longName;
            mVoiceOperatorAlphaShort = shortName;
```

```
            mVoiceOperatorNumeric = numeric;
            mDataOperatorAlphaLong = longName;
            mDataOperatorAlphaShort = shortName;
            mDataOperatorNumeric = numeric;
}
```

4. EVENT_POLL_STATE_NETWORK_SELECTION_MODE

EVENT_POLL_STATE_NETWORK_SELECTION_MODE 逻辑分支将处理查询网络选择类型是手动还是自动，其代码详情如下。

```
ints = (int[])ar.result;
mNewSS.setIsManualSelection(ints[0] == 1); //设置手动选网还是自动选网状态
if ((ints[0] == 1) && (mPhone.shouldForceAutoNetworkSelect())) {
            /*
             * Modem 当前是手动选网模式，但当前不允许手动选网，所以切换到自动选网模式
             */
            mPhone.setNetworkSelectionModeAutomatic (null);
}
break;
```

handlePollStateResultMessage 方法中四种类型的 Handler 消息处理逻辑，重点关注以下两点。

- mNewSS 对象保存当前最新的网络服务信息，如：mVoiceRegState、mDataRegState、mRilVoiceRadioTechnology、mRilDataRadioTechnology、mIsManualNetworkSelection 等属性。
- 更新 ServiceStateTracker 对象的 mEmergencyOnly、mGsmRoaming 和 mDataRoaming 状态。

注意

GsmCdmaPhone 对象通过 getServiceState 方法获取 ServiceStateTracker 对象的 mSS 属性，即当前网络服务信息，handlePollStateResultMessage 方法根据查询的网络服务信息更新 mNewSS 属性。那么，在 mNewSS 对象更新后，什么时候将最新的网络服务信息更新到 ServiceStateTracker 对象的 mSS 属性呢？那就是在 pollStateDone 方法中完成的。

7.3.3 继续更新 mNewSS

在 handlePollStateResult 方法中通过调用 pollStateDone 方法来完成查询网络服务信息结果的收尾工作，需要一个前提条件，就是 EVENT_POLL_STATE_XXX 类型的四个查询网络服务信息必须全部处理完成，其关键代码逻辑详情如下。

```
mPollingContext[0]--;//计数器减 1 操作
if (mPollingContext[0] == 0) {//计数器为 0 说明当前查询网络服务信息工作已经全部处理完成
    if (mPhone.isPhoneTypeGsm()) {//GSM
        updateRoamingState();//更新 mNewSS 漫游标志
        mNewSS.setEmergencyOnly(mEmergencyOnly);//更新 mNewSS 紧急呼救状态
    } else {......}
    pollStateDone();
}
```

updateRoamingState 方法中的主要逻辑是根据 ServiceStateTracker 对象的 mGsmRoaming 和 mDataRoaming 属性，获取漫游标志 roaming，最后调用 mNewSS.setVoiceRoaming(roaming)和 mNewSS.setDataRoaming(roaming)完成 mNewSS 对象的更新，注意两个 Roaming 状态保持一致。

在 mNewSS 对象更新了 Roaming 和 mIsEmergencyOnly 状态后，调用 pollStateDone 方法完成收尾工作。

7.3.4 完成收尾工作

pollStateDone 方法的主要处理逻辑是完成查询网络服务信息结果的收尾工作，分为三个主要的处理逻辑。

- 对比 mSS 和 mNewSS 两个 ServiceState 对象的网络服务信息，获取 hasXXX 网络服务信息更新标志。
- 更新 mSS、mCellLoc 对象信息，同时更新 mRejectCode、mMaxDataCalls、mReasonDataDenied 等属性。
- 发出网络服务变化的消息通知。

1. 如何获取服务信息改变

获取多种网络服务信息变化的标志如表 7-6 所示。

表 7-6　网络服务信息更新标志汇总表

网络服务信息更新标志	备注
hasRegistered/hasDeregistered	语音网络驻网成功/失败标志
hasDataAttached/hasDataDetached	移动数据网络驻网成功/失败标志
hasVoiceRegStateChanged hasDataRegStateChanged	语音网络/移动数据网络驻网状态变化标志
hasRilVoiceRadioTechnologyChanged hasRilDataRadioTechnologyChanged	语音网络/移动数据网络 RadioTechnology 变化标志
hasLocationChanged	小区信息变化标志
hasChanged	网络服务信息变化标志
hasVoiceRoamingOn hasVoiceRoamingOff	语音网络漫游状态标志
hasDataRoamingOn hasDataRoamingOff	数据网络漫游状态标志

表 7-6 总结了 pollStateDone 方法中比较关键的 12 个网络服务更新标志（总共 18 个更新标志），这些网络服务信息变化标志都是通过比较 ServiceStateTracker 对象的 mSS 和 mNewSS 对象的相关属性，从而得出服务信息变化的标志。

注意

请读者根据表 7-6，再结合 ServiceStateTracker 类 pollStateDone 方法中的代码逻辑，验证和思考这些处理逻辑，将有助于我们加深对 ServiceState 的学习和理解。

2. 更新 mSS、mCellLoc 对象

更新 mSS、mCellLoc 对象的主要处理逻辑详情如下：

```
if (mPhone.isPhoneTypeGsm()) {
    mReasonDataDenied = mNewReasonDataDenied;
    mMaxDataCalls = mNewMaxDataCalls;  //更新最大的移动数据连接数量
    mRejectCode = mNewRejectCode;
}
```

```
// 交换 mSS 和 mNewSS，并更新 mSS 对象的属性
ServiceState tss = mSS;
mSS = mNewSS;//更新 mSS
mNewSS = tss;
//为下一次使用，重置 nNewSS 属性
mNewSS.setStateOutOfService();//重置 mNewSS 对象状态

//交换 mCellLoc 和 mNewCellLoc，并更新 mCellLoc 对象的属性
CellLocation tcl = mCellLoc;
mCellLoc = mNewCellLoc;//更新 mCellLoc
mNewCellLoc = tcl;
```

在 ServiceStateTracker 与 RILJ 对象的交互过程中，首先更新 ServiceStateTracker 对象的 mNewXXX 属性，在调用 pollStateDone 方法进行收尾工作时，更新 ServiceStateTracker 对象中对应的 mXXX 属性，即当前网络服务信息对象或属性，这些对象或属性的对应关系如表 7-7 所示。

表 7-7　ServiceStateTracker 网络服务信息属性对应关系表

当前信息	更新的信息
mSS	mNewSS
mCellLoc	mNewCellLoc
mReasonDataDenied	mNewReasonDataDenied
mMaxDataCalls	mNewMaxDataCalls
mRejectCode	mNewRejectCode

3. 发出网络服务信息变化的消息通知

在 pollStateDone 方法中，一共判断出 18 个网络服务信息变化标志。这里重点介绍注册的通话服务状态信息变化后的处理逻辑，即 hasChanged 为 true 的情况。

```
if (hasChanged) {
    updateSpnDisplay();//更新 spn 显示
    //通过 TelephonyManager 设置运营商信息和漫游标志
    tm.setNetworkOperatorNameForPhone(mPhone.getPhoneId(), mSS.getOperatorAlpha());
    tm.setNetworkOperatorNumericForPhone(mPhone.getPhoneId(), operatorNumeric);
    tm.setNetworkRoamingForPhone(mPhone.getPhoneId(),
            mPhone.isPhoneTypeGsm() ? mSS.getVoiceRoaming() :
                    (mSS.getVoiceRoaming() || mSS.getDataRoaming()));

    setRoamingType(mSS);//更新漫游标志
    mPhone.notifyServiceStateChanged(mSS);//使用 GsmCdmaPhone 对象发出网络服务变化消息，
重点关注 mSS ServiceState 对象作为参数时的情况

    // 通过 ServiceStateProvider 记录数据库，将触发 JobScheduler 从而唤醒对应的应用程序
    mPhone.getContext().getContentResolver()//记录当前最新网络服务信息
            .insert(getUriForSubscriptionId(mPhone.getSubId()),
                    getContentValuesForServiceState(mSS));
}
```

在网络服务信息发生了变化以后，主要有两方面的处理。
- 调用 updateSpnDisplay 方法更新 SPN 的显示。
- 通过 TelephonyManager 设置驻网的运营商的名称和编号、是否漫游标志。

TelephonyManager 对应的处理逻辑是设置对应的 SystemProperties，读者可通过输入命令 adb shell getprop 查看。

```
$ adb shell getprop|grep gsm.operator.
[gsm.operator.alpha]: [CHN-UNICOM]
[gsm.operator.iso-country]: [cn]
[gsm.operator.isroaming]: [false]
[gsm.operator.numeric]: [46001]
```

com.android.internal.telephony.TelephonyProperties 对应的 Properties 定义如下：

```
/** 驻网运营商网络名称
 *
 * 可用性：成功驻网，CDMA 网络类型的运营网络名称或许不可用
 */
static final String PROPERTY_OPERATOR_ALPHA = "gsm.operator.alpha";

/** 运营商编号 (MCC 移动国家码+MNC 移动网络码)
 *
 * 可用性：成功驻网，CDMA 网络类型的运营商网络编号或许不可用
 */
static final String PROPERTY_OPERATOR_NUMERIC = "gsm.operator.numeric";

/** 移动设备驻网 GSM 漫游网络，取值为 true
 * 可用性：成功驻网
 */
static final String PROPERTY_OPERATOR_ISROAMING = "gsm.operator.isroaming";

/** ISO 国家代码，与当前驻网获取的运营商编号中的 MCC 国家代码相同
 *
 *
 * 可用性：成功驻网，CDMA 网络类型的 ISO 国家代码或许不可用
 */
static final String PROPERTY_OPERATOR_ISO_COUNTRY = "gsm.operator.iso-country";
```

- 通过 GsmCdmaPhone 对象的 notifyServiceStateChanged 方法调用，发出网络服务变化的通知，重点关注其参数 mSS，即当前网络服务信息 ServiceState 对象。

pollStateDone 方法的其他消息通知，绝大部分都是发起 mXXXRegistrants.notifyRegistrants 调用，请读者自行阅读和学习。

7.4 *#*#4636#*#*测试工具

在手机拨号盘输入*#*#4636#*#*，将启动 Android 的测试工具，其中有三个子项目：
- Phone information
- Usage statistics
- WiFi information

选择进入 Phone information，可发现 Telephony ServiceState 业务的主要信息，详情如图 7-4 所示。

通过查找 Android Events 日志，可以发现用于展示 Phone information 界面的 Activity 是 com.android.settings/.RadioInfo，对应的 Java 程序是 packages/apps/Settings/src/com/android/settings/RadioInfo.java。在 Setting 应用的配置文件 AndroidManifest.xml 中可以找到有关 Activity 的定义，详情如下：

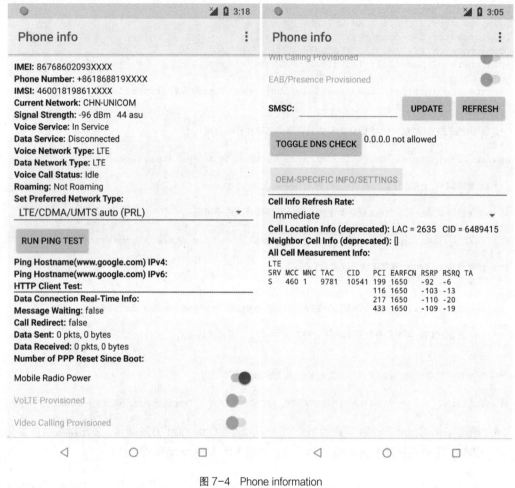

图 7-4 Phone information

```
<activity android:name="RadioInfo" android:label="@string/phone_info_label"
    android:process="com.android.phone">
    <intent-filter>
        <action android:name="android.intent.action.MAIN" />
        <category android:name="android.intent.category.DEVELOPMENT_PREFERENCE" />
    </intent-filter>
</activity>
```

注意 RadioInfo Activity 界面运行在 com.android.phone 进程空间，请读者思考为什么会有这样的设计和实现。

如图 7-4 所示的 Phone information 界面主要分为两个部分：
- 网络服务信息
- 小区信息

7.4.1 网络服务信息

汇总 Phone information 界面展示的网络服务信息，如表 7-8 所示。

表 7-8　网络服务信息汇总表

网络服务信息	View 控件	获取方法
IMEI	mDeviceId	phone.getDeviceId()
Phone Number	number	phone.getLine1Number()
IMSI	mSubscriberId	phone.getSubscriberId
Current Network	operatorName	ITelephonyRegistry
Signal Strength	dBm	ITelephonyRegistry
Voice Service	gsmState	ITelephonyRegistry
Data Service	gprsState	TelephonyManager.getDataState
Voice Network Type	voiceNetwork	phone.getServiceState().getRilVoiceRadio-Technology
Data Network Type	dataNetwork	phone.getServiceState().getRilDataRadioT-echnology
Voice Call Status	callState	ITelephonyRegistry
Roaming	roamingState	ITelephonyRegistry
Preferred Network Type	preferredNetworkType	phone.getPreferredNetworkType
Mobile Radio Power	radioPowerOnSwitch	phone.getServiceState().getState

表 7-8 所示的网络服务信息有两个获取通道：

- 通过 GsmCdmaPhone 对象直接获取
- 通过 ITelephonyRegistry 回调

现在就可以回答为什么 RadioInfo 运行在 com.android.phone 进程空间。因为可方便地获取 GsmCdmaPhone 对象，获取方式在 RadioInfo 类的 onCreate 方法中：phone = PhoneFactory.get-DefaultPhone()。

ITelephonyRegistry 消息的注册是在 RadioInfo 类的 onResume 方法中完成的，详情如下：

```
mTelephonyManager.listen(mPhoneStateListener,
        PhoneStateListener.LISTEN_CALL_STATE
      | PhoneStateListener.LISTEN_DATA_CONNECTION_STATE
      | PhoneStateListener.LISTEN_DATA_ACTIVITY
      | PhoneStateListener.LISTEN_CELL_LOCATION
      | PhoneStateListener.LISTEN_MESSAGE_WAITING_INDICATOR
      | PhoneStateListener.LISTEN_CALL_FORWARDING_INDICATOR
      | PhoneStateListener.LISTEN_CELL_INFO
      | PhoneStateListener.LISTEN_SERVICE_STATE
      | PhoneStateListener.LISTEN_SIGNAL_STRENGTHS
      | PhoneStateListener.LISTEN_DATA_CONNECTION_REAL_TIME_INFO);

private final PhoneStateListener mPhoneStateListener = new PhoneStateListener() {
    @Override
    public void onXXXChanged(int state) {
        updateXXX();//更新界面显示
        ......//updateXXX 调用
    }
    ......//重写父类 PhoneStateListener onXXXChanged 方法
}
```

上面的代码使用 TelephonyManager 向 ITelephonyRegistry 服务监听了 10 个 PhoneState 变化的消息，Callback 消息回调是 mPhoneStateListener，它继承自 PhoneStateListener 类，重写了父类的 11 个方法。

注意

PhoneStateListener 类为了实现跨进程的 Callback 调用，有一个 IPhoneStateListenerStub 类型的内部类对象 callback，继承了 IPhoneStateListener.Stub，其中的方法逻辑是通过 Handler 消息转化为对主类方法的调用，从而实现了模板方法。

7.4.2 扩展 ITelephonyRegistry

TelephonyManager 类中的 getTelephonyRegistry 获取 ITelephonyRegistry 系统服务调用 Binder 对象，可反推出此系统服务的服务名为"telephony.registry"。

ITelephonyRegistry 系统服务的 AIDL 接口定义文件主要在 frameworks/base/telephony/java/com/android/internal/telephony 目录下，有三个关键接口定义文件：ITelephonyRegistry.aidl、IPhoneStateListener.aidl 和 IOnSubscriptionsChangedListener.aidl。

ITelephonyRegistry 接口实现的 Java 程序为：frameworks/base/services/core/java/com/android/server/TelephonyRegistry.java，关键的代码信息详情如下：

```
class TelephonyRegistry extends ITelephonyRegistry.Stub
//服务启动过程
SystemServer.startOtherServices
telephonyRegistry = new TelephonyRegistry(context);
ServiceManager.addService("telephony.registry", telephonyRegistry);
```

ITelephonyRegistry 系统服务运行在 system_server 进程空间，加载 Android 系统的过程中将同步加入到系统服务中，服务名为"telephony.registry"。

总结 ITelephonyRegistry 服务的核心类图，如图 7-5 所示。

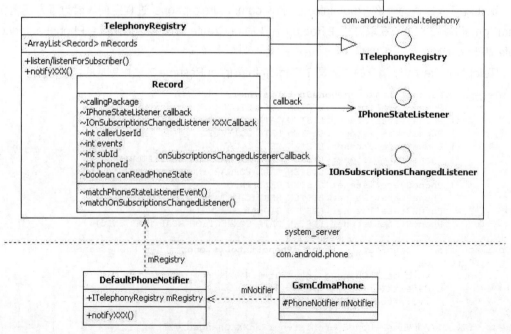

图 7-5　ITelephonyRegistry 核心类图

对于图 7-5 所示的 ITelephonyRegistry 核心类图，我们重点关注以下几点：
- TelephonyRegistry 实现了 ITelephonyRegistry，提供的系统服务主要有两个类型：listen/listenForSubscriber 对 mRecords 进行管理（Callback 列表）。
- TelephonyRegistry 的内部类 Record 将记录 Callback 信息。
- ITelephonyRegistry 系统服务运行在 system_server 进程空间。
- GsmCdmaPhone 对象触发 PhoneState 变化消息 Callback 调用。在创建 GsmCdmaPhone 对象时，将同步创建 DefaultPhoneNotifier 对象并保存 mNotifier 引用。在 DefaultPhoneNotifier 的构造方法中，将获取 ITelephonyRegistry 服务的 Binder 对象。因此，GsmCdmaPhone 通过 mNotifier 发起 PhoneState 变化消息通知，通过第一次的跨进程调用（com.android.phone→system_server）触发 TelephonyRegistry 提供的消息通知接口。
- TelephonyRegistry 接收 notifyXXX 调用，遍历 mRecords，发出 callback 调用，其处理逻辑框架代码详情如下：

```
public void notifyXXX(int phoneId, int subId, ServiceState state) {
    if (!checkNotifyPermission("notifyXXX()")){//权限检查
        return;
    }
    synchronized (mRecords) {//加同步锁
        if (validatePhoneId(phoneId)) {//验证 phoneId
            for (Record r : mRecords) {//循环 mRecords,遍历 Observer
                if (r.matchPhoneStateListenerEvent(PhoneStateListener.LISTEN_XXX) &&
                        idMatch(r.subId, subId, phoneId)) {//匹配监听消息
                    try {//回调
                        r.callback.onServiceStateChanged(new ServiceState(state));
                    } catch (RemoteException ex) {
                        mRemoveList.add(r.binder);
                    }
                }
            }
        }
    }
}
```

上面的代码框架中，请重点关注 r.callback，它是 IPhoneStateListener 类型，说明了 TelephonyRegistry 服务发出的消息回调具备跨进程调用的能力，即完成了第二次的跨进程消息传递：system_server→监听 PhoneState 变化的应用进程。

ITelephonyRegistry 系统服务的运行机制主要是两次跨进程接口调用：第一次跨进程接口调用是 GsmCdmaPhone 对象调用 ITelephonyRegistry 系统服务的 notifyXXX，消息源头是 com.android.phone 进程中的 GsmCdmaPhone 对象；第二次跨进程接口调用，是通过已经注册监听的 mRecords 列表通过 IPhoneStateListener 发起 Callback 回调，消息的处理终点在监听 PhoneState 变化的应用进程中。

7.4.3 展示小区信息

Phone information 界面涉及的小区信息共有两处：
- 设置小区信息更新频率
- 展示小区信息

设置小区信息更新频率的下拉列表默认值是 Disabled，对应的 View 控件是 cellInfoRefresh-RateSpinner，在 RadioInfo 类的 onCreate 和 onResume 方法中的关键处理逻辑如下：

```
//获取控件
cellInfoRefreshRateSpinner = (Spinner) findViewById(R.id.cell_info_rate_select);
ArrayAdapter<String> cellInfoAdapter = new ArrayAdapter<String>(this,
        android.R.layout.simple_spinner_item, mCellInfoRefreshRateLabels);//填充 List 选项
cellInfoRefreshRateSpinner.setAdapter(cellInfoAdapter);//设置控件适配器
cellInfoRefreshRateSpinner.setOnItemSelectedListener(mCellInfoRefreshRateHandler);

AdapterView.OnItemSelectedListener mCellInfoRefreshRateHandler =
        new AdapterView.OnItemSelectedListener() {
    public void onItemSelected(AdapterView parent, View v, int pos, long id) {
        mCellInfoRefreshRateIndex = pos;
        mTelephonyManager.setCellInfoListRate(mCellInfoRefreshRates[pos]);
        updateAllCellInfo();
    }......
};
```

上面的代码主要是针对 cellInfoRefreshRateSpinner 控件的处理，手动选择小区信息更新频率后，将通过 TelephonyManager 发起 setCellInfoListRate 调用，最终在 GsmCdmaPhone 对象中发起 mCi.setCellInfoListRate 调用来设置 Modem 小区信息的更新频率；

关注 RadioInfo 类的 mCellInfoRefreshRateLabels 和 mCellInfoRefreshRates 两个数组的对应关系，作为设置小区信息更新频率的参数。

RadioInfo 类的 mPhoneStateListener 在显示 Phone information 界面时，向 ITelephonyRegistry 服务注册 PhoneStateListener.LISTEN_CELL_LOCATION 和 PhoneStateListener.LISTEN_CELL_INFO 两个关于小区信息的监听回调，小区信息变化后的 Callback 响应方法分别是 onCellLocationChanged 和 onCellInfoChanged，其中，onCellLocationChanged 方法更新的小区信息已弃用，mPhoneStateListener.onCellInfoChanged 方法的响应逻辑的详情如下：

```
@Override
public void onCellInfoChanged(List<CellInfo> arrayCi) {
    mCellInfoResult = arrayCi;
    updateCellInfo(mCellInfoResult);
}
private final void updateCellInfo(List<CellInfo> arrayCi) {
    mCellInfo.setText(buildCellInfoString(arrayCi));//更新界面展示
}
```

重点关注 mPhoneStateListener.onCellInfoChanged 方法接收小区信息变化的 Callback 调用，其参数是 List<CellInfo>，而 buildCellInfoString 方法将小区信息列表转换为界面显示的小区信息（String 类型）。

CellInfo 作为抽象类，有四个子类：CellInfoLte、CellInfoWcdma、CellInfoGsm 和 CellInfoCdma，分别代表 4G、3G 和 2G（GSM）和 2G（CDMA）网络下的小区信息。

7.4.4　小区信息更新源头

RadioInfo 界面通过向 ITelephonyRegistry 系统服务注册小区信息变化的监听 mPhoneStateListener.onCellInfoChanged 接收到 Callback 回调，获取当前最新的小区信息列表从而在 Phone-information 界面显示。

通过对 ITelephonyRegistry 系统服务的学习，我们知道 PhoneState 状态变化的消息源是 com.android.phone 进程中的 GsmCdmaPhone 对象，反推 onCellInfoChanged，可追溯到 ServiceStateTracker 响应 RILJ 对象发出的 EVENT_UNSOL_CELL_INFO_LIST 消息回调。经过 mPhone.notifyCellInfo→mNotifier.notifyCellInfo→mRegistry.notifyCellInfoForSubscriber 调用过程，ItelephonyRegistry 系统服务响应 notifyCellInfoForSubscriber 服务接口调用，最终遍历 mRecords 发出 onCellInfoChanged 调用，整个过程将传递 CellInfo 列表对象。

在 ServiceStateTracker 的构造方法中，调用 mCi.registerForCellInfoList 方法向 RILJ 对象发起 EVENT_UNSOL_CELL_INFO_LIST 类型的 Handler Message 消息注册。

7.4.5 信号强度实时变化

Phone information 界面的 dBm 控件展示了当前网络的信号强度，我们可通过它直接观察到信号强度的实时变化。信号强度的实时更新实现机制与 mCellInfo 实时展示小区信息的原理是相似的，通过 TelephonyManager 注册 PhoneStateListener.LISTEN_SIGNAL_STRENGTHS 类型的监听完成信号强度更新的消息回调，主要的代码逻辑详情如下：

```java
private final PhoneStateListener mPhoneStateListener = new PhoneStateListener() {
    @Override
    public void onSignalStrengthsChanged(SignalStrength signalStrength) {//监听回调
        updateSignalStrength(signalStrength);
    }
}
//更新 dBm 界面显示
private final void updateSignalStrength(SignalStrength signalStrength) {
    Resources r = getResources();
    int signalDbm = signalStrength.getDbm();
    int signalAsu = signalStrength.getAsuLevel();
    if (-1 == signalAsu) signalAsu = 0;
    dBm.setText(String.valueOf(signalDbm) + " "
        + r.getString(R.string.radioInfo_display_dbm) + "    "
        + String.valueOf(signalAsu) + " "
        + r.getString(R.string.radioInfo_display_asu));//更新信号强度显示
}
```

通过对 ITelephonyRegistry 系统服务的学习，我们知道 PhoneState 状态变化的消息源是 com.android.phone 进程中的 GsmCdmaPhone 对象，反推 onSignalStrengthsChanged，可追溯到 ServiceStateTracker 响应 RILJ 对象发出的 EVENT_SIGNAL_STRENGTH_UPDATE 和 EVENT_GET_SIGNAL_STRENGTH 消息回调。经过 ServiceStateTracker.onSignalStrengthResult→mPhone.notifySignalStrength→mNotifier.notifySignalStrength→mRegistry.notifySignalStrengthForPhoneId 调用过程，ITelephonyRegistry 系统服务响应 notifySignalStrengthForPhoneId 服务接口调用，最终遍历 mRecords 发出 onSignalStrengthsChanged 调用，整个过程将传递 SignalStrength 对象。

关于 EVENT_SIGNAL_STRENGTH_UPDATE 和 EVENT_GET_SIGNAL_STRENGTH 两个 HandlerMessage 消息在 ServiceStateTracker 类中的核心处理流程，总结如图 7-6 所示。

对于图 7-6 所示的 SignalStrength 信号强度更新流程，我们重点关注以下几点：

● 设置 EVENT_SIGNAL_STRENGTH_UPDATE 信号强度更新回调

在步骤 2 的 ServiceStateTracker 构造方法中，通过 mCi.setOnSignalStrengthUpdate 调用设置信号强度更新的回调 Message 消息类型为 EVENT_SIGNAL_STRENGTH_UPDATE。

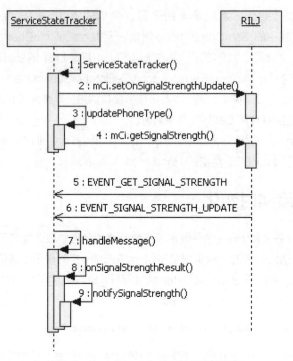

图 7-6 SignalStrength 信号强度更新流程

- 主动查询信号强度

在步骤 4 的 ServiceStateTracker 构造方法的调用过程中，通过 mCi.getSignalStrength 调用，查询当前最新信号强度，其回调的 Message 类型为 EVENT_GET_SIGNAL_STRENGTH。

- EVENT_SIGNAL_STRENGTH_UPDATE 和 EVENT_GET_SIGNAL_STRENGTH 响应过程

在步骤 5 和步骤 6 中，RILJ 对象发出 Message 消息回调。我们要注意虽然触发方式不同，但 ServiceStateTracker 中的响应逻辑相同（步骤 7、步骤 8 和步骤 9）。

- 发出信号变化消息通知

步骤 9 中将使用 GmsCdmaPhone 对象发起 mPhone.notifySignalStrength 调用，发出信号变化消息通知。

EVENT_GET_SIGNAL_STRENGTH 与 EVENT_SIGNAL_STRENGTH_UPDATE 消息处理逻辑的差异如下：EVENT_GET_SIGNAL_STRENGTH 消息是由 ServiceStateTracker 主动发起，将延迟 20 秒循环调用 queueNextSignalStrengthPoll 方法，并未在图 7-6 中表现出来，其处理逻辑的详情如下：

```
private void queueNextSignalStrengthPoll() {
    Message msg;
    msg = obtainMessage();
    msg.what = EVENT_POLL_SIGNAL_STRENGTH;
    long nextTime;
    //延时 20 秒 POLL_PERIOD_MILLIS 发送此 Handler 消息
    sendMessageDelayed(msg, POLL_PERIOD_MILLIS);
}
```

ServiceStateTracker 的本质是自定义的 Handler 消息处理类，因此，发起调用 sendMessage-Delayed 方法 20 秒后，ServiceStateTracker 对象的 handlerMessage 方法会接收和响应此消息，其处理逻辑详情如下：

```
case EVENT_POLL_SIGNAL_STRENGTH:
    // Just poll signal strength...not part of pollState()//向RILJ对象查询最新的网络信号量
    mCi.getSignalStrength(obtainMessage(EVENT_GET_SIGNAL_STRENGTH));
    break;
```

上面的代码首先创建了查询信号量成功后的 Callback 消息对象,其消息对象类型为 EVENT_GET_SIGNAL_STRENGTH;RILJ 对象通过 Callback 消息对象发起 Handler 消息回调,再一次回到 EVENT_GET_SIGNAL_STRENGTH 消息处理。

EVENT_GET_SIGNAL_STRENGTH 消息是由 ServiceStateTracker 每隔 20 秒循环调用 queueNextSignalStrengthPoll 方法来查询当前最新的网络信号强度。
EVENT_SIGNAL_STRENGTH_UPDATE 则是 RILJ 根据变化的网络信号强度主动上报。

7.5 飞行模式

手机在飞行模式开启的情况下,将关闭所有的无线网络,包括无线通信模块、WiFi 和蓝牙。本节我们将从两个方面解析 Android 平台飞行模式的实现机制。
- 飞行模式开启关闭入口逻辑
- 无线通信模块响应飞行模式开启关闭逻辑

7.5.1 飞行模式开启关闭入口逻辑

Android 手机设置飞行模式主要有两种方法:通知栏快捷按钮和网络设置,这里选择后者进行解析。

首先进入网络设置界面,其操作路径为:Settings→Network&Internet,可通过 Airplane mode 进行飞行模式的开关操作。开关飞行模式对应的代码是:packages/apps/Settings/src/com/android/settings/AirplaneModeEnabler.java,其关键处理逻辑的详情如下:

```
private void setAirplaneModeOn(boolean enabling) {
    //更改系统设置
    Settings.Global.putInt(mContext.getContentResolver(),
Settings.Global.AIRPLANE_MODE_ON, enabling ? 1 : 0);//设置飞行模式状态的系统变量
    //发送 Intent 对象
    Intent intent = new Intent(Intent.ACTION_AIRPLANE_MODE_CHANGED);
    intent.putExtra("state", enabling);//飞行模式开启或关闭的标志
    mContext.sendBroadcastAsUser(intent, UserHandle.ALL);//发出广播
}
```

通过上面的代码逻辑,可以发现开关飞行模式主要有两个操作:
- 更新 Settings.Global.AIRPLANE_MODE_ON 系统设置信息
- 发出 ACTION_AIRPLANE_MODE_CHANGED 的广播

frameworks/base/services/core/java/com/android/server/ConnectivityService.java 系统服务同样提供了对飞行模式的控制接口 setAirplaneMode,通知栏快捷按钮对飞行模式的控制便是调用此系统服务接口实现的,其处理逻辑与 AirplaneModeEnabler 中的 setAirplaneModeOn 方法相同。

7.5.2 Radio 模块开启关闭

查找 Android 源代码中的 Intent.ACTION_AIRPLANE_MODE_CHANGED 广播接收器，发现开启或关闭 Radio 无线通信模块的处理逻辑是在 PhoneGlobals 类中。PhoneAppBroadcast Receiver.onReceive 接收到该广播之后，经过 handleAirplaneModeChange→maybeTurnCellOff/ maybeTurnCellOn→setRadioPowerOff/setRadioPowerOn→PhoneUtils.setRadioPower 调用过程，最终调用到 GsmCdmaPhone 对象的 setRadioPower 方法，关键处理逻辑的详情如下：

```
static final void setRadioPower(boolean enabled) {
    for (Phone phone : PhoneFactory.getPhones()) {//遍历 GsmCdmaPhone 对象
        phone.setRadioPower(enabled);//调用 GsmCdmaPhone 对象的 setRadioPower 方法
    }
}
@Override//GsmCdmaPhone 的 setRadioPower 方法
public void setRadioPower(boolean power) {
    mSST.setRadioPower(power);//调用 ServiceStateTracker 对象的 setRadioPower 方法
}
```

上面的代码逻辑，最终是由 GsmCdmaPhone 对象的 mSST，即 ServiceStateTracker 向 RILJ 对象发起关闭或开启 Radio 无线通信模块的请求。

ServiceStateTracker 类的 setRadioPower 方法的关键流程可参考图 7-3。

注意

在开启飞行模式的过程中，不仅需要关闭 Radio 无线通信模块，还需要关闭 WiFi 和蓝牙的无线通信能力。

7.5.3 WiFi 模块开启关闭

在 frameworks/opt/net/wifi/service/java/com/android/server/wifi/WifiServiceImpl.java 代码中注册了 Intent.ACTION_AIRPLANE_MODE_CHANGED 类型的广播接收器，其处理逻辑的详情如下：

```
mContext.registerReceiver(
new BroadcastReceiver() {
    @Override
    public void onReceive(Context context, Intent intent) {
        ......//同步开关 WiFi
    }
}, new IntentFilter(Intent.ACTION_AIRPLANE_MODE_CHANGED));
```

7.5.4 蓝牙模块开启关闭

在 frameworks/base/services/core/java/com/android/server/BluetoothManagerService.java 代码中监听 Settings.Global.AIRPLANE_MODE_ON 数据变化，其处理逻辑的详情如下。

```
mContentResolver.registerContentObserver(
        Settings.Global.getUriFor(Settings.Global.AIRPLANE_MODE_ON),
        true, mAirplaneModeObserver);
//mAirplaneModeObserver 发起开关蓝牙模块
sendDisableMsg(REASON_AIRPLANE_MODE);
sendEnableMsg(mQuietEnableExternal, REASON_AIRPLANE_MODE);
```

7.6 扩展 SIM 卡业务

SIM（Subscriber Identification Module）卡也称为用户身份识别卡，由 CPU、ROM、RAM、EEPROM 和 I/O 电路组成，与 COS（Chip Operation System，芯片操作系统）一起构成一个完整的计算机系统，具有独立的数据处理能力；COS 控制 SIM 卡与外界交换信息，管理卡内存储器并在卡内完成各种命令的处理。

SIM 卡发展至今，主要在尺寸和容量两个方面有大的变化。目前市面上主要有三种尺寸的 SIM 卡：
- 标准卡（2FF）25mmx15mmx0.76mm
- Micro 卡（3FF）15mmx12mmx0.76mm
- Nano 卡（4FF）12.3mmx8.8mmx0.67mm

在功能机时代，因手机自身存储大小的限制，SIM 卡也会存储电话号码、短信等信息，因此 SIM 卡在容量方面不断提升，从 8KB 提升到 64KB 容量，甚至出现过 512KB 以上的大容量。

7.6.1 SIM 卡业务分析

SIM 卡业务主要体现在四个方面：
- 存储功能

保存电话号码、短信等信息。
- 安全管理

PIN、PIN2、PUK 等安全管理。
- 驻网鉴权

移动网络端匹配 SIM 卡内驻网鉴权的计算结果，保障只有运营商发出的合法 SIM 卡才能接入和使用其移动网络，达到身份验证的目的。
- STK

STK（SIM Tool Kit，SIM 卡应用工具包）运行卡内的小应用程序与手机用户进行交互，实现增值服务。

这四个方面的 SIM 卡业务都是通过 APDU（Application Protocol Data Unit）与手机进行数据交互实现的，APDU 可以是命令，也可以是命令的响应，读者可参考《GSM 11.11》和《ISO/IEC7816-4》规范，其中的几个关键点总结如下：
- APDU 命令

《GSM 11.11》9.1 An APDU can be a command APDU or a response APDU
- SIM 卡文件系统

《ISO/IEC7816-4》5.3 Structures for applications and data
- SIM 卡 APDU 命令响应

《ISO/IEC7816-4》5.1 Command-response pairs

随着智能手机的不断发展，SIM 卡的存储功能、安全管理和 STK 已基本不再使用，仅承担驻网鉴权业务完成身份验证。

那么 SIM 卡发展至今，在业务形态方面有什么变化吗？主要有两个分支：

- eSIM
- SoftSim

eSIM（Embedded-SIM，嵌入式 SIM）卡直接嵌入到移动设备主板，使用软件从网络端下载 SIM 卡信息并写入硬件，完成空中发卡；其业务流程也有非常大的改变，用户不用去运营商的营业厅办理 SIM 卡，不受时间和空间的限制，通过网络即可完成发卡业务，最终实现接入和使用运营商的移动网络。

eSIM 标准由 GSMA（Global System for Mobile Communications Assembly）GSM 协会主导制订，共有两套 eSIM 标准：

- 面向物联网 M2M（Machine to Machine）的 eSIM 规范
- 面向消费者可穿戴设备的 eSIM 规范

SoftSim，顾名思义，即用软件模拟 SIM 卡业务。因没有物理卡存在，与 eSIM 相似均采用空中发卡机制，由软件模拟 SIM 卡 COS 系统中的 APDU 响应来完成 SIM 卡的驻网鉴权。

7.6.2　驻网过程分析

GSM 网络的鉴权采用 A3A8 算法，又称 Comp128 算法，3G、4G 网络的鉴权采用的则是 Milenage 算法。

在手机注册移动网络的时候，移动网络会产生一个 128 位 16 字节的随机数据 RAND 发送给手机，手机再将这个数据发给 SIM 卡，SIM 卡使用自己的密钥 K_i 和 RAND 做运算以后，生成一个 32 位 4 字节的应答 SRES 发回给手机，最后发送到移动网络；移动网络收到 SRES 后进行相同运算，比较手机端和本地的运算结果是否相同，相同就说明这个卡是合法的，准其注册和使用移动网络服务。这个算法在 GSM 规范里面叫作 A3，m = 128bit，k = 128bit，c = 32bit。很显然，这个算法要求已知 m 和 k 可以很简单地算出 c，但是已知 m 和 c 却很难算出 k。

SIM 卡计算出 SRES 后，还会使用 RAND 和 K_i 计算出通信过程中加密使用的密钥 K_c，密钥 K_c 的长度是 64 位 8 字节，生成密钥 K_c 的算法是 A8，因为 A3 和 A8 算法的输入参数完全相同，所以在实现的时候使用了同一个函数，同时生成 SRES 和 K_c，因此统一称为 A3A8 算法。

在通信过程中加密算法叫作 A5。

下面摘录 a3a8.txt 文件中有关 A3A8 和 A5 算法的关键代码，详情如下：

```
/* An implementation of the GSM A3A8 algorithm.  (Specifically, COMP128.)
 *
 * Copyright 1998, Marc Briceno, Ian Goldberg, and David Wagner.
 * All rights reserved.
 * ......
 */
void A3A8(/* in */ Byte rand[16], /* in */ Byte key[16],
         /* out */ Byte simoutput[12]) {
        Byte x[32], bit[128];
        int i, j, k, l, m, n, y, z, next_bit;

        /* 加载 rand 数组到 x 数组的最低 10 个元素 */
        for (i=16; i<32; i++)
                x[i] = rand[i-16];

        /*循环 8 次 */
        for (i=1; i<9; i++) {
                /*加载密钥到 x 数组的前 16 个元素 */
```

```
            for (j=0; j<16; j++)
                    x[j] = key[j];
        /*执行算法 */
        ......
}
for (i=0; i<4; i++)
        simoutput[i] = (x[2*i]<<4) | x[2*i+1];
for (i=0; i<6; i++)
        simoutput[4+i] = (x[2*i+18]<<6) | (x[2*i+18+1]<<2)
                       | (x[2*i+18+2]>>2);
simoutput[4+6] = (x[2*6+18]<<6) | (x[2*6+18+1]<<2);
simoutput[4+7] = 0;
}
```

上面的代码逻辑，可重点关注以下几点：

- 函数名 A3A8。
- rand[16]，传入 128 位网络端生成的随机数。
- key[16]，传入 128 位 SIM 卡内置的加密密钥 K_i。
- simoutput[12]，前 4 个字节是 key 和 rand 计算生成的应答 SRES，后 8 个字节是 key 和 rand 计算生成的密钥 K_c。

Milenage 驻网鉴权算法可参考《3GPP TS 35.206》规范文档，其中详细描述了 f1、f2、f3、f4 和 f5 的原理和实现代码，关键的信息如下：

```
OPC    a 128-bit value derived from OP and K and used within the computation of the
       functions.
RAND   a 128-bit random challenge that is an input to the functions f1, f1*, f2, f3,
       f4, f5 and f5*.
AK     a 48-bit anonymity key that is the output of either of the functions f5 and f5*.
AMF    a 16-bit authentication management field that is an input to the functions f1 and f1*.
CK     a 128-bit confidentiality key that is the output of the function f3.
IK     a 128-bit integrity key that is the output of the function f4.
RES    a 64-bit signed response that is the output of the function f2.
MAC-A  a 64-bit network authentication code that is the output of the function f1.
MAC-S  a 64-bit resynchronisation authentication code that is the output of the function f1*.
```

7.6.3 SoftSim 业务实现分析

基于 Android 平台实现 SoftSim 业务，可分为两个关键步骤：

- 模拟 SIM 卡 App 应用

模拟 COS（Chip Operation System，芯片操作系统）对 SIM 卡的管理，完成 APDU 命令的响应逻辑，主要集中在模拟 SIM 卡文件系统和 Milenage 驻网鉴权算法这两个方面。

- 打通与应用 APDU 通道

将 Modem 与实体 SIM 卡的 APDU 交互完成驻网鉴权，修改为与 AP 侧 SoftSim 应用 APDU 交互并扩展 RIL 接口，打通 BP 与 AP 侧 APDU 的数据交换通道。

eSIM 和 SoftSim 的核心思想是跨越时间和地域的限制完成空中发卡，从而快速接入移动网络。这两种技术需要从技术上保障空中发卡过程中、本地保存和使用的 SIM 卡信息不被泄露；将 SIM 卡 K_i、K_c 等敏感信息进行非对称加密传输和保存，并在 TEE（Trusted Execution Environment）、QSEE（Qualcomm Secure Execution Environment）等可信执行环境中解密及使用。

本 章 小 结

本章解析 Telephony ServiceState 网络服务业务，以 ServiceStateTracker 为核心围绕 handlePollStateResult 和 pollStateDone 两个关键方法管理 mSS、mNewSS、mLastCellInfoList、mSignalStrength 等网络服务信息，主要包括驻网信息（Voice Call 语音和 Data Call 移动数据驻网信息）、运营商信息、网络信号、小区信息等核心网络服务信息 ServiceState。

ServiceStateTracker 运行机制与 GsmCdmaCallTracker 运行机制非常相似，包括主动向 RILJ 发起 ServiceState 网络服务信息查询和管理请求，以及 ServiceStateTracker 对象被动接收 RILJ 发出的网络服务信息变化的消息回调，一切都以 Handler 消息处理机制为基础。

4636 测试工具和飞行模式作为 ServiceState 网络服务业务的扩展和实践，特别是在 4636 测试工具中展示网络服务信息的两种获取方式，是帮助我们认识和理解 ServiceState 网络服务业务的利器，也是掌握 GsmCdmaPhone 对象直接获取方式和 ITelephonyRegistry 注册监听/Callback 回调的参考实现。

了解 A3A8 和 Milenage 驻网鉴权算法的原理和相关规范，以及 SIM 卡业务扩展——eSIM 和 SoftSim。

第 8 章

Data Call 移动数据业务

学习目标

- 理解 DcTracker 初始化流程和机制。
- 掌握 StateMachine 状态机设计原理、实现方式和关键运行机制。
- 掌握开启和关闭默认移动数据业务的关键业务流程。
- 理解 DataConnection 消息处理机制及状态转换流程。
- 学习 tcpdump 和 Wireshark 工具的使用。

通过前面的学习了解到 Android Telephony 三大业务能力分别是：

- Voice Call 语音通话
- ServiceState 网络服务
- Data Call 移动数据业务

将前面两章，我们已经对 Voice Call 语音通话和 ServiceState 网络服务业务进行了详细的分析和总结，本章将围绕 DcTracker 类展开对 Data Call 移动数据业务的学习，通过解析 DcTracker 的核心处理机制和关键业务流程，学习和理解 Android 手机中移动数据业务的原理和关键流程。

8.1 DcTracker 初始化过程

TeleService 系统应用在加载 Telephony 业务模型的过程中，会同步创建 DcTracker 对象。DcTracker 对象构造方法中的业务逻辑，主要分为以下五部分：

- Handler 消息注册
- 初始化 ApnContext
- 创建 DcController 对象
- 注册 Observer
- 注册广播接收器

8.1.1 Handler 消息注册

在 DcTracker 的构造方法中体现了 Handler 消息注册的逻辑，主要体现为对 registerForAllEvents 和 registerServiceStateTrackerEvents 两个方法的调用，其代码逻辑的详情如下：

```
private void registerForAllEvents() {
    mPhone.mCi.registerForAvailable(this, DctConstants.EVENT_RADIO_AVAILABLE, null);
    mPhone.mCi.registerForOffOrNotAvailable(this,
        DctConstants.EVENT_RADIO_OFF_OR_NOT_AVAILABLE, null);
    mPhone.mCi.registerForDataCallListChanged(this,
        DctConstants.EVENT_DATA_STATE_CHANGED, null);
    mPhone.getCallTracker().registerForVoiceCallEnded(this,
        DctConstants.EVENT_VOICE_CALL_ENDED, null);
    mPhone.getCallTracker().registerForVoiceCallStarted(this,
        DctConstants.EVENT_VOICE_CALL_STARTED, null);
    registerServiceStateTrackerEvents();
    ......
}
public void registerServiceStateTrackerEvents() {
    mPhone.getServiceStateTracker().registerForDataConnectionAttached(this,
        DctConstants.EVENT_DATA_CONNECTION_ATTACHED, null);
    mPhone.getServiceStateTracker().registerForDataConnectionDetached(this,
        DctConstants.EVENT_DATA_CONNECTION_DETACHED, null);
     ......
}
```

上面的代码可重点关注以下几点：

- Handler 消息注册对象

DcTracker 主要向 RILJ、GsmCdmaCallTracker、ServiceStateTracker 三个对象发起 Handler-Message 消息注册。

- Handler 消息类型

DcTracker 注册的 Handler 消息类型，以 DctConstants. EVENT_XXX 为主。

- 消息响应

DcTracker 作为自定义的 Handler 对象，在发起 Handler Message 消息注册时，传入 this 对象，并在 handleMessage 方法中响应 Message 消息回调。

8.1.2 初始化 ApnContext

在 DcTracker 的构造方法中将调用 initApnContexts 方法来初始化 ApnContext，代码逻辑的详情如下：

```
private void initApnContexts() {
    String[] networkConfigStrings = mPhone.getContext().getResources().getStringArray(
        com.android.internal.R.array.networkAttributes);//xml 文件获取网络配置
    for (String networkConfigString : networkConfigStrings) {
        NetworkConfig networkConfig = new NetworkConfig(networkConfigString);
        ApnContext apnContext = null;

        switch (networkConfig.type) {
        case ConnectivityManager.TYPE_MOBILE:
            apnContext = addApnContext(//增加 ApnContext 配置
                PhoneConstants.APN_TYPE_DEFAULT, networkConfig);
```

```
                break;
            case ConnectivityManager.TYPE_MOBILE_MMS:
                apnContext = addApnContext(
                            PhoneConstants.APN_TYPE_MMS, networkConfig);
                break;
            ......
        }
    }
}
```

上面的代码逻辑可重点关注以下四点。

- networkConfigStrings 内容

networkConfigStrings 数组的内容是从 frameworks/base/core/res/res/values/config.xml 中获取的，共有 12 个网络配置信息，详情如下：

```
<string-array translatable="false" name="networkAttributes">
    <item>"wifi,1,1,1,-1,true"</item>
    <item>"mobile,0,0,0,-1,true"</item>
    <item>"mobile_mms,2,0,2,60000,true"</item>
    <item>"mobile_supl,3,0,2,60000,true"</item>
    <item>"mobile_dun,4,0,2,60000,true"</item>
    <item>"mobile_hipri,5,0,3,60000,true"</item>
    <item>"mobile_fota,10,0,2,60000,true"</item>
    <item>"mobile_ims,11,0,2,60000,true"</item>
    <item>"mobile_cbs,12,0,2,60000,true"</item>
    <item>"wifi_p2p,13,1,0,-1,true"</item>
    <item>"mobile_ia,14,0,2,-1,true"</item>
    <item>"mobile_emergency,15,0,2,-1,true"</item>
</string-array>
```

重点关注 mobile 移动数据上网和 mobile_mms 彩信移动数据上网这两种类型的网络配置信息。

- 创建 NetworkConfig 对象

循环 networkConfigStrings 数组，首先根据网络配置的 String 字符串信息创建 NetworkConfig 对象，然后使用 NetworkConfig 对象构造 ApnContext 对象。最后保存 ApnContext 对象。主要的处理逻辑详情如下：

```
public NetworkConfig(String init) {
    String fragments[] = init.split(",");//分割 xml 配置文件中的 String
    name = fragments[0].trim().toLowerCase(Locale.ROOT);
    type = Integer.parseInt(fragments[1]);//与 ConnectivityManager.TYPE_XXX 定义一致
    radio = Integer.parseInt(fragments[2]);
    priority = Integer.parseInt(fragments[3]);
    restoreTime = Integer.parseInt(fragments[4]);
    dependencyMet = Boolean.parseBoolean(fragments[5]);//默认为 true
}
```

- 创建并保存 ApnContext 对象到三个集合列表中

```
private ApnContext addApnContext(String type, NetworkConfig networkConfig) {
    ApnContext apnContext = new ApnContext(mPhone, type, LOG_TAG, networkConfig, this);
    mApnContexts.put(type, apnContext);
    mApnContextsById.put(ApnContext.apnIdForApnName(type), apnContext);
    mPrioritySortedApnContexts.add(apnContext);
    return apnContext;
}
public ApnContext(Phone phone, String apnType, String logTag, NetworkConfig config,
        DcTracker tracker) {
    mPhone = phone;
```

```
    mApnType = apnType;// ConnectivityManager.TYPE_XXX
    mState = DctConstants.State.IDLE;
    setReason(Phone.REASON_DATA_ENABLED);
    mDataEnabled = new AtomicBoolean(false);
    mDependencyMet = new AtomicBoolean(config.dependencyMet);//默认为 true
    priority = config.priority;
    mDcTracker = tracker;
    mRetryManager = new RetryManager(phone, apnType);
}
```

- NetworkConfig 和 ApnContext 的关系

从 config.xml 中解析出网络配置信息从而创建了 12 个 NetworkConfig 对象，共匹配了 10 个 ApnContext 对象，匹配关系是：NetworkConfig 对象的 type 属性（config.xml 配置字符串中的第二个配置项）为 ConnectivityManager.TYPE_MOBILE_XXX，对应关系可参考 initApnContexts 的 switch 分支处理逻辑。因此，3 个集合 mApnContexts、mApnContextsById 和 mPrioritySortedApnContexts 中均保存了 10 个 ApnContext 对象列表。

注意

mPrioritySortedApnContexts 列表中将根据 ApnContext 对象的 priority 属性（即 config.xml 配置信息的倒数第二个选项：网络配置的优先级）排序；mPrioritySortedApnContexts 列表中的第一个 ApnContext 对象的 mApnType 是 ConnectivityManager.TYPE_MOBILE_HIPRI 类型，priority 属性为 3；最后一个 ApnContext 对象的 mApnType 是 ConnectivityManager.TYPE_MOBILE_EMERGENCY 类型，priority 属性为 -1。

优先级从高到低依次是：hipri、mms、supl、cbs、dun、fota、ims、default、ia、emergency。手机上网将建立 default 类型的数据连接，当手机接收到一条彩信时，因为彩信建立的数据连接是 mms 类型，mms 比 default 的数据连接优先级要高，所以会断开 default 数据连接而创建 mms 数据连接，从而能快速接收彩信。因此，在发送和接收彩信的同时不能上网。

8.1.3 认识 APN

APN（Access Point Name）是 Android 手机实现移动数据上网业务必须配置的参数，用来决定手机通过哪种接入方式来访问网络，其配置信息全部记录在 telephony.db 的 SQLite 数据库中名为 carriers 的表中。

首先将此数据库文件 pull 到本地，然后可通过 SQLite 数据库工具（SqliteDev 或 SQLite Expert Professional）查看 carriers 表的结构和其中的 APN 配置信息数据，pull 操作详情如下。

```
$ adb pull data/user_de/0/com.android.providers.telephony/databases/telephony.db .
```

1. APN 配置关键字段

基于 Android 8.1 版本的 Nexus 6P 手机，其 telephony.db 数据库文件中的 carriers 表共有 34 个字段、2942 条 APN 配置信息，关键字段如表 8-1 所示。

表 8-1　APN 关键字段

字段	说明
name	APN 配置名称
numeric	运营商编号
apn	APN 接入点，中国移动 cmwap 和 cmnet

续表

字段	说明
Proxy	代理服务器地址
port	端口号
mmsproxy	彩信代理服务器地址
mmsport	彩信代理服务器端口号
mmsc	彩信接入服务地址
type	APN 接入类型

不同运营商的 APN 配置信息都不一样，可上网查询或直接找运营商咨询。

2. 增加 APN 配置

手机中的 APN 配置信息非常关键和重要，一旦有误，将导致手机不能上网。在 telephony.db 的 SQLite 数据库的 carriers 表中，可以找到 Android 源代码中默认的一些 APN 配置信息，但这些信息仍然不够，例如要做国内手机的 Android 定制开发，或是国外手机的定制开发；那么该如何新增 APN 配置呢？共有两个方法。

- 修改配置文件
- APN 配置管理界面

（1）修改配置文件

现在的手机开发过程中有一个非常重要的步骤，那就是运营商认证。如果发布到市场上的手机还需要用户去手动配置上网参数，首先就完成不了运营商认证，并且用户的使用体验度也不好，因此在定制开发的过程中，要尽量将 APN 配置信息预置在手机中。

Nexus 6P 手机中预置的 APN 信息，对应的 xml 配置文件在 Android 8.1 源码中的路径是：device/huawei/angler/apns-full-conf.xml。

由 packages/providers/TelephonyProvider/src/com/android/providers/telephony/TelephonyProvider.java 中的 initDatabase 方法将 apns-full-conf.xml 文件中的 APN 信息插入到 carriers 数据库表中的处理逻辑请读者自行了解。

要增加 APN 的配置信息，仅需修改 device/huawei/angler/apns-full-conf.xml 文件即可。

中国移动部分 APN 配置信息的详情如下。

```
<apn carrier="中国移动 (China Mobile) GPRS"
  mcc="460"
  mnc="00"
  apn="cmnet"
  type="default,supl"
/>
<apn carrier="中国移动 (China Mobile) WAP"
  mcc="460"
  mnc="00"
  apn="cmwap"
  proxy="10.0.0.172"
  port="80"
  type="default,supl"
/>
......
```

修改后，将重新生成的 system.img 镜像文件写入 Nexus 6P 手机中即可完成默认 APN 配置信息的新增或修改。

如果没有生效，需要执行以下命令来删除数据库文件：

```
adb shell rm data/user_de/0/com.android.providers.telephony/databases/telephony.db
```

再次启动 Nexus 6P 手机，才会加载新的配置文件。

在接收或发送彩信的过程中，会使用到 MMS 彩信数据连接并产生相应的流量，但是，运营商只会收取彩信的发送费用，并不会记录和收取数据流量产生的费用。这是因为不同运营商的彩信 APN 配置中有与 MMSC 彩信服务器相关的配置信息，手机发送和接收彩信访问彩信服务器时产生的流量是免费的，并不记录其访问明细。

（2）APN 配置管理界面

APN 配置管理界面的进入路径是 Settings→Network&Internet→Mobile network→Access Point Names，在该界面中可手动修改 APN 配置，同时，提供的 Reset to default 功能可重置为默认配置信息。对应的 Java 文件是：packages/apps/Messaging/src/com/android/messaging/ui/appsettings/ApnSettingsActivity.java。

APN 配置管理界面仅展现了当前驻网运营商对应的 APN 配置列表。ApnSettingsActivity 文件中的关键逻辑详情如下：

```
//获取运营商编号 mcc 和 mnc
final String mccMnc = PhoneUtils.getMccMncString(PhoneUtils.get(mSubId).getMccMnc())
;
new AsyncTask<Void, Void, Cursor>() {//查询 Carriers 表
    @Override
    protected Cursor doInBackground(Void... params) {
        String selection = Telephony.Carriers.NUMERIC + " =?";//查询条件
        String[] selectionArgs = new String[]{ mccMnc };//匹配运营商编号
        final Cursor cursor = mDatabase.query(
                ApnDatabase.APN_TABLE, APN_PROJECTION,
                selection, selectionArgs, null, null, null);//关注查询条件
        return cursor;
    }
    ......
}
```

8.1.4　创建 DcController

DcTracker 的 mDcc 属性是 DcController 类型。DcController 构造方法中的主要处理逻辑详情如下：

```
HandlerThread dcHandlerThread = new HandlerThread("DcHandlerThread");
dcHandlerThread.start();
Handler dcHandler = new Handler(dcHandlerThread.getLooper());
mDcc = DcController.makeDcc(mPhone, this, dcHandler);//创建 DcController 对象
public static DcController makeDcc(Phone phone, DcTracker dct, Handler handler) {
    DcController dcc = new DcController("Dcc", phone, dct, handler);
    dcc.start();//启动 StateMachine
    return dcc;
}
private DcController(String name, Phone phone, DcTracker dct,
        Handler handler) {//private 私有构造方法
    super(name, handler);// StateMachine 构造方法调用
```

```
    mPhone = phone;
    mDct = dct;
    addState(mDccDefaultState);//初始化状态
    setInitialState(mDccDefaultState);
    ......
}
```

原来 DcController 类继承自 StateMachine 类，其有一个内部类 DccDefaultState 对象 mDcc-DefaultState，继承自 State 类。

StateMachine 实现了 State 状态模式，将在后面的章节中重点解析其设计原理和实现机制。

8.1.5　注册 Observer

在 DcTracker 的构造方法中完成了 mApnObserver 和 mSettingsObserver 两个对象的创建，并且完成了 Observer 的注册。

mApnObserver 监听 Telephony.Carriers.CONTENT_URI，即 Telephony 数据库 Carriers 表中保存的 APN 信息是否发生改变；ApnObserver 的 onChange 方法响应 APN 配置信息的变化，通过 sendMessage(obtainMessage(DctConstants.EVENT_APN_CHANGED))发出 Handler 消息，最终由 DcTracker 的 handleMessage 方法响应。

mSettingsObserver 则监听 Settings.Global.DEVICE_PROVISIONING_MOBILE_DATA_ENABLED、Settings.Global.DATA_ROAMING 和 Settings.Global.DEVICE_PROVISIONED 数据库更新，DcTracker.handleMessage 响应的 Message 类型分别是：

DctConstants.EVENT_ROAMING_SETTING_CHANGE；

DctConstants.EVENT_DEVICE_PROVISIONING_CHANGE；

DctConstants.EVENT_DEVICE_PROVISIONED_CHANGE。

8.1.6　广播接收器

mIntentReceiver 作为 DcTracker 的匿名内部类对象，继承自 BroadcastReceiver 类，其注册广播的主要逻辑详情如下：

```
IntentFilter filter = new IntentFilter();
filter.addAction(Intent.ACTION_SCREEN_ON);
filter.addAction(Intent.ACTION_SCREEN_OFF);
filter.addAction(WifiManager.NETWORK_STATE_CHANGED_ACTION);
filter.addAction(WifiManager.WIFI_STATE_CHANGED_ACTION);
filter.addAction(INTENT_DATA_STALL_ALARM);
filter.addAction(INTENT_PROVISIONING_APN_ALARM);
filter.addAction(CarrierConfigManager.ACTION_CARRIER_CONFIG_CHANGED);
mPhone.getContext().registerReceiver(mIntentReceiver, filter, null, mPhone);
```

8.1.7　加载 ApnSetting

DcTracker 的 mAllApnSettings 属性将保存当前运营商的 APN 配置信息列表，即 ApnSetting 对象列表。加载过程的关键逻辑在 createAllApnList 方法中，详情如下：

```
String selection = Telephony.Carriers.NUMERIC + " = '" + operator + "'";
Cursor cursor = mPhone.getContext().getContentResolver().query(
```

```
Telephony.Carriers.CONTENT_URI, null, selection, null, Telephony.Carriers._ID);
```

与 APN 配置管理界面的处理一致，先查询 Carriers 数据库表，匹配出当前运营商的 APN 配置信息；然后创建 ApnSetting 对象，保存在 mAllApnSettings 列表中。

ApnSetting 类使用 types、carrier、apn、proxy、port、mmsc、mmsProxy、mmsPort 等 28 个属性来保存 APN 配置信息，可与表 8-1 中的数据进行对应。

 DcTracker 管理 ApnContext 和 ApnSetting 两个对象的集合，都与 APN 相关。ApnContext 的 mApnSetting 是 ApnSetting 对象，提供 setApnSetting 方法来设置 mApnSetting 属性。

在 Nexus 6P 手机中插入中国联通的 4G 电话卡，加载 mAllApnSettings 的 radio 日志，详情如下：

```
D/DCT    ( 1125): [0]onRecordsLoadedOrSubIdChanged: createAllApnList
D/DCT    ( 1125): [0]createAllApnList: selection=numeric = '46001'
D/DCT    ( 1125): [0]createApnList: X result=[
[ApnSettingV3] 沃 3G 连接互联网 (China Unicom), 2273, 46001, 3gnet, , , , , -1, default |
supl, IP, IP, true, 0, 0, 0, false, 0, 0, 0, 0, , , false,
[ApnSettingV3] 沃 3G 手机上网 (China Unicom), 2274, 46001, 3gwap, 10.0.0.172, , , 80, -1,
default | supl, IP, IP, true, 0, 0, 0, false, 0, 0, 0, 0, , , false,
[ApnSettingV3] 联通 2G 上网 (China Unicom), 2275, 46001, uninet, , , , , -1, default |
supl, IP, IP, true, 0, 0, 0, false, 0, 0, 0, 0, , , false,
[ApnSettingV3] 联通彩信 (China Unicom), 2276, 46001, 3gwap, , http://mmsc.myuni.com.cn,
10.0.0.172, 80, , -1, mms, IP, IP, true, 0, 0, 0, false, 0, 0, 0, 0, , , false,
[ApnSettingV3] 联通 2G 彩信 (China Unicom), 2277, 46001, uniwap, , http://mmsc.myuni.com.cn,
10.0.0.172, 80, , -1, mms, IP, IP, true, 0, 0, 0, false, 0, 0, 0, 0, , , false]
```

numeric = '46001'作为查询数据库表 Carriers 的查询条件，46001 是中国联通的运营商编号。mAllApnSettings 列表中保存了五个 ApnSetting 对象。

总结 DcTracker 类的关系，如图 8-1 所示。

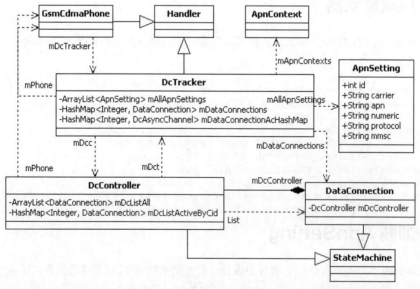

图 8-1　DcTracker 关键类图

图 8-1 所示的 DcTracker 关键类图，我们重点关注以下几点：

- DcTracker 的本质是自定义的 Handler 消息处理类,与 GsmCdmaPhone 对象相互引用。
- mApnContexts、mApnContextsById 和 mPrioritySortedApnContexts 三个关键列表中分别保存了 10 个 ApnContext 对象,可重点关注 mPrioritySortedApnContexts 列表中根据 ApnContext 的优先级属性排序。
- 加载当前运营商 APN:mAllApnSettings 保存了当前运营商的 APN 配置信息。
- DcController 和 DataConnection 都继承于 StateMachine,它们之间有组合和相互引用的关系。

8.2 解析 StateMachine

StateMachine 状态机类实现了 State 设计模式,可以查找 Android 源码中 StateMachine 的子类,主要体现在蓝牙、Wifi 和 Telephony 三个模块中。

在解析 StateMachine 状态机类之前,首先要弄清楚什么是 State 设计模式。

8.2.1 State 设计模式

State 设计模式将对象的状态封装成一个对象,在不同状态下,同样的调用将执行不同的操作,如图 8-2 所示。

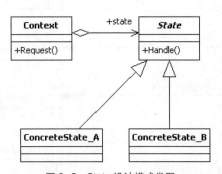

图 8-2 State 设计模式类图

Context 可维护多个 State 子类的实例,每个实例代表一种状态。State 定义一个接口,以封装与 Context 的一个特定状态相关的行为,即 Context 只需固定调用 State→Handle 方法,即可将与状态相关的请求委托给当前 ConcreteState 对象的 Handle 方法处理。

Context 是客户端使用的主要接口,客户端并不需要直接与 ConcreteState 状态对象进行交互。所有 ConcreteState 状态对象的交互和管理都由 Context 来实现,并且 ConcreteState 状态对象之间有一定的关系。

8.2.2 StateMachine 核心类

StateMachine 核心类如图 8-3 所示。

StateMachine 类对外提供状态相关操作的接口方法,而 SmHandler 类则是作为 StateMachine 状态机的核心,负责 Handler 消息的发送和接收,用来管理和更新 State 对象。

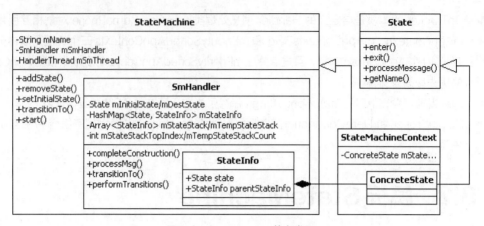

图 8-3　StateMachine 核心类图

8.2.3　初始化流程

总结 StateMachine 初始化流程如图 8-4 所示。

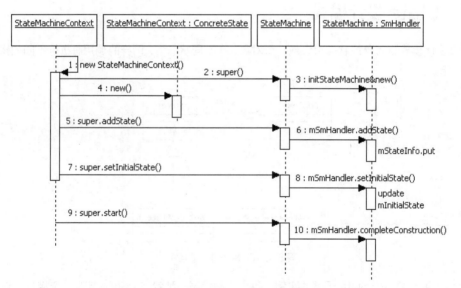

图 8-4　StateMachine 初始化流程

图 8-4 所示的 StateMachine 初始化流程可归纳为四个关键过程：

- 隐藏对象的创建过程。StateMachineContext 作为 StateMachine 的子类，在创建过程中，步骤 3 同步创建 SmHandler 对象，步骤 4 创建 StateMachineContext 管理的 ConcreteState 对象。
- addState 增加管理的状态对象。步骤 5 和步骤 6 中的 SmHandler 对象的 mStateInfo 保存 State 列表。
- setInitialState 设置状态的初始化对象，步骤 7 和步骤 8 中的 SmHandler 对象更新 mInitialState 属性。
- start 开始运行，步骤 9 和步骤 10 中，SmHandler 对象的 completeConstruction 方法逻辑主要是根据 mInitialState 初始化状态来初始化 mStateStack 数组，此数组将保存当前状态的

State 对象。步骤 5、步骤 7 和步骤 9 中的三个方法是 StateMachine 对象对外提供的接口方法，经过转换内部方法的调用，最终交给 SmHandler 对象中对应的方法来处理。addState 增加管理的状态对象，这些状态对象之间存在着树状关系。SmHandler 对象的 addState 方法定义如下。

```
private final StateInfo addState(State state, State parent)
```

8.2.4 运行流程

总结 StateMachine 状态机的运行流程，如图 8-5 所示。

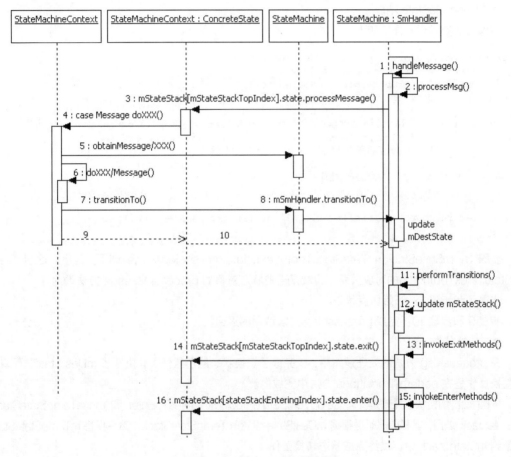

图 8-5　StateMachine 运行流程

图 8-5 所示的 StateMachine 运行流程可重点关注以下几点：

- 消息入口

SmHandler 对象的 handleMessage 方法作为 StateMachine 的消息入口。下面以 DcController 类中注册的 Handler 消息为例，代码逻辑的详情如下：

```
mPhone.mCi.registerForDataCallListChanged(getHandler(),
        DataConnection.EVENT_DATA_STATE_CHANGED, null);
```

DcController 作为 StateMachine 的子类，调用 getHandler 方法将返回 mSmHandler 对象，因此，

mCi 即 RILJ 对象发起 EVENT_DATA_STATE_CHANGED 的 Message 消息回调时，由 SmHandler 对象的 handleMessage 方法进行响应。

- handleMessage 消息主要有两个处理逻辑

步骤 2 和步骤 11 分别对应 processMsg 和 performTransitions。

handleMessage 处理逻辑简化后的代码详情如下：

```
@Override
public final void handleMessage(Message msg) {
    processMsg (msg);
    performTransitions ();
}
```

- ConcreteState 响应

步骤 2 的处理逻辑详情如下：

```
private final State processMsg(Message msg) {
    //这里非常关键，用于获取当前状态对象
    StateInfo curStateInfo = mStateStack[mStateStackTopIndex];
    if (isQuit(msg)) {
        transitionTo(mQuittingState);
    } else {//实现 State 设计模式的关键所在，所有 State 对象都重写了 processMessage 方法
        while (!curStateInfo.state.processMessage(msg)) {
            //更新职责链到父状态对象，进入下一次循环
            curStateInfo = curStateInfo.parentStateInfo;
            ......//异常处理
        }
    }
    return (curStateInfo != null) ? curStateInfo.state : null;
}
```

步骤 3：mStateStack[mStateStackTopIndex].state.processMessage 调用。注意 mStateStack[mStateStackTopIndex] 的对应关系，也就是当前状态对象的 processMessage 方法调用。

步骤 7 和步骤 8：开始切换状态。

步骤 9 和步骤 10：完成 processMsg 方法调用的返回。

- 切换状态

在 processMsg 方法的处理逻辑中，仅完成了切换状态的前置任务，更新了 mDestState，而真正的切换操作是在 performTransitions 方法中完成的。

步骤 12：切换状态操作是发起 setupTempStateStackWithStatesToEnter 和 moveTempStateStackToStateStack 调用；其核心思想是根据 mDestState 设置 mTempStateStack，然后再将 mTempStateStack 更新到 mStateStack 中，最终完成状态切换工作。

- ConcreteState 对象的 exit 和 enter 响应过程

performTransitions 将更新当前状态，因此涉及新旧状态 ConcreteState 对象的响应，步骤 14 调用旧状态 ConcreteState 对象的 exit 方法，退出当前状态；步骤 15 调用新状态 ConcreteState 对象的 enter 方法，进入当前状态。

exit 和 enter 调用过程均在 SmHandler 中定义了调用模板。

8.2.5 小结

StateMachine 实现的状态机关键业务流程，可参考图 8-4 和图 8-5。Android 平台实现的 State

设计模式 StateMachine 具有以下三个特点。

- StateMachine 类对外提供对 State 对象操作的相关接口，如 addState、setInitialState 和 start 这三个启动状态机的基本方法。
- SmHandler 内部类作为自定义的 Handler 消息处理类，通过发送、接收和处理 Handler 消息来管理和更新所有的 State 对象。重点关注 processMsg 方法，其用来处理所有 State 对象接收到的 Handler 消息。
- StateMachine 管理的所有 State 状态对象之间具有树状关系，一个 State 对象自己不能处理的消息，会交给其父节点处理，如果父节点仍不能处理，又交给上一个父节点处理，直到根节点。这里涉及另外一个面向对象的设计模式——职责链模式（Chain of Responsibility），感兴趣的读者可自行学习。

8.3　DataConnection

DataConnection（数据连接）类在 Telephony 业务模型中管理移动数据业务，一个 DataConnection 对象代表手机移动数据业务的一个数据连接。Android 8.1 最多支持 10 种移动数据连接类型，总结 DataConnection 关键类图，如图 8-6 所示。

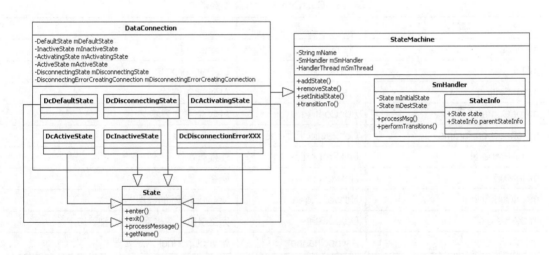

图 8-6　DataConnection 关键类图

DataConnection 继承于 StateMachine 类，也是 Data Call 移动数据业务涉及的第二个 StateMachine 状态机；它定义了 DcDefaultState、DcInactiveState、DcActivatingState、DcActiveState、DcDisconnectingState 和 DcDisconnectionErrorCreatingConnection 共六个私有内部类，全部继承自 State 类。

简化后的 DataConnection 代码框架详情如下：

```
//六个 DcXXXState 私有内部类全部重写了父类的 processMessage 方法
private class DcXXXState extends State {
    @Override
    public void enter() {......}//进入此状态处理逻辑响应
    @Override
```

```
            public void exit() {......}//退出此状态处理逻辑响应
    @Override//根据 Message 消息类型进入不同分支处理
            public boolean processMessage(Message msg) {......}
}
```

总结 DataConnection 类的六个私有 State 子类，如表 8-2 所示。

表 8-2　DataConnection 类的 State 子类

State 子类	说明
DcDefaultState	默认状态
DcInactiveState	不活动状态
DcActivatingState	正在激活状态
DcActiveState	激活状态
DcDisconnectingState	断开中状态
DcDisconnectionErrorCreatingConnection	断开失败并且正在创建状态

8.3.1　关键属性

总结 DataConnection 类的关键属性，如表 8-3 所示。

表 8-3　DataConnection 类关键属性

属性	类型	描述
mPhone	Phone	Phone 对象
mDct	DcTracker	DcTracker 对象
mApnSetting	ApnSetting	APN 配置信息
mApnContexts	ApnContext	APN 上下文列表
mConnectionParams	ConnectionParams	连接参数
mLinkProperties	LinkProperties	连接属性
mNetworkInfo	NetworkInfo	网络信息
mNetworkAgent	NetworkAgent	NetworkInfo 的代理
mXXXState	DcXXXState	完整覆盖了数据连接的六种状态
mAc	AsyncChannel	AsyncChannel 消息处理

表 8-3 列出的 DataConnection 类关键属性，重点关注以下几点。
- 在 DataConnection 切换过程中或是不同的连接状态下，分别由这六个 mXXXState 对象的 enter、exit 和 processMessage 三个方法响应不同的业务逻辑，减少了对状态的烦琐判断，这也正是使用 State 设计模式所带来的优势。
- 具有 GsmCdmaPhone、DcTracker 对象的引用。
- 保存 ApnSetting 和 ApnContext 配置信息。

8.3.2　关键方法

总结 DataConnection 类的关键方法，如表 8-4 所示。

表 8-4 DataConnection 类关键方法

分类	方法	说明
对象创建	makeDataConnection	创建 DataConnection 并启动 StateMachine
Data Call 操作	onConnect	发起 setupDataCall 请求
	tearDownData	发起 deactivateDataCall 请求
	onSetupConnectionCompleted	setupDataCall 请求的回调响应逻辑
消息通知	notifyAllOfConnected	根据 mApnContexts 发出消息通知
	notifyAllDisconnectCompleted	
	notifyConnectCompleted	通过 DisconnectParams 发出消息通知
	notifyDisconnectCompleted	

表 8-4 列出的 DataConnection 类关键方法，重点关注以下两点。

- DataConnection 对象必须通过 public static 类型的 makeDataConnection 方法来创建，它的构造方法是 private 类型的。
- Data Call 操作和消息通知对应的方法都是私有的，这些方法的调用方都是六大 State 对象。

8.3.3 StateMachine 初始化流程

DataConnection 类继承自 StateMachine 状态机类，StateMachine 的初始化逻辑详情如下。

```
public static DataConnection makeDataConnection(Phone phone, int id,
        DcTracker dct, DcTesterFailBringUpAll failBringUpAll,
        DcController dcc) {
    DataConnection dc = new DataConnection(phone,
            "DC-" + mInstanceNumber.incrementAndGet(), id, dct, failBringUpAll, dcc);
    dc.start();//启动 StateMachine
    return dc;
}
private DataConnection(Phone phone, String name, int id,
        DcTracker dct, DcTesterFailBringUpAll failBringUpAll,
        DcController dcc) {
    super(name, dcc.getHandler());
    mPhone = phone;
    mDct = dct;
    mDcController = dcc;
    mId = id;
    ......
    //addState 六个 State 状态对象
    addState(mDefaultState);
    addState(mInactiveState, mDefaultState);
    addState(mActivatingState, mDefaultState);
    addState(mActiveState, mDefaultState);
    addState(mDisconnectingState, mDefaultState);
    addState(mDisconnectingErrorCreatingConnection, mDefaultState);
    setInitialState(mInactiveState);//设置初始化状态 mInactiveState

    mApnContexts = new HashMap<ApnContext, ConnectionParams>();
}
```

DataConnection 类的 mName 属性的命名规则是 "DC-数字"。

六个 State 状态对象之间有一个简单的树状关系，从代码的排列上可以明显看出，mDefaultState

作为根节点,而其他五个 State 对象作为它的子节点,即叶子节点。

DataConnection 类的 makeDataConnection 方法的逻辑中,完成了 StateMachine 初始化,有四大步骤:
- 创建对象(State 和 SmHandler)
- addState
- setInitialState
- start

8.4 开启移动数据业务

Nexus 6P 手机开启或关闭默认的移动数据业务时有两个交互界面:
- 通知栏 Mobile data 快捷控制
- Mobile network 设置界面

响应这两个开关操作的调用过程总结如下:

TelephonyManager.setDataEnabled→ITelephony(PhoneInterfaceManager).setDataEnabled→phone.setDataEnabled→mDcTracker.setDataEnabled

DcTracker 对象的 setDataEnabled 方法用来响应移动数据业务的开关操作。

8.4.1 流程分析

setDataEnabled 方法通过 CMD_SET_USER_DATA_ENABLE 类型的 Handler 消息,转换成 onSetUserDataEnabled 方法调用,其关键逻辑详情如下:

```
private void onSetUserDataEnabled(boolean enabled) {
    synchronized (mDataEnabledSettings) {//同步锁
        if (mDataEnabledSettings.isUserDataEnabled() != enabled) {//开关状态发生改变
            mDataEnabledSettings.setUserDataEnabled(enabled);//更新设置
            //更新数据库设置,注意双卡模式下的异同
            Settings.Global.putInt(mResolver, Settings.Global.MOBILE_DATA, enabled ? 1 : 0);
            if (enabled) {//开启移动数据业务
                reevaluateDataConnections();
                onTrySetupData(Phone.REASON_DATA_ENABLED);
            } else {//关闭移动数据业务
                onCleanUpAllConnections(Phone.REASON_DATA_SPECIFIC_DISABLED);
            }
        }
    }
}
```

onSetUserDataEnabled 方法有两个逻辑处理分支:开启移动数据业务和关闭移动数据业务。继续关注开启移动数据业务 onTrySetupData 的处理分支,会接着调用 setupDataOnConnectableApns 方法,此方法的主要逻辑详情如下:

```
for (ApnContext apnContext : mPrioritySortedApnContexts) {//遍历 ApnContext 列表
    if (apnContext.isConnectable()) {//可连接的 ApnContext
        apnContext.setReason(reason);//设置开启移动数据业务的原因
        trySetupData(apnContext);//继续调用,关注传入的 apnContext 参数
    }
}
```

上面的代码逻辑，我们重点关注以下两点：

- 从优先级高的 ApnContext 对象开始遍历 mPrioritySortedApnContexts 列表
- 可连接的 ApnContext 判断条件

isConnectable 用于判断当前 ApnContext 是否可以发起连接，后面再详细分析其判断逻辑。继续关注 trySetupData 方法，传入的是 ApnContext 对象，前置条件是 isConnectable。trySetupData 的主要处理逻辑详情如下：

```
trySetupData(ApnContext apnContext) {
    DataConnectionReasons dataConnectionReasons = new DataConnectionReasons();
    boolean isDataAllowed = isDataAllowed(apnContext, dataConnectionReasons);
    if (isDataAllowed) {
            //获取驻网的移动数据业务 RadioTechnology
            int radioTech = mPhone.getServiceState().getRilDataRadioTechnology();
            apnContext.setConcurrentVoiceAndDataAllowed(mPhone.getServiceStateTracker()
                    .isConcurrentVoiceAndDataAllowed());
            if (apnContext.getState() == DctConstants.State.IDLE) {
                ArrayList<ApnSetting> waitingApns =
                        buildWaitingApns(apnContext.getApnType(), radioTech);
                if (waitingApns.isEmpty()) {......return false; }//未找到对应 APN 配置信息
                else {
                    apnContext.setWaitingApns(waitingApns);
                }
            }
            boolean retValue = setupData(apnContext, radioTech, dataConnectionReasons.contains(
                    DataAllowedReasonType.UNMETERED_APN));
            notifyOffApnsOfAvailability(apnContext.getReason());
            return retValue;
    } else {
        ......
        return false;
    }
}
```

上面的代码逻辑中有两个重要的方法调用：isDataAllowed 和 buildWaitingApns，后面再详细分析其处理逻辑。

继续分析 setupData 的处理逻辑，其详情如下：

```
private boolean setupData(ApnContext apnContext, int radioTech, boolean unmeteredUseOnly) {
    ApnSetting apnSetting;
    DcAsyncChannel dcac = null;
    apnSetting = apnContext.getNextApnSetting();//RetryManager
    if (dcac == null) {
        dcac = findFreeDataConnection();//Cache 中的可用 DataConnection 对象
        if (dcac == null) {
            dcac = createDataConnection();
        }
    }
    final int generation = apnContext.incAndGetConnectionGeneration();//计数器
    apnContext.setDataConnectionAc(dcac);
    apnContext.setApnSetting(apnSetting);//ApnContext 与 ApnSetting 关系
    apnContext.setState(DctConstants.State.CONNECTING);//设置连接中状态
    //GsmCdmaPhone 对象发出 DataConnection 变化消息通知
    mPhone.notifyDataConnection(apnContext.getReason(), apnContext.getApnType());

    Message msg = obtainMessage();//注意 DcTracker 创建的 Message 对象
    msg.what = DctConstants.EVENT_DATA_SETUP_COMPLETE;
    msg.obj = new Pair<ApnContext, Integer>(apnContext, generation);
```

```
        dcac.bringUp(apnContext, profileId, radioTech, unmeteredUseOnly, msg, generation);
        return true;
}
```

上面的代码逻辑，我们重点关注以下几点：

- 使用 createDataConnection 方法创建 DataConnection 对象，并建立两个 Handler 消息传递通道 AsyncChannel。
- ApnContext 和 ApnSetting 对象的最后更新并建立关联。
- 使用 GsmCdmaPhone 对象发出 DataConnection 变化消息通知。
- 使用 bringUp 激活移动数据业务。

接着，继续解析 createDataConnection 方法，其主要逻辑详情如下：

```
private DcAsyncChannel createDataConnection() {
    int id = mUniqueIdGenerator.getAndIncrement();//DataConnection mId
    DataConnection conn = DataConnection.makeDataConnection(mPhone, id,
                    this, mDcTesterFailBringUpAll, mDcc);
    mDataConnections.put(id, conn);//保存 DataConnection 对象
    DcAsyncChannel dcac = new DcAsyncChannel(conn, LOG_TAG);//自定义 AsyncChannel
    //full 类型的 AsyncChannel 连接
    int status = dcac.fullyConnectSync(mPhone.getContext(), this, conn.getHandler());
    if (status == AsyncChannel.STATUS_SUCCESSFUL) {//AsyncChannel 连接成功
            mDataConnectionAcHashMap.put(dcac.getDataConnectionIdSync(), dcac);
    } else {......}
    return dcac;
}
```

上面的代码逻辑，我们重点关注以下几点。

- DataConnection.makeDataConnection 创建 DataConnection 对象并启动 StateMachine。
- DcTracker 对象的 mDataConnections 属性保存了 DataConnection 列表。
- DcAsyncChannel 打通了 DcTracker 和 DataConnection 的两个 Handler 消息传递通道。

最后，分析 dcac.bringUp 的业务逻辑，其详情如下：

```
public void bringUp(ApnContext apnContext, int profileId, int rilRadioTechnology,
                    boolean unmeteredUseOnly, Message onCompletedMsg,
                    int connectionGeneration) {
    mLastConnectionParams = new ConnectionParams(apnContext, profileId,
            rilRadioTechnology, unmeteredUseOnly, onCompletedMsg, connectionGeneration);
    sendMessage(DataConnection.EVENT_CONNECT, mLastConnectionParams);
}
```

首先根据传入的参数创建 ConnectionParams 对象，然后通过 sendMessage 方法发送 DataConnection.EVENT_CONNECT 类型的消息。

结合 AsyncChannel 的处理机制思考一下 DcAsyncChannel 对象中调用 sendMessage 方法发送的 Message 在什么地方接收呢？那就是由 DataConnection 对象 SmHandler 的 handleMessage 方法响应该消息。

再次结合前面学习的 StateMachine 的运行机制和业务流程，可以确定 DataConnection.mInactiveState 对象的 processMessage 方法将响应并处理此 Message 消息，业务逻辑的详情如下：

```
case EVENT_CONNECT:
    ConnectionParams cp = (ConnectionParams) msg.obj;//获取参数对象
    if (initConnection(cp)) {//更新 mApnSetting、mApnContexts 等属性
```

```
            onConnect(mConnectionParams);//调用主类的 onConnect 方法
            transitionTo(mActivatingState);//更新状态为 mActivatingState 激活中
    } else {......}//异常处理
    retVal = HANDLED;
    break;
```

上面代码中最关键的就是对 onConnect 和 transitionTo 这两个方法的调用。

- onConnect 方法是 DataConnection 主类中实现的私有方法。
- 根据 StateMachine 的学习,可知 transitionTo 方法主要完成 DataConnection 状态切换。

DataConnection 对象的 onConnect 方法的主要处理逻辑详情如下:

```
private void onConnect(ConnectionParams cp) {
    mCreateTime = -1;
    mLastFailTime = -1;
    mLastFailCause = DcFailCause.NONE;
    //注意 SmHandler 创建的 Message 对象
    Message msg = obtainMessage(EVENT_SETUP_DATA_CONNECTION_DONE, cp);
    msg.obj = cp;

    DataProfile dp = new DataProfile(mApnSetting, cp.mProfileId);
    boolean isModemRoaming = mPhone.getServiceState().getDataRoamingFromRegistration();
    boolean allowRoaming = mPhone.getDataRoamingEnabled()
            || (isModemRoaming && !mPhone.getServiceState().getDataRoaming());

    mPhone.mCi.setupDataCall(cp.mRilRat, dp, isModemRoaming, allowRoaming, msg);
}
```

上面代码中最关键的是对 mPhone.mCi.setupDataCall 方法的调用,mCi 即 RILJ 对象,最终由 RIL 完成 Data Call 移动数据业务的处理。

摘录激活 Data Call 移动数据业务的 RIL 关键日志详情如下:

```
[3887]> SETUP_DATA_CALL,radioTechnology=14,isRoaming=false,allowRoaming=true,
DataProfile=0/3gnet/IP/0////0/0/0/0/true/5/IP/0/0///false [SUB0]
//已经获取到了 ip 地址、网关、dns
[3887]< SETUP_DATA_CALL DataCallResponse: { status=0 retry=-1 cid=0 active=2 type=IP
 ifname=rmnet_data0 mtu=1500 addresses=[10.54.188.70/30] dnses=[116.116.116.116,221.
5.88.88] gateways=[10.54.188.69] pcscf=[]} [SUB0]
```

另外,需要关注 msg 对象,它作为 setupDataCall 方法调用的参数之一,也就是说 RIL 完成激活 Data Call 移动数据业务处理之后,会使用此 Message 对象发起 Callback 调用。

DataConnection.SmHandler 对象的 handleMessage 方法进行响应,StateMachine 的状态已经切换到 mActivatingState,因此 mActivatingState 对象的 processMessage 方法将响应此消息回调,其处理逻辑详情如下:

```
case EVENT_SETUP_DATA_CONNECTION_DONE:
ar = (AsyncResult) msg.obj;
cp = (ConnectionParams) ar.userObj;

DataCallResponse.SetupResult result = onSetupConnectionCompleted(ar);
switch (result) {
    case SUCCESS: //成功激活 Data Call 移动数据业务
        mDcFailCause = DcFailCause.NONE;
        transitionTo(mActiveState);//转换为激活状态 mActiveState
        break;
    case ERR_BadCommand:
        ......//异常处理
}
```

```
retVal = HANDLED;
break;
```

上面的代码主要分为两个处理逻辑,详情如下。
- onSetupConnectionCompleted 根据 RIL 激活 Data Call 移动数据业务的返回信息,判断并获取激活 Data Call 移动数据业务的处理结果。
- 根据处理结果切换数据连接 DataConnection 状态。

注意

现在是否理解了 State 设计模式,以及 StateMachine 状态机在 DataConnection 中的运行机制?其实在 DataConnection 中对连接状态有大量的判断,Android 使用 StateMachine 状态机把对状态的判断逻辑转移到了表示不同状态的六个 State 对象中,使复杂的状态判断逻辑得到较大的简化。

总结开启默认移动数据业务的关键流程,如图 8-7 所示。

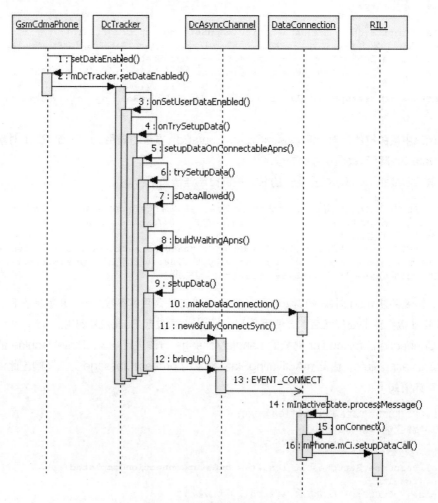

图 8-7 开启移动数据业务流程

图 8-7 所示的开启移动数据业务流程中,我们重点关注以下几点
- PhoneInterfaceManager 服务提供 setDataEnabled 方法,最后由 DcTracker 的 setData

Enabled 方法响应开启移动数据业务。

- 步骤 3 中的同步调用通过发送 CMD_SET_USER_DATA_ENABLE 消息，由 DcTracker 的 onSetUserDataEnabled 方法进行响应。
- 步骤 4 中的 onSetUserDataEnabled 方法有开启和关闭移动数据业务两个处理分支。
- 步骤 5 遍历 mPrioritySortedApnContexts，找到 isConnectable 的 ApnContext 对象，作为参数继续发起对 trySetupData 的调用，进入步骤 6。
- 步骤 7 和步骤 8 在下一节详细解析。
- 在 DcTracker 中，关注步骤 2、步骤 3、步骤 4、步骤 5、步骤 6 和步骤 9 中方法名的变化和规律：setDataEnabled、onSetUserDataEnabled、onTrySetupData、setupDataOnConnectableApns、trySetupData 和 setupData。
- 步骤 10 和步骤 11 开启 DataConnection StateMachine 状态机运行机制。
- 步骤 12 发送 Message 消息触发开启移动数据业务，DataConnection 对象的 SmHandler 响应此消息。
- 步骤 16 激活 Data Call 移动数据业务，向 RILJ 对象发起 setupDataCall 请求。

8.4.2 前置条件分析

8.4.1 节有关激活 Data Call 默认移动数据业务的流程分析中，涉及的 isConnectable、isDataAllowed 和 buildWaitingApns 三个关键方法的业务处理逻辑在本节进行详细解析。

1. isConnectable

在 setupDataOnConnectableApns 方法中遍历 mPrioritySortedApnContexts，找出 isConnectable 的 ApnContext 对象来激活移动数据业务，其判断逻辑的详情如下：

```
public boolean isConnectable() {
    return isReady() && ((mState == DctConstants.State.IDLE)
                        || (mState == DctConstants.State.SCANNING)
                        || (mState == DctConstants.State.RETRYING)
                        || (mState == DctConstants.State.FAILED));
}
public boolean isReady() {
    return mDataEnabled.get() && mDependencyMet.get();
}
```

首先关注 mState 的状态，根据 DctConstants.State 的七个常量定义，发现并不是 CONNECTING、CONNECTED 和 DISCONNECTING 这三个状态，因此都返回 true，这几个状态不难理解。

接着是 isReady 的条件判断，根据配置文件信息，mDependencyMet 都默认为 true。因此，在 isConnectable 的条件判断中，最关键的就是 mDataEnabled，它在构造方法中默认为 false。改变 mDataEnabled 属性可追溯到 TelephonyNetworkFactory 类，它继承自 NetworkFactory，PhoneFactory.makeDefaultPhone 在创建 GsmCdmaPhone 对象时，同步创建了 TelephonyNetworkFactory 对象。总结初始化的业务流程如图 8-8 所示。

图 8-8 所示的 TelephonyNetworkFactory 初始化业务流程，需要重点关注以下几点。

- 父类 NetworkFactory 实现的方法

步骤 4 和步骤 6 中发起的 setCapabilityFilter、setScoreFilter 和 register 三个调用，都是父类 NetworkFactory 实现的方法。

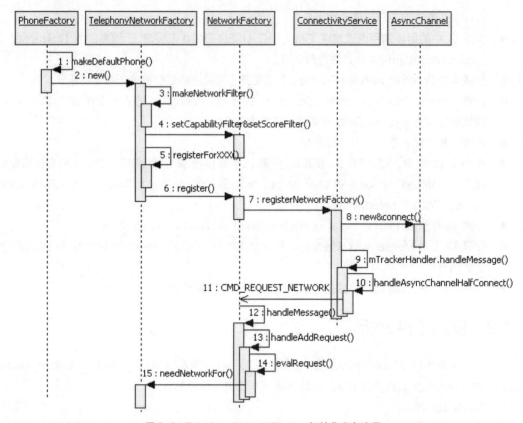

图 8-8 TelephonyNetworkFactory 初始化业务流程

- 创建 NetworkCapabilities

步骤 3 调用 makeNetworkFilter 创建 NetworkCapabilities，即 Telephony 上网能力，主要是 NetworkCapabilities.NET_CAPABILITY_XXX 等，可与 ApnContext 列表中的 apnType 类型相匹配。

- 注册 NetworkFactory

步骤 6、步骤 7 和步骤 8 中，ConnectivityService 系统服务创建通道 AsyncChannel，建立 NetworkFactory 与 ConnectivityService 之间的 mTrackerHandler 消息通道。

- 建立 AsyncChannel 响应过程

步骤 9 和步骤 10，由 ConnectivityService 中的 mTrackerHandler 响应 AsyncChannel.CMD_CHANNEL_HALF_CONNECTED 消息，向 NetworkFactory 发出 NetworkFactory.CMD_REQUEST_NETWORK 消息，再经过 NetworkFactory 的消息处理，最终由 TelephonyNetworkFactory 的 needNetworkFor 方法接收 ConnectivityService 管理的 NetworkRequest 对象。

- 匹配 NetworkCapabilities

步骤 14：evalRequest 方法中的关键逻辑详情如下：

```
if (n.requested == false && n.score < mScore &&
        n.request.networkCapabilities.satisfiedByNetworkCapabilities(
        mCapabilityFilter) && acceptRequest(n.request, n.score)) {
    needNetworkFor(n.request, n.score);//调用子类重写的方法 needNetworkFor
    n.requested = true;
}
```

n.score < mScore 是评分的判断条件，satisfiedByNetworkCapabilities 将匹配 setCapabilityFilter

设置的网络能力，NetworkFactory 则根据 mScore 和 mCapabilityFilter 过滤 NetworkRequest 对象。

 ConnectivityService 管理的 NetworkRequest 对象可参考 mDefaultRequest 和 mDefault MobileDataRequest 对象，与 TelephonyNetworkFactory 获取的 NetworkRequest 对象是匹配的。

在 TelephonyNetworkFactory 接收到的 NetworkRequest 对象保存在 mDefaultRequests 列表中；在 Telephony ServiceState 网络服务驻网成功后，TelephonyNetworkFactory 响应 EVENT_ACTIVE_PHONE_SWITCH 和 EVENT_SUBSCRIPTION_CHANGED 类型的消息回调，最终将通过 applyRequests 方法调用向 DcTracker 发起 requestNetwork 请求，更新 ApnContext 对象的 mDataEnabled 属性，如图 8-9 所示。

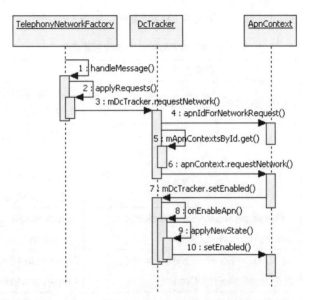

图 8-9　TelephonyNetworkFactory requestNetwork 流程总结

图 8-9 所示的 TelephonyNetworkFactory requestNetwork 流程中，重点关注以下几点。

- 匹配 APN ID

步骤 4：ApnContext 提供的静态方法 apnIdForNetworkRequest 的关键逻辑详情如下：

```
public static int apnIdForNetworkRequest(NetworkRequest nr) {
    NetworkCapabilities nc = nr.networkCapabilities;
    if (nc.hasCapability(NetworkCapabilities.NET_CAPABILITY_INTERNET)) {
        apnId = DctConstants.APN_DEFAULT_ID;
    }
    if (nc.hasCapability(NetworkCapabilities.NET_CAPABILITY_MMS)) {
        if (apnId != DctConstants.APN_INVALID_ID) error = true;
        apnId = DctConstants.APN_MMS_ID;
    }......
    return apnId;
}
```

NetworkCapabilities.NET_CAPABILITY_XXX 匹配的顺序依优先级从低到高，并且 if 逻辑中高级别的能力值可替换低级别的能力值。

如果我们接收到既有 INTERNET 也有 MMS 能力的 requestNetwork 对象，最终获取的 APN ID 是 DctConstants.APN_MMS_ID。

- 获取 ApnContext 对象

步骤 5 中使用 APN ID 在 mApnContextsById 列表中获取 ApnContext 对象。

- 更新 ApnContext

经过步骤 7、步骤 8、步骤 9 和步骤 10，最终调用 ApnContext 的 setEnabled 方法更新的 mDataEnabled 属性。

注意：因为 TelephonyNetworkFactory 接收的是默认数据连接 NetworkRequest 对象，因此，DcTracker 更新的是 ApnContext 列表中 APN 类型为 Default 的 ApnContext 对象，其 mDataEnabled 属性为 true。

2．isDataAllowed

总结 DcTracker 的 isDataAllowed 方法中的判断条件，如表 8-5 所示。

表 8-5 isDataAllowed 方法的判断条件

标志	获取方法	说明
internalDataEnabled	mDataEnabledSettings.isInternalDataEnabled	默认 true，紧急呼救 GsmCdmaCallTracker 会设置为 false
attachedState	DcTracker.mAttached	移动数据网络注册状态改变后，ServiceStateTracker 消息回调 EVENT_DATA_CONNECTION_ATTACHED/EVENT_DATA_CONNECTION_DETACHED
desiredPowerState	mPhone.getServiceStateTracker().getDesiredPowerState()	ServiceStateTracker 的 mDesiredPowerState 属性
radioStateFromCarrier	mPhone.getServiceStateTracker().getPowerStateFromCarrier()	ServiceStateTracker 的 mRadioDisabledByCarrier 属性
recordsLoaded	mIccRecords.get().getRecordsLoaded()	SIM 卡加载完成状态
defaultDataSelected	SubscriptionManager.isValidSubscriptionId	双卡模式下上网卡的匹配状态

根据表 8-5 所示的 isDataAllowed 关键判断条件，返回 true 值的主要条件如下：

- 不在紧急呼救业务中，internalDataEnabled 为 true。
- DcTracker.mAttached 为 true，Data Reg 移动数据网络注册成功。
- Radio Power 为开启状态 desiredPowerState 和 radioStateFromCarrier。
- SIM 卡加载完成，recordsLoaded 为 true。
- 默认上网卡与当前 DcTracker 保持一致，为 defaultDataSelected。

3．buildWaitingApns

DcTracker 的 buildWaitingApns 方法传递的参数是 ApnContext 的 requestedApnType 和 radioTech，返回 ApnSetting 对象列表，其有两个逻辑处理分支。

- 调用 getPreferredApn 方法获取 ApnSetting 对象列表。
- 直接使用 mAllApnSettings 列表匹配 ApnSetting 对象。

getPreferredApn 方法获取 ApnSetting 对象，通过缓存的 id 在 mAllApnSettings 中进行匹配，缓存过程是调用 setPreferredApn 方法实现的。其中，最关键的是两次匹配过程，其关键业务逻辑详情如下：

```
for (ApnSetting apn : mAllApnSettings) {
    if (apn.canHandleType(requestedApnType)) {//第一次匹配
        if (ServiceState.bitmaskHasTech(apn.bearerBitmask, radioTech)) {//第二次匹配
            apnList.add(apn);
}}}......
public boolean canHandleType(String type) {
    if (!carrierEnabled) return false;
    boolean wildcardable = true;
    for (String t : types) {
        if (t.equalsIgnoreCase(type) ||APN 类型相同，不区分大小写
                    (wildcardable && t.equalsIgnoreCase(PhoneConstants.APN_TYPE_ALL)) ||
                    (t.equalsIgnoreCase(PhoneConstants.APN_TYPE_DEFAULT) &&
                    type.equalsIgnoreCase(PhoneConstants.APN_TYPE_HIPRI))) {
            return true;
        }
    }
    return false;
}
public static boolean bitmaskHasTech(int bearerBitmask, int radioTech) {
    if (bearerBitmask == 0) {//普通情况下 bearerBitmask 设置为 0
        return true;
    } else if (radioTech >= 1) {
        return ((bearerBitmask & (1 << (radioTech - 1))) != 0);//位操作进行 radioTech 匹配
    }
    return false;
}
```

第一次匹配是 ApnContext 的 mApnType 与 ApnSetting 的 String 类型数组中的信息进行匹配；第二次匹配是注册数据网络 radioTech 与 ApnSetting 配置的 bearerBitmask 进行匹配。

最终返回 APN 类型匹配、radioTech 匹配的 ApnSetting 列表。

8.4.3 DcActiveState 收尾工作

DataConnection 与 RILJ 交互，RIL 的 setupDataCall 成功之后，状态 mActivatingState 将转换为 mActiveState，根据 StateMachine 的处理机制，将调用此状态的 enter 方法，其处理逻辑可分为以下几点。

- 更新 mNetworkInfo。调用 NetworkInfo 的 setSubtype、setRoaming、setDetailedState 和 setExtraInfo 方法来更新 mNetworkInfo。
- 注册消息。通过 mPhone.getCallTracker().registerForXXX 调用，向 GsmCdmaCallTracker 注册 EVENT_DATA_CONNECTION_VOICE_CALL_STARTED 和 EVENT_DATA_CONNECTION_VOICE_CALL_ENDED 消息，注意 Handler 对象是 SmHandler。
- 保存 DataConnection 对象到 DcController.mDcListActiveByCid 列表。
- 创建 DcNetworkAgent 对象。

我们重点解析创建 DcNetworkAgent 对象的业务逻辑，DcNetworkAgent 是 DataConnection 的内部私有类，其代码逻辑的详情如下：

```
private class DcNetworkAgent extends NetworkAgent {
    public DcNetworkAgent(Looper l, Context c, String TAG, NetworkInfo ni,
            NetworkCapabilities nc, LinkProperties lp, int score, NetworkMisc misc) {
        super(l, c, TAG, ni, nc, lp, score, misc);
    }
}
```

NetworkAgent 构造方法中的处理逻辑，将建立 NetworkAgent 与 ConnectivityService 之间的 mTrackerHandler AsyncChannel 消息通道。

DataConnection 的 DcNetworkAgent 对象究竟有什么作用呢？主要是在网络状态和能力信息发生改变时，使用 mNetworkAgent.sendLinkProperties、mNetworkAgent.sendNetworkCapabilities 和 mNetworkAgent.sendNetworkInfo 调用，通过 AsyncChannel 消息通道向 ConnectivityService 发送网络信息变化的最新信息。

ConnectivityService 不在本书的介绍范围，感兴趣的读者可自行学习。

8.4.4　Suspend 挂起状态

在 DcActiveState 的 enter 处理逻辑中，向 GsmCdmaCallTracker 注册 registerForVoiceCallStarted 和 VoiceCallEnded 的消息；在开启移动数据的情况下，DcActiveState 将响应 VoiceCall 变化的消息回调，其处理逻辑的详情如下：

```
case EVENT_DATA_CONNECTION_VOICE_CALL_STARTED:
case EVENT_DATA_CONNECTION_VOICE_CALL_ENDED: {
    if (updateNetworkInfoSuspendState() && mNetworkAgent != null) {
        mNetworkAgent.sendNetworkInfo(mNetworkInfo);//发送更新后的 NetworkInfo 对象
    }
    retVal = HANDLED;
    break;
}
```

updateNetworkInfoSuspendState 中的方法通过 GsmCdmaCallTracker 获取当前通话状态，将设置两个状态。

- 通话非 IDLE 状态设置为 NetworkInfo.DetailedState.SUSPENDED。
- IDLE 状态设置为 NetworkInfo.DetailedState.CONNECTED。

8.4.5　查看手机上网基本信息

Nexus 6P 手机成功开启移动数据业务后可查看上网的基本信息，详情如下：

```
$ adb shell ifconfig
rmnet_data0 Link encap:UNSPEC
            inet addr:10.71.177.232  Mask:255.255.255.240
            inet6 addr: fe80::d653:279c:46a1:e7d/64 Scope: Link
            UP RUNNING  MTU:1500  Metric:1
            RX packets:56 errors:0 dropped:0 overruns:0 frame:0
            TX packets:108 errors:0 dropped:0 overruns:0 carrier:0
            collisions:0 txqueuelen:1000
            RX bytes:41855 TX bytes:9708
lo          Link encap:Local Loopback
            inet addr:127.0.0.1  Mask:255.0.0.0
            inet6 addr: ::1/128 Scope: Host
            UP LOOPBACK RUNNING  MTU:65536  Metric:1
            RX packets:0 errors:0 dropped:0 overruns:0 frame:0
            TX packets:0 errors:0 dropped:0 overruns:0 carrier:0
```

```
            collisions:0 txqueuelen:0
            RX bytes:0 TX bytes:0
$ adb shell ip route
10.71.177.224/28 dev rmnet_data0  proto kernel  scope link  src 10.71.177.232
```

8.5 关闭移动数据业务

关闭移动数据业务的逻辑相对开启过程要简单一些，总结如图 8-10 所示。

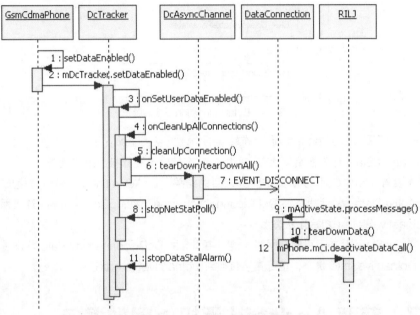

图 8-10　关闭移动数据业务流程

在图 8-10 所示的关闭移动数据业务流程中，重点关注以下几点。
- 步骤 4：onCleanUpAllConnections 方法响应移动数据业务关闭请求。
- 步骤 6：循环 mApnContext 通过 ApnContext 对象获取到对应的 DcAsyncChannel，发起 tearDown 或 tearDownAll 请求。
- 步骤 7：发出 EVENT_DISCONNECT 或 EVENT_DISCONNECT_ALL 消息请求。
- 步骤 9：DataConnection SmHandler 对象响应此消息，交给当前状态 mActiveState 处理。
- 步骤 10：DataConnection.tearDownData 方法中的关键处理逻辑是调用 mPhone.mCi.deactivate-DataCall，回调 Message 消息类型是 EVENT_DEACTIVATE_DONE。

8.6　DataConnection 状态转换

DataConnection 类中定义了六个状态的内部类，并分别创建了六个 State 对象。在数据连接状态转换的过程中，仅限 DcDefaultState 的五个子 State 对象之间的转换，总结如图 8-11 所示。

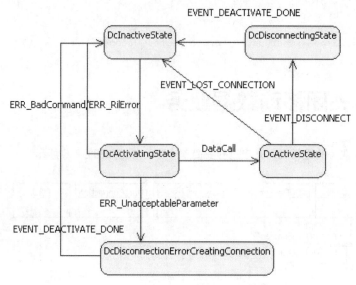

图 8-11　DataConnection 状态转换图

在图 8-11 所示的 DataConnection 状态转换图中，重点关注以下几点。
- DcInactiveState 作为初始化状态，所有的状态转换最终都将回到此状态。
- 根据设置 Data Call 移动数据业务返回的结果。DcActivatingState 可转换成三个状态：DcActiveState（激活状态）、DcDisconnectionErrorCreatingConnection（连接网络失败）和 DcInactiveState（初始化状态）。
- DcActiveState 状态下可断开网络连接，中间会由 DcDisconnectingState 状态过渡，最终回到 DcInactiveState 状态，也可直接转换到 DcInactiveState 状态。

8.7　获取 Android 手机上网数据包

Android 平台是基于并运行在 Linux Kernel 之上的操作系统平台，因此在开发、调试手机上网的过程中，tcpdump 工具的使用以及分析上网数据包的过程是必不可少的。那么，在 Android 平台下是如何操作的呢？可分为以下两个操作步骤。
- 获取上网数据包。
- Wireshark 分析数据。

8.7.1　使用 tcpdump 工具抓取 TCP/IP 数据包

要抓取 TCP/IP 数据包，当然离不开 tcpdump 工具，此工具是基于 Linux 平台的网络抓包工具，可以将网络中传送的数据包"头"完全截获下来并提供分析。

抓取 TCP/IP 数据包的过程非常简单，直接使用 Android 提供的操作命令即可完成，操作详情如下。

```
//抓取 TCP 数据包，将数据文件保存在虚拟设备的/sdcard/目录下的 tcp.pcap 文件中，可修改数据文件保存的目录
$ adb shell tcpdump -i any -p -s 0 -w /sdcard/tcp.pcap
//将抓取的 TCP 数据包路径复制到计算机当前目录下
$ adb pull /sdcard/ tcp.pcap .
```

可通过帮助或上网查找相关资料了解 tcpdump 相关参数。在执行此命令后，可以使用 Nexus 6P 手机访问百度网站，访问网络的 TCP/IP 数据包将保存在/sdcard/tcp.pcap 文件中。

8.7.2 使用 Wireshark 软件分析 TCP/IP 数据包

前面已经抓取了 Nexus 6P 手机访问网络的 TCP 数据包文件 tcp.pcap，用普通的文本编辑器打开此文件，会显示乱码，但可以使用 Wireshark 软件打开 tcp.pcap 文件。软件的安装命令如下。

```
//安装 Wireshark 分析工具
$ sudo apt-get install Wireshark
```

使用 Nexus 6P 手机访问百度网站，抓取的 TCP/IP 数据包中的关键信息如图 8-12 所示。

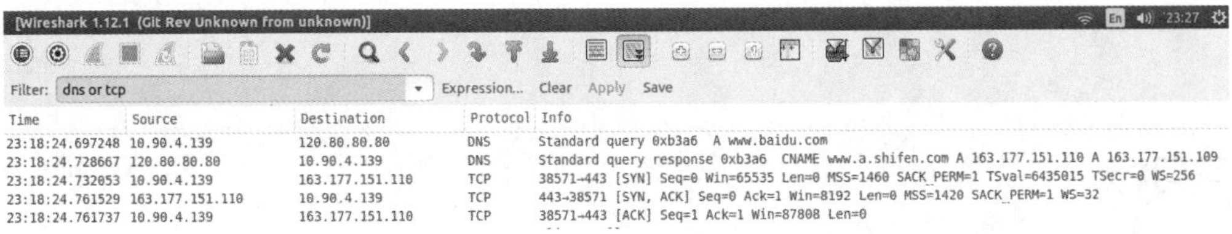

图 8-12　Wireshark 解析出的 TCP/IP 关键信息

从图 8-12 所示的 Wireshark 解析出的 TCP/IP 关键信息，可知访问百度网站分成如下两个步骤。

- 解析域名

手机端的本地 IP 地址是 10.90.4.13，DNS 服务器解析出百度网站的域名有两个 IP 地址：163.177.151.110 和 163.177.151.109。

- TCP/IP 三次握手

10.90.4.13 与 163.177.151.110（本地与远端服务器）之间的 TCP/IP 三次握手连接状态：SYN、SYN&ACK 和 ACK，完成建立 Socket 连接操作。

本 章 小 结

本章主要解析 Data Call 移动数据业务，请读者掌握以下几点。

- StateMachine 状态机的运行原理和机制，以及在 DataConnection 类中的应用。其中重点理解 State 设计模式和 Chain of Responsibility 职责链模式，以及 SmHandler 消息转换和处理机制。
- 开启或关闭 Data Call 移动数据业务的关键业务流程和运行机制。
- DcTracker 的运行机制与 GsmCdmaCallTracker 和 ServiceStateTracker 的运行机制非常相似，它们之间的最大不同在于，DcTracker 能直接接收和响应 GsmCdmaCallTracker 和 ServiceStateTracker 发出的 Handler 消息，例如移动数据连接挂起状态。

第 9 章

SMS&MMS 业务

学习目标

- 掌握短信发送和接收流程。
- 理解短信应用中 Action 原理及运行机制。
- 掌握 MMS Data Call 移动数据业务流程。
- 掌握彩信发送和接收流程。

前面已经重点分析和学习了 Telephony 业务模型中的三大业务：Voice Call 语音通话、ServiceState 网络服务、Data Call 移动数据业务，Telephony 业务模型中还包括使用非常多的短消息业务（Short Messaging Service，SMS）和彩信业务（Multimedia Messaging Service，MMS）。

本章将从这两个业务的流程入手，重点解析 SMS 和 MMS 的实现机制及关键业务流程。

- 发送 MO（Mobile Originate）
- 接收 MT（Mobile Terminate）

9.1 短信发送流程

短信业务由发送短信和接收短信两个最常见的业务场景组成。本节将解析发送短信的业务流程，通过阅读和分析源代码并结合 Nexus 6P 手机使用过程中对应的日志输出，重点解析短信发送业务的关键流程。

9.1.1 进入短信应用

首先，需要找到短信应用的代码入口，打开 Nexus 6P 手机，在主界面点击短信图标，便可打开短信应用进入到短信会话列表展示界面，同时查看 Android 系统事件 event 日志，便可找到进入短信应用对应的代码，系统事件 event 日志详情如下。

```
I/am_activity_launch_time(807):[0,172132693,com.android.messaging/.ui.conversationlist.
ConversationListActivity,362,362]
```

通过上面的日志信息可以非常直观地看出，打开短信应用进入的短信会话列表界面是 ConversationListActivity，对应的 Java 代码是 ConversationListActivity.java，短信应用 package 包路径为 com.android.messaging。

查看 ConversationListActivity.java 文件的路径，可以发现短信应用的代码库是：packages/apps/Messaging/，短信会话列表界面在 AndroidManifest.xml 中的配置信息详情如下：

```
<activity
    android:name=".ui.conversationlist.ConversationListActivity"
    android:configChanges="orientation|screenSize|keyboardHidden"
    android:screenOrientation="user"
    android:label="@string/app_name"
    android:theme="@style/BugleTheme.ConversationListActivity">
    <intent-filter>
        <action android:name="android.intent.action.MAIN" />
        <category android:name="android.intent.category.LAUNCHER" />
        <category android:name="android.intent.category.DEFAULT" />
        <category android:name="android.intent.category.APP_MESSAGING" />
    </intent-filter>
</activity>
```

这里配置的 Activity 名称是.ui.conversationlist.ConversationListActivity，再结合 package 包定义：package="com.android.messaging"，可知与 ConversationListActivity 类的 package 包路径完全匹配：com.android.messaging.ui.conversationlist.ConversationListActivity。

注意

ConversationListActivity.java 显示了短信会话列表界面，Nexus 6P 手机默认是没有一条短信的，可试着接收几条短信，便可看见此界面 Activity 的运行效果。

9.1.2 短信编辑界面

打开 ConversationListActivity.java 短信会话列表界面，点击屏幕右下角的新建短信按钮，便可进入短信编辑界面，输入短信的接收方电话号码以及短信的内容，点击右下角的发送按钮即可发出短信，短信编辑界面的主要日志和代码详情如下：

```
I/am_activity_launch_time(807):0,96371999,com.android.messaging/.ui.conversation.
ConversationActivity,642,642]

conversationFragment = new ConversationFragment();
fragmentTransaction.add(R.id.conversation_fragment_container,
        conversationFragment, ConversationFragment.FRAGMENT_TAG);
```

短信编辑界面是 ConversationActivity，在其 onCreate 方法中调用 updateUiState 来创建 ConversationFragment；在其对应的 layout 文件 conversation_fragment.xml 中并未找到发送短信按钮相关 View。分析其处理逻辑发现 onCreateView 方法中 mComposeMessageView 的构造逻辑如下：

```
mComposeMessageView = (ComposeMessageView)
        view.findViewById(R.id.message_compose_view_container);
mComposeMessageView.bind(DataModel.get().createDraftMessageData(
        mBinding.getData().getConversationId()), this);
```

在 ConversationMessageView 中可找到发送短信按钮 View，代码详情如下：

```
mSendButton = (ImageButton) findViewById(R.id.send_message_button);
```

```java
mSendButton.setOnClickListener(new OnClickListener() {
    @Override
    public void onClick(final View clickView) {
        sendMessageInternal(true /* checkMessageSize */);
    }
});
```

点击发送短信按钮,其响应事件是调用 ConversationMessageView 类的 sendMessageInternal 方法,关键处理逻辑的详情如下:

```java
final String messageToSend = mComposeEditText.getText().toString();
mBinding.getData().setMessageText(messageToSend);
final String subject = mComposeSubjectText.getText().toString();
mBinding.getData().setMessageSubject(subject);
mBinding.getData().checkDraftForAction(checkMessageSize,mHost.getConversationSelfSubId(),
    new CheckDraftTaskCallback() {
        @Override
        public void onDraftChecked(DraftMessageData data, int result) {
            mBinding.ensureBound(data);
            switch (result) {
                case CheckDraftForSendTask.RESULT_PASSED:
                    final MessageData message = mBinding.getData()
                            .prepareMessageForSending(mBinding);
                    if (message != null && message.hasContent()) {
                        playSentSound();
                        mHost.sendMessage(message);
                        ......
                    }
                    break; //省略检查未通过的异常处理
                case CheckDraftForSendTask.RESULT_XXX;break;
            }
        }
    }, mBinding);
```

上面的代码逻辑,可重点关注以下几点:

- mBinding.getData()获取 DraftMessageData 对象。
- 更新 DraftMessageData 对象的 mMessageText 和 mMessageSubject 属性,即短信内容和短信接收者信息。
- CheckDraftTaskCallback 内部匿名类对象实现了回调的接口 onDraftChecked。
- checkDraftForAction 逻辑。首先创建 DraftMessageData 类的内部类对象 CheckDraftForSendTask,它继承了 SafeAsync Task;接着调用此对象的 executeOnThreadPool 方法触发重写父类的三个方法调用 onPreExecute、doInBackgroundTimed 和 onPostExecute,这几个方法的处理逻辑是发送短信的前置条件判断,最终通过 mCallback.onDraftChecked 调用将判断结果发送给 CheckDraftTaskCallback 对象。
- onDraftChecked 逻辑。首先调用 prepareMessageForSending 创建 MessageData 对象,然后调用 mHost.sendMessage (message),即 ConversationFragment 类的 sendMessage 方法开始发送 Message。

跟踪 ConversationFragment 类的 sendMessage 方法的处理逻辑,将继续调用 mBinding.getData().sendMessage(mBinding, message)发送 Message。再次进入 ConversationData 的 sendMessage 方法,其关键处理逻辑详情如下:

```java
InsertNewMessageAction.insertNewMessage(message, systemDefaultSubId);
```

InsertNewMessageAction 继承自 Action 类，先解析 Messaging 应用中 Action 的运行机制后，再接着分析短信发送流程。

9.1.3　Action 处理机制

在 Messaging 代码库中，package com.android.messaging.datamodel.action 中有 37 个与 Action 相关的类，总结 Action 核心类如图 9-1 所示。

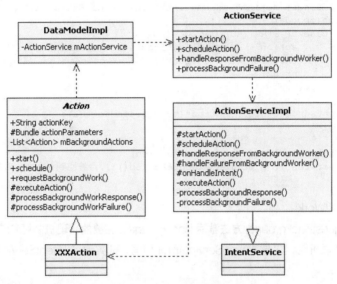

图 9-1　Action 关键类图

图 9-1 所示的 Action 关键类图，需要重点关注以下几点：
- Action 抽象类

Action 抽象类实现了 Parcelable 接口，有 28 个子类，提供的 start 和 schedule 方法将调用 DataModel 的静态方法 startActionService 和 scheduleAction，传入的参数是 Action 对象。
- DataModel

DataModelImpl 作为普通 Java 类，继承自 DataModel 抽象类，并实现了其抽象方法；其 startAction-Service 和 scheduleAction 方法逻辑通过 mActionService 发起 startAction 和 scheduleAction 调用。
- ActionService

ActionService 作为普通 Java 类，提供了五个 ActionServiceImpl 静态方法的代理调用。
- ActionServiceImpl

ActionServiceImpl 继承自 IntentService 类，其提供的五个静态方法中的主要处理逻辑是启动 ActionServiceImpl 服务，onHandleIntent 则返回后台执行耗时的异步任务。以 startAction 为例，其处理逻辑的详情如下：

```
protected static void startAction(final Action action) {
    //创建 Intent，关注 Action 和 EXTRA_OP_CODE
    final Intent intent = makeIntent(OP_START_ACTION);
    actionBundle.putParcelable(BUNDLE_ACTION, action);//保存 Action 对象
    intent.putExtra(EXTRA_ACTION_BUNDLE, actionBundle);
    startServiceWithIntent(intent);
}
```

```java
private static void startServiceWithIntent(final Intent intent) {
    intent.setClass(context, ActionServiceImpl.class);//指定启动 IntentService 为当前类
    if (context.startService(intent) == null) {//启动 IntentService
        sWakeLock.release(intent, opcode);
    }
}
@Override
protected void onHandleIntent(final Intent intent) {//IntentService 异步响应
    final int opcode = intent.getIntExtra(EXTRA_OP_CODE, 0);//OP_START_ACTION
    try {
        Action action;
        final Bundle actionBundle = intent.getBundleExtra(EXTRA_ACTION_BUNDLE);
        switch(opcode) {
            case OP_START_ACTION: {
                action = (Action) actionBundle.getParcelable(BUNDLE_ACTION);
                executeAction(action);//调用 executeAction，传入 Action 对象
                break;
            }......
        }
        action.sendBackgroundActions(mBackgroundWorker);
    }
}
private void executeAction(final Action action) {
    final Object result = action.executeAction();//最终回到 Action 对象的 executeAction
}
```

- BackgroundWork

Action 通过 requestBackground 方法激活 Background，主要的实现机制可参考 WorkBackgroundWorkerService 的处理机制，它与 ActionServiceImpl 相同，都实现了 IntentService。请读者自行分析其处理逻辑。

总结 Action 的 start 处理逻辑和关键流程，如图 9-2 所示。

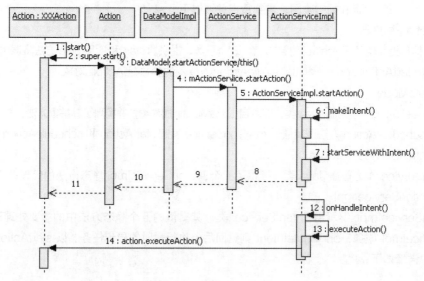

图 9-2 Action start 流程

图 9-2 所示的 Action start 流程，可重点关注以下两点。

- start 方法调用过程。总结 Action 对象的 start 方法调用过程，详情如下：Action.start → DataModel.startAction Service → ActionService.startAction → ActionServiceImpl.startAction →

startServiceWithIntent→startService→onHandleIntent→executeAction→Action.executeAction。
- 从步骤 1 到步骤 14，Action.start 调用最终回到 Action.executeAction 调用，为什么会有这样的设计呢？

Action.executeAction 是运行在后台的异步任务，使用了 Android IntentService 的运行机制，而 Messaging 应用使用 Action，并不关心如何异步以及后台调度等处理机制。

9.1.4 继续跟进短信发送流程

继续分析 InsertNewMessageAction 对象的 insertNewMessage 方法，其中关键的业务逻辑详情如下。

```
public static void insertNewMessage(final MessageData message, final int subId) {
    Assert.isFalse(subId == ParticipantData.DEFAULT_SELF_SUB_ID);
    final InsertNewMessageAction action = new InsertNewMessageAction(message, subId);
    action.start();
}
```

上面的代码逻辑创建了 InsertNewMessageAction 对象并调用其 start 方法。根据 Action 的处理机制，将以后台异步方式调用 InsertNewMessageAction 对象的 executeAction 方法，主要有两个处理逻辑：

- 将 message 信息插入数据库表中。
- 调用 ProcessPendingMessagesAction.scheduleProcessPendingMessagesAction。

出现另外一个 Action 对象：ProcessPendingMessagesAction，继续分析其静态方法 scheduleProcessPendingMessagesAction，关键的处理逻辑详情如下：

```
final ProcessPendingMessagesAction action = new ProcessPendingMessagesAction();
if (action.queueActions(processingAction)) {
    ......return;
}
private boolean queueActions(final Action processingAction) {
    final String toSendMessageId = findNextMessageToSend(db, now);
    final String toDownloadMessageId = findNextMessageToDownload(db, now);
    if (toSendMessageId != null) {
        if (!SendMessageAction.queueForSendInBackground(
                toSendMessageId, processingAction)) {
            succeeded = false;
        }
    }......
}
```

上面的代码逻辑，可重点关注以下几点：

- 在 ProcessPendingMessagesAction 的静态方法中创建 ProcessPendingMessagesAction 并调用其 queueActions 方法，并未触发 Action 异步执行后台任务。
- 第三个 Action 类：SendMessageAction。

SendMessageAction 类的静态方法 queueForSendInBackground 将创建 SendMessageAction 对象并调用 queueAction 方法，与 ProcessPendingMessagesAction.scheduleProcessPendingMessagesAction 的调用过程非常相似。

SendMessageAction 对象的 queueAction 方法有以下两个主要处理逻辑：

- 从数据库读取 MessageData
- processingAction.requestBackgroundWork(this)调用

SendMessageAction 对象的 doBackgroundWork 方法在后台执行耗时的异步任务，关键的处理逻辑详情如下：

```
status = MmsUtils.sendSmsMessage(recipient, messageText, messageUri, subId,
            smsServiceCenter, deliveryReportRequired);
```

短信发送流程跟踪到此，做一下阶段性总结，如图 9-3 所示。

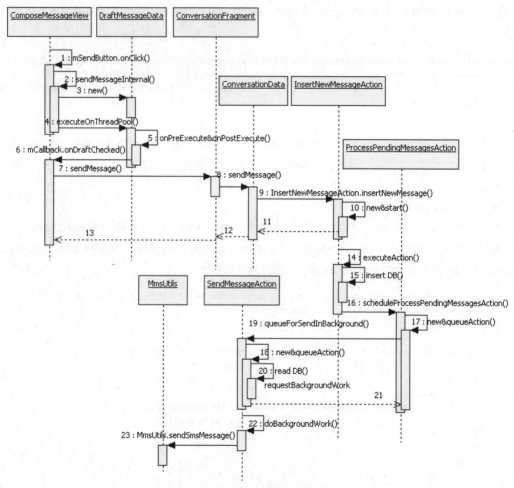

图 9-3　短信发送流程阶段性总结

图 9-3 所示的短信发送阶段性流程总结，可重点关注以下几点：

- 步骤 5：短信发送的前置条件判断。
- 步骤 10：InsertNewMessageAction 触发 IntentService Action 处理，步骤 14 在后台异步响应 Action 请求。
- 步骤 16 到步骤 17：ProcessPendingMessagesAction 中的 Action 处理都是同步的方法调用。
- 步骤 22：在后台异步响应 Action 请求，最终发起 MmsUtils.sendSmsMessage 调用（步骤 23）。

MmsUtils.sendSmsMessage 在 Messaging 应用中的处理流程，总结如图 9-4 所示。

第 9 章 SMS&MMS 业务

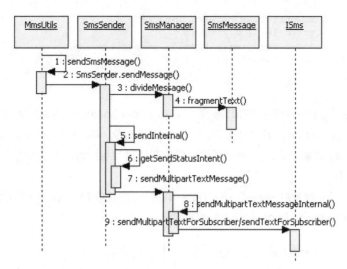

图 9-4 MmsUtils.sendSmsMessage 流程总结

图 9-4 所示的 MmsUtils.sendSmsMessage 流程总结，可重点关注以下几点。

- 拆分长短信

如果短信中的内容超过 160 个字节，将拆分成多条短信进行发送，步骤 3 和步骤 4 的返回值是 ArrayList<String>，即长短信分割后的短信内容列表。

感兴趣的读者可测试步骤 4 中 fragmentText 的短信拆分逻辑，拆分后的第一条短信长度没有达到一条短信的最大长度，这是由于拆分后的短信头中需要保存其短信的序列号，因此短信头增加了长度。而且长短信不能按照普通短信发送，否则，短信的接收端无法将长短信按照拆分后的短信序列号顺序重新拼接成为一条长短信。

- 创建回调 PendingIntent

步骤 6：创建 MESSAGE_DELIVERED_ACTION 和 MESSAGE_SENT_ACTION 两个 Intent 对象，class 为 SendStatusReceiver，最后添加到 deliveryIntents 和 sentIntents PendingIntent 中。

- 短信和长短信两个处理分支

步骤 9：调用 ISms 服务的 sendMultipartTextForSubscriber 接口发送长短信，调用 sendTextForSubscriber 接口发送普通短信，判断条件是步骤 3 返回 ArrayList<String>的 size，大于 1 发送长短信，否则发送正常短信。

- SmsManager 代码空间和运行空间

frameworks/base/telephony/java/android/telephony/SmsManager.java 在 framework 代码库中，供所有 Android 应用使用，此处运行在 Messaging 应用空间。它提供了短信拆分、短信发送、将短信复制到 SIM 卡上、从 SIM 卡上删除短信和小区广播等操作接口。

9.1.5 phone 进程中的短信发送流程

ISms 系统服务运行在 com.android.phone 进程空间，其接口定义文件是：frameworks/base/telephony/java/com/android/internal/telephony/ISms.aidl。

frameworks/opt/telephony/src/java/com/android/internal/telephony/UiccSmsController.java 实现了 ISms 接口，代码逻辑详情如下：

```
public class UiccSmsController extends ISms.Stub {
protected UiccSmsController() {//构造方法
    if (ServiceManager.getService("isms") == null) {
        ServiceManager.addService("isms", this);//增加服务名为 isms 的系统服务
    }......
}
```

isms 系统服务是在什么时候完成加载的呢？是在 PhoneFactory.makeDefaultPhone 逻辑中，通过 ProxyController.getInstance 调用创建 ProxyController 对象，再在 ProxyController 的构造方法中创建 UiccSmsController 对象。因此，isms 系统服务是在加载 Telephony 业务模型时，同步完成的初始化和系统服务的发布。

UiccSmsController 对象的 sendTextForSubscriber 方法响应 Messaging 应用发起的发送短信请求调用。总结 com.android.phone 进程中的响应逻辑，详情如图 9-5 所示。

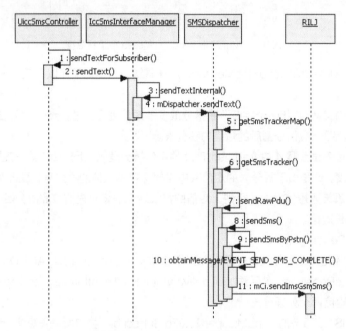

图 9-5　phone 进程短信发送流程

图 9-5 所示的 phone 进程短信发送流程，可重点关注以下几点：

● SMSDispatcher

SMSDispatcher 是抽象类，主要有 getFormat、sendText、sendSms 和 sendSmsByPstn 等抽象方法，其三个子类分别是 CdmaSMSDispatcher、GsmSMSDispatcher 和 ImsSMSDispatcher，通过名字便能知道，它们分别作为 Cdma、Gsm 和 Ims 类型的短信分发器。

SMSDispatcher 的处理步骤，除了步骤 4、步骤 8 和步骤 9 是在子类中完成调用逻辑，其他都是由父类中提供的方法实现。

● 创建 SmsTracker

步骤 5 和步骤 6 创建 SmsTracker 对象，重点关注 mData、mSentIntent、mDeliveryIntent 等属性。mData 是步骤 5 根据 destAddr、scAddr、text 和 pdu（Protocol Data Unit，短信协议数据单元）等参数，创建的 HashMap；mSentIntent 和 mDeliveryIntent 则是 Messaging 应用传过来的回调 PendingIntent 对象。

SmsTracker 跟踪短信发送情况，其注释为 "Keeps track of an SMS that has been sent to the RIL"。
- sendSmsByPstn

sendSmsByPstn 作为发送短信流程中 SMSDispatcher 对象最后调用的方法，主要有两个处理逻辑，分别在步骤 10 和步骤 11：创建 EVENT_SEND_SMS_COMPLETE 类型的 Message 对象；mCi.sendSMS 向 RILJ 发起发送短信接口调用。

至此，发送短信的业务流程已经跟踪到 RILJ 对象，其业务流程和 Voice Call 语音通话非常相似，短信发送请求从应用层分多个步骤传递到 Telephony Framework 框架层，Telephony 框架层根据短信发送信息创建 PDU、SmsTracker 等对象，最终调用 RILJ 对象 sendSMS 方法，将发送短信请求转换成 RIL 的处理逻辑。

9.2 扩展短信发送业务

从 Messaging 应用层到 RIL，我们已经对短信发送流程做了详细的解析。本节将从以下三个方面扩展短信发送业务。
- 确认短信发送结果
- 重发机制
- 状态报告

9.2.1 确认短信发送结果

SMSDispatcher 调用 mCi.sendSMS 发送短信，回调的 Message 对象是 SMSDispatcher 创建的 EVENT_SEND_SMS_COMPLETE 类型 Message。

RIL 与 Modem 完成短信发送的交互，Modem 通过网络将短信发送给运营商的短信中心并将结果反馈给 RIL，RILJ 对象使用 Message 消息回调完成短信发送结果的返回。因此，SMSDispatcher 作为自定义 Handler 消息处理对象，在 handleMessage 方法中响应 EVENT_SEND_SMS_COMPLETE 类型 Message 消息，其处理逻辑的详情如下：

```
@Override
protected void sendSmsByPstn(SmsTracker tracker) {
    HashMap<String, Object> map = tracker.getData();
    byte smsc[] = (byte[]) map.get("smsc");
    byte[] pdu = (byte[]) map.get("pdu");
    Message reply = obtainMessage(EVENT_SEND_SMS_COMPLETE, tracker);
    //发送短信
    mCi.sendSMS(IccUtils.bytesToHexString(smsc),
                IccUtils.bytesToHexString(pdu), reply);
}
@Override
public void handleMessage(Message msg) {
    switch (msg.what) {
    case EVENT_SEND_SMS_COMPLETE://发送短信结果回调
        handleSendComplete((AsyncResult) msg.obj);
        break;
    ......
}
```

上面的代码逻辑中，重点关注创建 Message 对象的两个参数：EVENT_SEND_SMS_COMPLETE 和 SmsTracker 对象。在 SMSDispatcher 对象的 handleMessage 方法中调用 handleSendComplete 方法，传入的参数 msg.obj 对象中将保留 SmsTracker 对象。handleSendComplete 的处理逻辑详情如下：

```java
protected void handleSendComplete(AsyncResult ar) {
    SmsTracker tracker = (SmsTracker) ar.userObj;
    PendingIntent sentIntent = tracker.mSentIntent;

    if (ar.exception == null) {
        tracker.onSent(mContext);
    } else {......}
}
```

onSent 方法中的关键逻辑是发起 mSentIntent.send(context, Activity.RESULT_OK, fillIn) 调用，mSentIntent 作为 PendingIntent，将在 Messaging 应用中创建。关键逻辑详情如下：

```java
//创建 Intent 对象及相关参数
getSendStatusIntent(context, SendStatusReceiver.MESSAGE_SENT_ACTION,
        messageUri, partId, subId), 0)
private static Intent getSendStatusIntent(final Context context, final String action,
        final Uri requestUri, final int partId, final int subId) {
    final Intent intent = new Intent(action, requestUri, context, SendStatusReceiver.class);
    intent.putExtra(SendStatusReceiver.EXTRA_PART_ID, partId);
    intent.putExtra(SendStatusReceiver.EXTRA_SUB_ID, subId);
    return intent;
}
```

因此，mSentIntent.send 将在 Messaging 应用中由 SendStatusReceiver 接收短信发送成功消息。

9.2.2 重发机制

SMSDispatcher 对象的 handleSendComplete 方法响应 RILJ 返回的短信发送结果，该方法中有两个处理分支：

- 发送成功
- 发送异常

短信发送异常的处理逻辑详情如下：

```java
if (!isIms() && ss != ServiceState.STATE_IN_SERVICE) {
    tracker.onFailed(mContext, getNotInServiceError(ss), 0/*errorCode*/);
} else if ((((CommandException)(ar.exception)).getCommandError()
        == CommandException.Error.SMS_FAIL_RETRY) &&
        tracker.mRetryCount < MAX_SEND_RETRIES) {//重发次数未超过 3 次
    tracker.mRetryCount++;//增加重试次数
    Message retryMsg = obtainMessage(EVENT_SEND_RETRY, tracker);
    sendMessageDelayed(retryMsg, SEND_RETRY_DELAY);
} else {//超过重试次数，短信发送按失败处理
    tracker.onFailed(mContext, error, errorCode);
}
```

MAX_SEND_RETRIES 为 3，SEND_RETRY_DELAY 为 2000，即重发次数为 3，重发时间间隔为 2 秒。

总结 EVENT_SEND_RETRY 消息的响应逻辑如下：handleMessage.EVENT_SEND_RETRY → sendRetrySms → mImsSMSDispatcher.sendRetrySms → mGsmDispatcher.sendSms → sendSmsByPstn。

整个过程将传递 SmsTracker，最后调用 SMSDispatcher 的 sendSmsByPstn 方法重发短信。

9.2.3 状态报告

首先是手机通过移动网络将短信成功发送到运营商的短信中心，短信中心再经过短信路由将短信成功发送给短信接收方，并将短信成功发送到短信接收方的消息发给短信发送方，该消息就是短信发送状态报告。

因此，在 handleSendComplete 中仅能确认短信成功发送到运营商的短信中心，并无法确认对方是否已经收到；而短信发送状态报告更能真实地反映短信发送状态，表示对方已经成功接收短信。简单来说，短信发送状态报告的作用和目的是精确地标识短信已被对方正常接收。

Messaging 应用提供了短信发送状态报告的设置开关，此开关默认是关闭的，因此需要我们手动开启，操作入口是：Messaging→Settings→Advanced→SMS delivery reports。

短信发送状态报告是 MT 类型的消息，Modem 最先接收到，然后将消息分发给 RIL，再由 RILJ 对象发出 EVENT_NEW_SMS_STATUS_REPORT 类型的 Message 消息。该消息的注册和响应逻辑详情如下：

```
//RIL 消息日志
D/RILJ    ( 1525): [UNSL]< UNSOL_RESPONSE_NEW_SMS_STATUS_REPORT [SUB0]
D/RILJ    ( 1525): [3871]> SMS_ACKNOWLEDGE success = true cause = 1 [SUB0]
D/RILJ    ( 1525): [3871]< SMS_ACKNOWLEDGE    [SUB0]

//GsmSMSDispatcher 类构造方法注册 Handler 消息
mCi.setOnSmsStatus(this, EVENT_NEW_SMS_STATUS_REPORT, null);
//响应 Message 消息回调
case EVENT_NEW_SMS_STATUS_REPORT:
        handleStatusReport((AsyncResult) msg.obj);

// handleStatusReport 的关键处理逻辑
PendingIntent intent = tracker.mDeliveryIntent;
Intent fillIn = new Intent();
fillIn.putExtra("pdu", pdu);
fillIn.putExtra("format", getFormat());
try {
     intent.send(mContext, Activity.RESULT_OK, fillIn);
} catch (CanceledException ex) {}
mCi.acknowledgeLastIncomingGsmSms(true, Intents.RESULT_SMS_HANDLED, null);
```

因此，mDeliveryIntent.send 将在 Messaging 应用中由 SendStatusReceiver 接收短信状态报告，短信应用界面展示的短信发送状态将增加显示 "√"。

9.3 短信接收流程

上一节已经详细解析了短信的发送流程，本节继续解析 Android 8.1 中实现短信接收业务的关键流程。

9.3.1 RIL 接收短信消息

Nexus 6P 手机接收到一条短信，查看其 radio 日志的详情如下：

```
D/RILJ    ( 1525): [UNSL]< UNSOL_RESPONSE_NEW_SMS [SUB0]
```

RILJ 接收到 UNSOL_RESPONSE_NEW_SMS 类型的消息，找到 RadioIndication.java 中的 newSms 方法，将发出 mRil.mGsmSmsRegistrant.notifyRegistrant 消息通知。而 mGsmSmsRegistrant 的注册方为 GsmInboundSmsHandler，其构造方法中的逻辑详情如下：

```
phone.mCi.setOnNewGsmSms(getHandler(), EVENT_NEW_SMS, null);

public void setOnNewGsmSms(Handler h, int what, Object obj) {
    mGsmSmsRegistrant = new Registrant (h, what, obj);
}
```

9.3.2 GsmInboundSmsHandler

GsmInboundSmsHandler 继承自抽象类 InboundSmsHandler，而 InboundSmsHandler 又继承自 StateMachine 类。既然实现了 StateMachine 状态机，首先查看一下其初始化业务逻辑，详情如下：

```
protected InboundSmsHandler(String name, Context context, SmsStorageMonitor
storageMonitor, Phone phone, CellBroadcastHandler cellBroadcastHandler) {
    super(name);
    addState(mDefaultState);
    addState(mStartupState, mDefaultState);
    addState(mIdleState, mDefaultState);
    addState(mDeliveringState, mDefaultState);
    addState(mWaitingState, mDeliveringState);

    setInitialState(mStartupState);
}
```

共有五个状态对象：mDefaultState、mStartupState、mIdleState、mDeliveringState 和 mWaitingState，设置的初始化状态是 mStartupState。

总结 GsmInboundSmsHandler 的初始化过程，构造方法的调用过程如下：GsmCdma Phone→mTelephonyComponentFactory.makeIccSmsInterfaceManager→new→IccSmsInterfaceManager→new→ImsSMSDispatcher→GsmInboundSmsHandler.makeInboundSmsHandler→new GsmInbound-SmsHandler。

原来，在创建 GsmCdmaPhone 对象的过程中，同步创建了 GsmInboundSmsHandler 对象，完成了 EVENT_NEW_SMS 消息注册和 StateMachine 初始化操作。

注意

在 GsmInboundSmsHandler 的初始化过程中，ImsSMSDispatcher 构造方法中容易遗漏初始化调用 SmsBroadcastUndelivered.initialize，其关键的业务逻辑是发出 InboundSms-Handler.EVENT_START_ACCEPTING_SMS 消息，将 GsmInboundSmsHandler 的状态切换到 mIdleState 状态。

因此，mIdleState 的 processMessage 方法响应 EVENT_NEW_SMS 消息，执行 deferMessage 和 transitionTo(mDeliveringState)；接着 mDeliveringState 状态将响应 EVENT_NEW_SMS 消息，调用 handleNewSms 方法。

进入 handleNewSms 方法继续跟踪接收到短信的处理逻辑，可以发现以下四个连续调用：

- dispatchMessage

- dispatchMessageRadioSpecific
- dispatchNormalMessage
- addTrackerToRawTableAndSendMessage

其中比较关键的业务逻辑如下：在 dispatchNormalMessage 方法中创建 InboundSmsTracker 对象；在 addTrackerToRawTableAndSendMessage 方法中调用 addTrackerToRawTable 方法记录数据库并发送 EVENT_BROADCAST_SMS 消息，其参数是 InboundSmsTracker 对象。

mDeliveringState 继续处理 EVENT_BROADCAST_SMS 消息，处理逻辑的详情如下：

```
case EVENT_BROADCAST_SMS:
    InboundSmsTracker inboundSmsTracker = (InboundSmsTracker) msg.obj;
    if (processMessagePart(inboundSmsTracker)) {
        sendMessage(EVENT_UPDATE_TRACKER, inboundSmsTracker);
        transitionTo(mWaitingState);
    } else {......}
    return HANDLED;
```

上面的代码可重点关注以下两点：
- processMessagePart 发出新短信广播
- transitionTo(mWaitingState)

在 processMessagePart 中调用 filterSms 方法来过滤接收到的短信，再发起 dispatchSmsDeliveryIntent 调用，主要逻辑是创建 Intent 对象，详情如下：

```
Intent intent = new Intent();
intent.putExtra("pdus", pdus);//短信 pdus 通过 Intent 对象发出
intent.putExtra("format", format);

intent.setAction(Intents.SMS_DELIVER_ACTION)
dispatchIntent(intent, android.Manifest.permission.RECEIVE_SMS,
        AppOpsManager.OP_RECEIVE_SMS, options, resultReceiver, UserHandle.SYSTEM);
```

dispatchIntent 调用 sendOrderedBroadcastAsUser 发出广播。

mWaitingState 的主要作用是对新短信广播超时的处理，其处理逻辑的详情如下：

```
@Override
public void enter() {
    mTracker = null;//进入此状态后开始计时，30 秒时间
    sendMessageDelayed(EVENT_STATE_TIMEOUT, STATE_TIMEOUT);
}
@Override
public void exit() {//退出此状态，不做超时处理
    removeMessages(EVENT_STATE_TIMEOUT);
    removeMessages(EVENT_UPDATE_TRACKER);
}
@Override
public boolean processMessage(Message msg) {
    switch (msg.what) {
        case EVENT_BROADCAST_COMPLETE://广播已经完成
            sendMessage(EVENT_RETURN_TO_IDLE);
            transitionTo(mDeliveringState);//回到 mDeliveringState 状态
            return HANDLED;
        case EVENT_STATE_TIMEOUT://超时处理
            if (mTracker != null) {
                dropSms(new SmsBroadcastReceiver(mTracker));
            } else {
                sendMessage(EVENT_BROADCAST_COMPLETE);
```

```
            }
            return HANDLED;
......}}
```

SmsBroadcastReceiver 接收到新短信广播后,发出 EVENT_BROADCAST_COMPLETE 类型的消息,即 mWaitingState 退出短信广播发送超时处理机制。

GsmInboundSmsHandler 接收新短信的处理逻辑中最关键的是四个 State 的转换,总结如图 9-6 所示。

图 9-6 所示的 GsmInboundSmsHandler 接收短信状态图,可重点关注以下几点:

- mIdleState 作为接收广播的初始化状态,在 ImsSMSDispatcher 构造方法中完成状态切换。
- mDeliveringState 状态的两个关键处理消息:EVENT_NEW_SMS 消息调用 handleNewSms 方法,EVENT_BROADCAST_SMS 消息则调用 processMessagePart 方法发出新短信广播通知。
- mWaitingState 的超时处理机制。

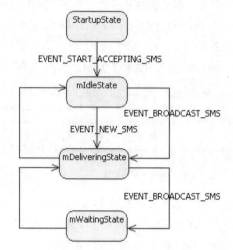

图 9-6 GsmInboundSmsHandler 接收短信状态图

注意 processMessagePart 方法经过 dispatchSmsDeliveryIntent→dispatchIntent 调用,最终通过 mContext.sendOrderedBroadcastAsUser 方式,发出 Action 为 Intents.SMS_DELIVER_ACTION 的广播,InboundSmsHandler 内部类 SmsBroadcastReceiver 对象接收此广播,并将 Action 转换为 Intents.SMS_RECEIVED_ACTION,继续调用 dispatchIntent 发出第二次广播,SmsBroadcastReceiver 对象接收到此广播后发出 EVENT_BROADCAST_COMPLETE 类型的消息,mWaitingState 退出短信广播发送超时处理机制。

因此,接收到新消息后,最终发出的广播 Action 为 Intents.SMS_RECEIVED_ACTION,即 android.provider.Telephony.SMS_RECEIVED。

9.3.3 Messaging 应用接收新短信

Messaging 应用接收到新短信后,主要有两个操作:

- 在通知栏展示接收到新短信
- 将短信内容保存到本地数据库中

Messaging 应用的 SmsReceiver 广播接收器将接收 GsmInboundSmsHandler 发出的新短信消息广播通知 android.provider.Telephony.SMS_RECEIVED。广播接收器 SmsReceiver 的定义如下:

```
<receiver android:name=".receiver.SmsReceiver"
          android:enabled="false"
          android:permission="android.permission.BROADCAST_SMS">
    <intent-filter android:priority="2147483647">
        <action android:name="android.provider.Telephony.SMS_RECEIVED" />
    </intent-filter>
    <intent-filter android:priority="2147483647">
```

```
            <action android:name="android.provider.Telephony.MMS_DOWNLOADED" />
        </intent-filter>
</receiver>
```

注意 android:enabled="false"配置，在 SmsReceiver 类中的 updateSmsReceiveHandler 方法可更新广播接收器的可用状态。

SmsReceiver 类的 onReceive 方法接收到广播后，首先调用 deliverSmsIntent 方法，接着调用 deliverSmsMessages 方法。在调用过程中，有两个关键处理逻辑：

- 解析短信内容

getMessagesFromIntent 方法首先在 intent 对象中获取 pdu 和 format 两个信息，最后调用 SmsMessage.createFromPdu(pdu, format)创建 SmsMessage 对象。

- 启动 ReceiveSmsMessageAction

ReceiveSmsMessageAction 作为 Messaging 应用中的 Action 子类，调用 start 方法，在后台激活 executeAction 异步任务，将短信保存到数据库中，并发起新短信的通知栏显示。

ReceiveSmsMessageAction 保存短信的主要逻辑详情如下：

```
db.beginTransaction();//开始事务
try {
    //插入新短信信息
    BugleDatabaseOperations.insertNewMessageInTransaction(db, message);
    //更新短信会话信息
    BugleDatabaseOperations.updateConversationMetadataInTransaction(db, conversationId,
            message.getMessageId(), message.getReceivedTimeStamp(), blocked,
            conversationServiceCenter, true /* shouldAutoSwitchSelfId */);
    db.setTransactionSuccessful();//设置事务成功处理
} finally {
    db.endTransaction();//结束事务
}
```

因为需要更新两张数据库表中的信息，所以增加了数据库事务处理机制。

接着，ReceiveSmsMessageAction 的 executeAction 方法调用 BugleNotifications.update 发起在状态栏显示新短信通知，都是在当前类中完成的操作，其调用过程的详情如下：update→createMessageNotification→processAndSend→sendNotification→fireOffNotification→doNotify。

最后，由 doNotify 方法调用 NotificationManager.notify 发出新短信的状态栏消息通知。

9.3.4 PDU

短信发送流程中，将发送的短信内容转换为 PDU 信息后再发送给 RILJ 对象，最终将短信发送给短信接收方。

短信接收流程中，RILJ 通过 Telephony Framework 框架层将接收到短信的 PDU 信息发送给 Messaging 应用，转换为接收到的短信内容后，再展示给手机用户；短信的发送和接收流程 PDU 的转换都是通过 SmsMessage 的两个静态方法完成的。

- getSubmitPdu 将文本信息转换为 pdu byte 数组。
- createFromPdu 将 pdu byte 数组转换为文本信息并创建 SmsMessage 对象。

Android 8.1 源码中主要有四个 Java 类 SmsMessage，它们之间的区别和联系如图 9-7 所示。我们可重点关注以下几点：

- 四个 SmsMessage 类，用包名来区分。

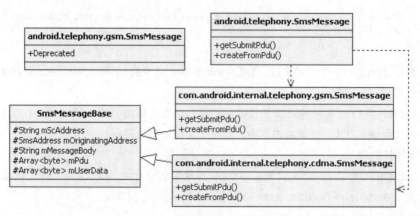

图 9-7 SmsMessage 类图

- 两个代码库。android.telephony.gsm.SmsMessage 类在 frameworks/opt/telephony 代码库中，并用@Deprecated 注解，表示已经不使用。其他三个类均在 frameworks/base/telephony/ 代码库中。
- 代理关系。根据 GSM 和 CDMA 等不同网络类型，android.telephony.SmsMessage 将作为 SmsMessageBase 两个子类的代理类。

注意

SmsMessage 类的 getSubmitPdu 和 createFromPdu 方法是按照 3GPP TS 23.040 规范实现的。感兴趣的读者可阅读《3GPP TS 23.040 V15.0.0 (2018-03)》文档。重点关注 9.2.3 Definition of the TPDU parameters。

对照《3GPP TS 23.040 V15.0.0 (2018-03)》规范，验证发送短信 PDU 第一个字节信息。

首先修改 com/android/internal/telephony/gsm/SmsMessage.java 代码中的方法 getSubmitPdu，在返回 SubmitPdu 对象前，打印其 encodedMessage 属性，它是 PDU 字节数组，可使用 IccUtils 的静态方法 bytesToHexString 将字节数组转换为十六进制字符串，并更新到 Nexus 6P 手机中。

使用 Nexus 6P 手机打开短信报告并发送一条短信，查看此短信 PDU 的第一个字节信息，十六进制数为 21，二进制数为 100001。

对应规范中最关键的两点如下：
- 9.2.3.1 SMS-SUBMIT (in the direction MS to SC)
- 9.2.3.5 A status report is requested

即发送短信类型和需要状态报告。

9.3.5 短信业务小结

前面三个小节完成了发送短信和接收短信业务的解析和关键流程的学习，总结 Telephony Framework 层短信业务的核心类，如图 9-8 所示。

图 9-8 所示的短信关键类图，可重点关注以下两点：

- SMSDispatcher 负责短信发送业务，是自定义的 Handler 类型，有三个子类分别承载 Gsm、Cdma 和 Ims 三种不同网络类型的短信发送业务。
- InboundSmsHandler 负责短信接收业务，虽然名字带有 Handler，但它实现了 StateMachine 状态机，由五个状态协作完成短信接收业务。

> > > > > > > > > > 第 9 章　SMS&MMS 业务

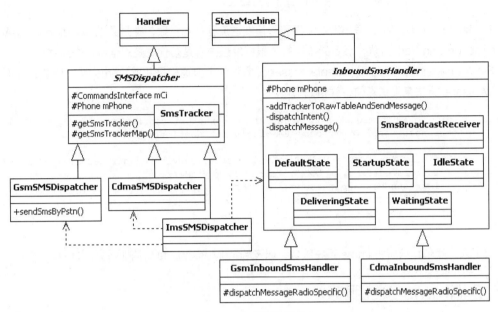

图 9-8　短信关键类图

9.4　彩信关键业务逻辑

短信业务只能收发文本信息，若要分享手机中的音乐、照片、视频等多媒体信息，可使用运营商提供的彩信业务来完成；彩信业务 MMS 可看作短信业务 SMS 的扩展和升级，它突破了短信业务仅能收发文本信息的限制，可发送和接收多媒体信息。

彩信业务是如何实现收发多媒体信息的呢？它会和短信业务一样先通过 RIL 再通过 Modem 接收和发送文本信息吗？彩信业务的设计基于 WAP 协议，作为 WAP 协议层上的网络应用，因此，在收发彩信的时候，可以传输多媒体信息。

9.4.1　彩信发送入口

SmsManager 提供了短信的发送入口，同样 sendMultimediaMessage 提供了彩信的发送入口，在此方法中打印 Java 调用堆栈并触发彩信发送业务，加入的代码及运行时 Java 的调用堆栈信息的详情如下：

```
android.util.Log.e("XXX", "sendMultimediaMessage", new RuntimeException());

sendMultimediaMessage
java.lang.RuntimeException
        at android.telephony.SmsManager.sendMultimediaMessage(SmsManager.java:1168)
        at android.support.v7.mms.MmsManager.sendMultimediaMessage(
                MmsManager.java:141)
        at com.android.messaging.sms.MmsSender.sendMms(MmsSender.java:169)
        at com.android.messaging.sms.MmsSender.sendMms(MmsSender.java:69)
        at com.android.messaging.sms.MmsUtils.sendMmsMessage(MmsUtils.java:2110)
```

```
        at com.android.messaging.datamodel.action.SendMessageAction.doBackgroundWork(
            SendMessageAction.java:250)
```

在 Messaging 应用中，短信发送流程和彩信发送流程基本一致，都是通过 SendMessageAction 的 doBackgroundWork 方法在后台异步执行，根据发送内容是短信或是彩信有两个处理分支。

SmsManager 的 sendMultimediaMessage 方法的主要逻辑是获取 imms 系统服务，并调用其 sendMessage 方法来发送彩信，详情如下：

```
public void sendMultimediaMessage(Context context, Uri contentUri, String locationUrl,
        Bundle configOverrides, PendingIntent sentIntent) {
    try {
        final IMms iMms = IMms.Stub.asInterface(ServiceManager.getService("imms"));
        android.util.Log.e("XXX", "sendMultimediaMessage", new RuntimeException());
        iMms.sendMessage(getSubscriptionId(), ActivityThread.currentPackageName(),
                contentUri, locationUrl, configOverrides, sentIntent);
    } catch (RemoteException e) {......}
}
```

接下来，我们继续解析 imms 系统服务，以及 sendMessage 接口的业务处理逻辑。

9.4.2 imms 系统服务

imms 系统服务的接口定义文件是 frameworks/base/telephony/java/com/android/internal/telephony/IMms.aidl，它的接口实现程序是 packages/services/Mms/src/com/android/mms/service/MmsService.java。在 packages/services/Mms 代码库中，Android.mk 编译脚本的关键信息详情如下：

```
include $(CLEAR_VARS)
LOCAL_PACKAGE_NAME := MmsService
LOCAL_PRIVILEGED_MODULE := true
LOCAL_JAVA_LIBRARIES := telephony-common okhttp
LOCAL_SRC_FILES := $(call all-java-files-under, src)
LOCAL_RESOURCE_DIR := $(LOCAL_PATH)/res
LOCAL_CERTIFICATE := platform
LOCAL_PRIVILEGED_MODULE := true
include $(BUILD_PACKAGE)
```

packages/services/Mms 代码库中将编译出系统签名的 MmsService.apk 系统应用，我们统一称为 MmsService。在 AndroidManifest.xml 中的定义如下：

```
<manifest xmlns:android="http://schemas.android.com/apk/res/android"
        xmlns:androidprv="http://schemas.android.com/apk/prv/res/android"
        package="com.android.mms.service"
        coreApp="true"
        android:sharedUserId="android.uid.phone">
    <uses-permission android:name="android.permission.RECEIVE_BOOT_COMPLETED" />
......
    <application android:label="MmsService"
            android:process="com.android.phone"
            android:usesCleartextTraffic="true">
        <service android:name=".MmsService"
            android:enabled="true"
            android:exported="true"/>
    </application>
</manifest>
```

上面的配置信息，我们重点关注以下几点：

- 包名 com.android.mms.service。
- 运行资源信息：运行在 com.android.phone 进程空间，用户 id 是 android.uid.phone，即 radio 用户。
- MmsService 服务：MmsService 应用中仅有 com.android.mms.service.MmsService 服务，并无其他 Activity、Broadcast 等定义，其运行代码框架总结如下：

```
public class MmsService extends Service implements MmsRequest.RequestManager {
    @Override
    public IBinder onBind(Intent intent) {
        return mStub;
    }
    private IMms.Stub mStub = new IMms.Stub() {
        @Override
        public void sendMessage() {}
        ......
    }
}
```

IMms 接口实现了两个类：MmsServiceBroker 和 MmsService。imms 系统服务是 MmsServiceBroker 服务在 system_server 系统进程启动过程中加载的，在 onStart 方法中调用 publishBinderService("imms", new BinderService())增加了名字为 imms 的系统服务，BinderService 实现了 IMms.Stub,实现的接口处理逻辑是代理 MmsService 服务对应的接口。首先绑定 MmsService 服务获取 Client Binder 对象，再通过 Binder 对象调用代理 MmsService 服务提供的服务接口。

感兴趣的读者可自行学习 MmsServiceBroker 代理 MmsService 服务的实现方式。

9.4.3 彩信发送流程

MmsService 服务接口 sendMessage 响应发送彩信调用，主要有两个处理逻辑：

- 创建 SendRequest
- addSimRequest→addToRunningRequestQueueSynchronized

addToRunningRequestQueueSynchronized 的调用逻辑的详情如下：

```
private void addToRunningRequestQueueSynchronized(final MmsRequest request) {
    mRunningRequestCount++;
    mRunningRequestExecutors[queue].execute(new Runnable() {
        @Override
        public void run() {
            try {
                request.execute(MmsService.this, getNetworkManager(request.getSubId()));
            } finally {......}
        }
    });
}
```

上面的代码，我们重点关注以下几点：

- mRunningRequestExecutors

它是 ExecutorService 对象数组，数组长度固定为 2，下标 QUEUE_INDEX_SEND 和 QUEUE_INDEX_DOWNLOAD 分别代表发送彩信和接收彩信的类型。

- Runnable

重写了 run 方法并调用 MmsRequest 对象的 execute 方法。

- 初始化 MmsNetworkManager 对象

调用 getNetworkManager 方法将创建 MmsNetworkManager 对象，并保存在 mNetworkManager Cache 缓存列表中。在创建 MmsNetworkManager 对象的逻辑中，最关键的是创建 NetworkRequest 对象，详情如下：

```
mNetworkRequest = new NetworkRequest.Builder()
        .addTransportType(NetworkCapabilities.TRANSPORT_CELLULAR)
        .addCapability(NetworkCapabilities.NET_CAPABILITY_MMS)
        .setNetworkSpecifier(Integer.toString(mSubId))
        .build();
```

接下来，我们继续分析 MmsRequest 对象的 execute 方法中的业务逻辑，详情如下：

```
for (int i = 0; i < RETRY_TIMES; i++) {//重试三次发送彩信
    try {
        networkManager.acquireNetwork(requestId);//请求网络
        final String apnName = networkManager.getApnName();
        try {
            ApnSettings apn = null;
            try {//加载彩信 ApnSetting 信息
                apn = ApnSettings.load(context, apnName, mSubId, requestId);
            } catch (ApnException e) {......}
            response = doHttp(context, networkManager, apn);//http 发送彩信内容
            result = Activity.RESULT_OK;
            break;
        } finally {//释放网络
            networkManager.releaseNetwork(requestId, this instanceof DownloadRequest);
        }
    } catch (ApnException e) {......}
}
```

上面的代码逻辑定义了彩信的发送业务流程，可分为以下四个步骤。

（1）acquireNetwork 请求 MMS 移动数据业务。

（2）ApnSettings.load 加载 MMS 类型的 APN 设置信息。

（3）doHttp 创建 http 连接发送彩信内容。

（4）releaseNetwork 释放网络资源。

接一来，重点分析 acquireNetwork 和 doHttp 处理逻辑。

9.4.4　Data Call

继续跟踪 acquireNetwork 请求 MMS 类型的 Data Call 移动数据业务，处理逻辑的详情如下：

```
public void acquireNetwork(final String requestId) throws MmsNetworkException {
    synchronized (this) {//同步锁
        if (mNetworkCallback == null) {
            startNewNetworkRequestLocked();//开始请求网络
        }
        while (waitTime > 0) {
            try {
                this.wait(waitTime);//等待
            } catch (InterruptedException e) {}
            if (mNetwork != null) {//已经获取 MMS 数据连接
```

```
                    // Success
                    return;//成功返回
                }
            }
            // Timed out, so release the request and fail
            releaseRequestLocked(mNetworkCallback);//超时处理
            throw new MmsNetworkException("Acquiring network timed out");
        }
    }
```

上面的代码逻辑中最核心的是 startNewNetworkRequestLocked 调用，请求 MMS 类型的 Data Call 移动数据业务，this.wait 将等待数据连接的返回，处理逻辑的详情如下：

```
private void startNewNetworkRequestLocked() {
    final ConnectivityManager connectivityManager = getConnectivityManager();
    mNetworkCallback = new NetworkRequestCallback();
    connectivityManager.requestNetwork(mNetworkRequest, mNetworkCallback,
            NETWORK_REQUEST_TIMEOUT_MILLIS);
}

private class NetworkRequestCallback extends ConnectivityManager.NetworkCallback {
    @Override
    public void onAvailable(Network network) {
        super.onAvailable(network);
        synchronized (MmsNetworkManager.this) {
            mNetwork = network;
            MmsNetworkManager.this.notifyAll();
        }
    }......
}
```

对 connectivityManager.requestNetwork 的调用，可重点关注以下几点：

- ConnectivityService 系统服务响应 requestNetwork 接口调用。
- 参数 mNetworkRequest 请求 NetworkCapabilities.NET_CAPABILITY_MMS 类型的数据连接。
- MMS Data Call 的执行。

ConnectivityService 接收到 mNetworkRequest 请求后，调用 sendUpdatedScoreToFactories 方法，遍历 mNetworkFactoryInfos 列表发送网络请求，其代码逻辑的详情如下：

```
for (NetworkFactoryInfo nfi : mNetworkFactoryInfos.values()) {
    nfi.asyncChannel.sendMessage(android.net.NetworkFactory.CMD_REQUEST_NETWORK,
            score, 0, networkRequest);
}
```

请读者思考，nfi.asyncChannel.sendMessage 使用 AsyncChannel 发送的 Handler 消息，在什么地方被接收呢？

答案是 com.android.phone 进程中 TelephonyNetworkFactory 对象的 needNetworkFor 方法响应。根据上一章学习的内容，发起 DcTracker 对象的 requestNetwork 调用，从而更新 MMS 类型 ApnContext 对象的可用状态为 true，最终调用 trySetupData，创建 MMS 类型的 Data Call 移动数据业务。

DataConnection 成功发起 MMS 类型 Data Call 进入 DcActiveState 状态，创建 DcNetworkAgent 对象，向 ConnectivityService 发起 registerNetworkAgent 请求，经过 ConnectivityService 服务的处理，最终调用 notifyNetworkCallbacks，由 NetworkRequestCallback 对象接收网络请求的结果。

NetworkRequestCallback 响应网络请求回调。更新 mNetwork 并发出 MmsNetworkManager.

this.notifyAll，结束 MmsNetworkManager 等待。

MmsNetworkManager 对象的 acquireNetwork 调用有两种处理结果：MMS 数据连接成功建立，返回 void；抛出异常。

总结 acquireNetwork 请求 MMS 类型的 Data Call 移动数据业务的核心处理流程，如图 9-9 所示。

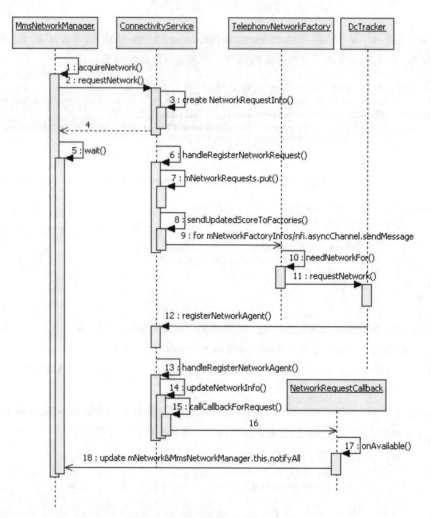

图 9-9　acquireNetwork 核心流程

图 9-9 所示的 acquireNetwork 核心流程，我们重点关注以下几点：

- 步骤 4、步骤 5、步骤 18 的同步处理机制

步骤 2：发起 MMS 类型的 Data Call 移动数据业务请求后，ConnectivityService 经过内部 Handler 消息，转换为异步处理，步骤 4 返回请求调用，进入步骤 5 等待状态。

- 创建 Data Call

步骤 11：DcTracker 首先更新类型为 MMS 的 ApnContext 状态为可用，接着调用 setupDataCall 向 RILJ 发起创建 MMS 类型的 Data Call 移动数据业务请求。

- 成功创建 MMS 数据连接

DataConnection 进入 DcActiveState 状态，通过创建 DcNetworkAgent 对象发起步骤 12 的调用。

- ConnectivityService 发起 requestNetwork 请求回调

步骤 14、步骤 15：使用 mNetworkRequests 列表，匹配出 requestNetwork 时的请求 Message 信息，发出 CallBack 调用。

注意，requestNetwork 请求时将保存 NetworkRequestInfo 对象到 mNetworkRequests 列表中（步骤 7）。

- ConnectivityService 交互

MmsNetworkManager 与 ConnectivityService 的交互，中间经过了 ConnectivityManager 的中转，请读者关注 HashMap<NetworkRequest, NetworkCallback> sCallbacks 中间的消息转换。

- NetworkRequestCallback 回调

NetworkRequestCallback 作为 MmsNetworkManager 的内部私有类，在步骤 18 中更新 NetworkRequestCallback 对象的 mNetwork 属性，并发起 MmsNetworkManager.this.notifyAll()调用；最终结束步骤 5 的等待，完成 acquireNetwork 调用返回。

9.4.5 doHttp

MmsRequest 作为抽象类，它的两个子类 SendRequest 和 DownloadRequest 分别实现了 doHttp 抽象方法，这两个子类分别承载发送彩信和接收彩信的处理逻辑。

SendRequest 类的 doHttp 方法将创建 MmsHttpClient 对象并执行其 execute 方法，该方法中的关键逻辑详情如下：

```
public byte[] execute(String urlString, byte[] pdu, String method, boolean isProxySet,
        String proxyHost, int proxyPort, Bundle mmsConfig, int subId, String requestId)
            throws MmsHttpException {
    HttpURLConnection connection = null;
    try {
        final URL url = new URL(urlString);//彩信 APN 配置的 URL 地址
        connection = (HttpURLConnection) mNetwork.openConnection(url, proxy);
        if (METHOD_POST.equals(method)) {
            final OutputStream out =
                    new BufferedOutputStream(connection.getOutputStream());
            out.write(pdu);//向彩信服务器地址通过 http 写入发送的彩信内容
            out.flush();
            out.close();
        } else if (METHOD_GET.equals(method)) ......//彩信接收处理逻辑
    } catch ()......
    } finally {
        if (connection != null) {
            connection.disconnect();
        }
    }
}
```

上面的代码框架是一个非常标准的基于 Http 网络的消息处理机制，同时也说明了彩信业务是基于 MMS 类型的 Data Call 移动数据业务的网络应用。

9.4.6 接收彩信

向 Nexus 6P 手机发送一条彩信，InboundSmsHandler 的 DeliveringState 将首先接收到一条比较特殊的短信，在调用 processMessagePart 方法时，若判断接收的端口号是 2948，则按照 WAP Push

的方式处理。

 注意 WAP Push 叫作服务信息或推入信息，是一种特殊格式的短信，其类型和功能比较多，如 SI、SL 类型的 WAP Push，或者 OMA OTA（通过 WAP Push 更新本地 APN 配置）等功能。

接着调用 WapPushOverSms.dispatchWapPdu 方法，将创建 Action 为 Intents.WAP_PUSH_DELIVER_ACTION 类型的 Intent 对象，再次调用 InboundSmsHandler 对象的 dispatchIntent 发出广播通知，SmsBroadcastReceiver 接收到此广播，将 Action 转换为 Intents.WAP_PUSH_RECEIVED_ACTION 类型消息，第二次调用 dispatchIntent 发出接收到 WAP Push 消息的广播通知。即 android.provider.Telephony.WAP_PUSH_RECEIVED 类型的广播，根据短信接收流程的解析，可知 Messaging 应用中的 MmsWapPushReceiver 将接收此广播，并激活 ReceiveSmsMessageAction 的 executeAction 后台异步处理机制。MmsWapPushReceiver 的配置信息详情如下：

```
<receiver android:name=".receiver.MmsWapPushReceiver"
          android:enabled="false"
          android:permission="android.permission.BROADCAST_WAP_PUSH">
    <intent-filter android:priority="2147483647">
        <action android:name="android.provider.Telephony.WAP_PUSH_RECEIVED" />
        <data android:mimeType="application/vnd.wap.mms-message" />
    </intent-filter>
</receiver>
```

此广播接收器默认为不可用状态，在 SmsReceiver 类的 updateSmsReceiveHandler 静态方法中，可更改其为可用状态。

ReceiveSmsMessageAction 中的 executeAction 处理逻辑主要由两个部分构成。
- 记录数据库
- 调用 ProcessPendingMessagesAction.scheduleProcessPendingMessagesAction

在 Messaging 应用的设置中，可开启或关闭自动下载新彩信选项。Messaging 应用接收到一条新的彩信后，首先检查此配置信息，有两个处理分支分别负责自动下载和手动下载彩信内容；因此在 scheduleProcessPendingMessagesAction 方法中也对应有两个处理分支。
- 显示彩信基本信息，提供界面让用户手动下载彩信内容。
- 激活 DownloadMmsAction 下载彩信内容。

两个处理分支只是激活方式不同，最终的结果都是调用 imms 系统服务的 downloadMessage 进入运营商的彩信服务器，通过 Http 下载彩信内容。

9.4.7 MmsService 小结

MmsService 应用运行在 com.android.phone 进程空间，承载着彩信发送和接收业务，是基于网络的应用，其核心类总结如图 9-10 所示。

图 9-10 所示的 MmsService 关键类图，可重点关注以下几点：
- MmsService。提供了发送和接收彩信的接口，imms 服务的实现是 MmsServiceBroker，它代理 MmsService 服务。
- MmsRequest 模板类。实现了发送或接收彩信关键四个步骤的调度模板，首先请求 MMS 类型的 Data Call 移动数据业务，接着加载 ApnSetting，然后通过 http 下载或发送彩信内容，有 Send 和 Download 两个子类，最后由 releaseNetwork 释放 Data Call 移动数据业务。

第 9 章 SMS&MMS 业务

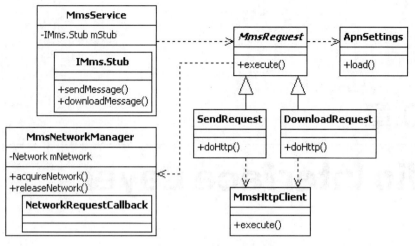

图 9-10 MmsService 关键类图

- MmsNetworkManager 负责请求或释放 MMS 类型的 Data Call 移动数据业务。
- MmsHttpClient 负责 http 数据传输，实现了数据的上传和下载。

本 章 小 结

本章主要解析短信和彩信的实现机制及关键业务流程，请读者掌握以下几点。

- 短信的发送和接收流程跨越了 Messaging 应用、Telephony Framework 和 RIL，最后由 RIL 与 Modem 交互完成短信的发送和接收。
- 掌握 SMSDispatcher 发送短信的处理流程和 InboundSmsHandler 状态机接收短信的处理流程。
- 掌握 Messaging 应用中的 Action 处理机制。
- 彩信是基于 Data Call 移动数据业务的网络应用，掌握 ConnectivityService 的处理机制，完成彩信移动数据连接的建立以及 RequestNetwork 的消息回调。

第 10 章
Radio Interface Layer

学习目标

- 掌握 RILJ 运行机制。
- 掌握 RILC 加载流程和关键运行机制。
- 掌握 RILJ 与 RILC 的交互机制。
- 理解 Solicited 和 UnSolicited 消息关键处理机制。
- 理解 RIL 系统架构。

Radio Interface Layer（无线通信接口层，RIL）运行在 Android 硬件抽象层（HAL）之上，其实现的源代码可分为两大部分。

- Telephony RILJ
- HAL 层中的 C/C++ 程序统一称为 RILC

RILC 的代码在 hardware/ril/目录下的关键代码信息，如图 10-1 所示。

如图 10-1 所示的 RILC 代码结构，可分为四个部分：

- 头文件

在 ril.h 头文件中，主要定义了以 RIL_ 开头的 RIL_Init、RIL_register、RIL_onRequestComplete 等函数和 RIL_Env、RIL_RadioFunctions 等结构体，以及 RIL_REQUEST_XXX 和 RIL_UNSOL_XXX 等 RIL 消息。

- libril

查看 Android.mk 编译脚本，将编译输出 libril.so 动态链接库，其中主要有 ril_commands.h 和 ril_unsol_commands.h 头文件，以及 ril.cpp 和 ril_service.cpp 两个重要的 C++代码文件。

- reference-ril

根据 Android.mk 编译脚本，将编译输出 libreference-ril.so 动态库，其依赖 libril。作为 Android 源码中 ril 的参考实现，实现了 RIL AT 命令的交互机制；该目录下的 ril.h 头文件，是一个链接文件。

- rild

rild.c 文件编译出 rild 可执行文件，其依赖 libril。

第 10 章 Radio Interface Layer

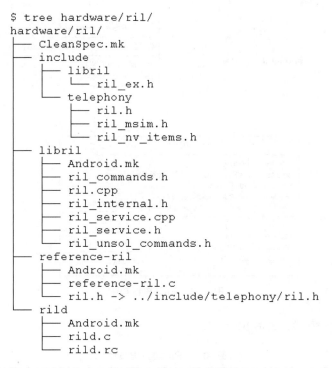

图 10-1 RILC 代码结构

RILC 将编译出 rild 可执行文件和 libril.so 和 libreference-ril.so 两个动态库。RILC 这三个部分的代码根据输出结果，本书统一称为 rild、libril 和 reference-ril。

10.1 解析 RILJ

结过前面几章的学习可知，RILJ 对象与 Telephony 业务模型产生最直接的交互，完成 Telephony 核心业务；作为 RIL 的消息入口，首先解析和学习 RILJ 的运行机制和关键业务模型。

10.1.1 认识 RIL 类

与 RILJ 相关的代码共有三个，分别是 CommandsInterface.java、BaseCommands.java 和 RIL.java，package 包路径统一是 com.android.internal.telephony，如图 10-2 所示。

图 10-2 所示的 RILJ 核心类图，我们重点关注以下几点：

- RIL 类继承自 BaseCommands 抽象类并实现了 CommandsInterface 接口。RIL 类的对象统一称为 RILJ 对象，Telephony 业务模型中三大 Tracker 和 SMSDispatcher 等对象的 mCi 属性，就是 RILJ 对象，ci 即 CommandsInterface 的缩写。
- BaseCommands 抽象类 mXXXRegistrants 和 mXXXRegistrant 保存监听 CallBack 的消息对象，提供注册和取消注册的管理方法 registerXXX 和 unregisterXXX。
- RIL 类有一个内部类 RILRequest，作为 RIL 请求对象的封装类，重点关注 mSerial 请求序列号、mRequest 请求类型和 mResult 消息回调对象等属性，提供静态方法 obtain 来统一创建 RILRequest 对象。

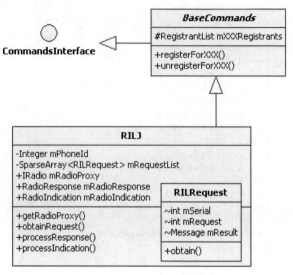

图 10-2　RILJ 核心类图

总结 RIL 类的关键属性，如表 10-1 所示。

表 10-1　RIL 类关键属性

属性	类型	描述
mPhoneId	Integer	PhoneId
mRilVersion	int	RIL 版本号
mState	RadioState	RADIO_OFF RADIO_UNAVAILABLE RADIO_ON
mXXXRegistrants	RegistrantList	39 个 mXXXRegistrants
mXXXRegistrant	Registrant	20 个 Registrant
mRequestList	SparseArray<RILRequest>	RILRequest RIL 请求对象列表
mRadioProxy	IRadio	IRadio 服务 Client
mRadioResponse	RadioResponse	RadioResponse 服务对象
mRadioIndication	RadioIndication	RadioResponse 服务对象

表 10-1 所示的 RIL 类关键属性，我们重点关注以下几点：

- mRequestList 列表保存内部类 RILRequest 对象。
- RILJ 的消息处理主要集中在 39 个 RegistrantList 对象和 20 个 Registrant 对象，其实现逻辑在父类 BaseCommands 中；RILJ 对象在接收到 RILC 发出的消息后，上述对象负责发出相应的 Handler 消息通知。
- RadioState 枚举类型在 CommandsInterface 接口中定义，共有三种状态的定义。
- IRadio、RadioResponse 和 RadioIndication 都是基于 Service 服务的处理框架，涉及跨进程的服务访问。

总结 RIL 类的关键方法，如表 10-2 所示。

第 10 章 Radio Interface Layer

表 10-2 RIL 类关键方法

分类	接口信息	描述
控制接口	dial/acceptCall/rejectCall/switchWaitingOrHoldingAndActive	通话控制相关接口
	setupDataCall/deactivateDataCall	移动数据业务控制接口
	setRadioPower/startNetworkScan	网络服务控制
	sendSMS/sendCdmaSms	发送短信接口
查询接口	getBasebandVersion	获取基带版本
	getCurrentCalls	查询 Voice Call 列表
	getDataCallList	查询 Data Call 列表
	getDataRegistrationState	获取数据网络驻网状态
	getSignalStrength	查询信号
消息处理	getRadioProxy	初始化 IRadio 相关服务
	obtainRequest	创建 RILRequest 对象
	addRequest	保存 RILRequest 对象
	processResponse	处理 RIL 请求返回结果
	processIndication	处理 RIL 上报信息

表 10-2 所示的 RIL 类关键方法，我们关注以下几点：

- RILJ 提供的接口主要由控制、查询和消息处理三大类组成，表中仅列出了较常见的接口。
- RILRequest 对象处理的相关接口：obtainRequest 封装了创建 RILRequest 对象的过程，addRequest 将保存 RILRequest 对象到 mRequestList 列表中，还有 findAndRemoveRequestFromList、clearRequestList 等方法也是用来处理 RILRequest 对象的。

10.1.2 RILRequest

RILJ 接收到 Telephony 业务模型发出的接口调用后将请求封装成 RILRequest 对象，主要保存请求消息类型以及 Callback 回调的 Message 消息对象等信息，可见 RILRequest 对象的重要程度。

总结 RILRequest 类的关键属性，代码详情如下。

```
static Random sRandom = new Random();
static AtomicInteger sNextSerial = new AtomicInteger(0);//下一个 RILRequest 对象编号

int mSerial;//RILRequest 对象唯一编号
int mRequest;//RIL 请求类型
Message mResult;//RIL 请求回调的 Message 对象
RILRequest mNext; //下一个 RILRequest 处理对象
```

RILRequest 类的关键属性可分为两大部分：

- 静态属性，为所有 RILRequest 对象共享。
- 普通成员变量，为 RIL 类请求对象独享。

重点掌握 RIL 请求类型 mRequest 属性和 RIL 请求回调的 Message 对象的 mResult 属性。Message 对象保存了请求参数，RIL 处理完 RIL 请求后，使用此对象发出 Message 消息通知，RIL 请求方即可接收到发出请求的响应结果。

RILRequest 类有两个非常关键的方法：obtain 和 onError。

RILRequest 类中重载了 obtain 方法，一个是 private static obtain 方法，将创建 RILRequest 对象；另一个是 package static obtain 方法，作为 private static obtain 方法的代理，提供给 RILJ 对象调用。

创建 RILRequest 对象的处理逻辑，其代码详情如下。

```
private static RILRequest obtain(int request, Message result) {
    RILRequest rr = null;
    synchronized(sPoolSync) {//同步锁
        if (sPool != null) {
            rr = sPool; //获取缓冲池中的RILRequest对象
            sPool = rr.mNext;
            rr.mNext = null;
            sPoolSize--;
        }
    }
    if (rr == null) {
        rr = new RILRequest();//如果没有缓冲对象时，创建RILRequest对象
    }
    rr.mSerial = sNextSerial.getAndIncrement();//AtomicInteger自动加1
    rr.mRequest = request;//保存RIL请求类型
    rr.mResult = result; //保存Message对象
    ......
    return rr;
}
```

上面的代码逻辑中，RILRequest 类的私有静态方法 obtain 用来创建或获取 RILRequest 对象，重点关注以下三点：

- synchronized(sPoolSync)同步机制保障了 sPool 对象处理的正确性。
- mSerial 标识唯一的 RILRequest 对象，使用 AtomicInteger 自动加 1 取值。
- mRequest 和 mResult 作为 RILRequest 对象非常关键的属性，通过 obtain 方法传入的参数进行初始化操作。

RIL 请求返回异常或失败，将调用 RILRequest 对象的 onError 方法，其处理逻辑的详情如下。

```
void onError(int error, Object ret) {
    CommandException ex;
    ex = CommandException.fromRilErrno(error);//通过异常编号找到CommandException

    if (mResult != null) {
        AsyncResult.forMessage(mResult, ret, ex);//创建Message对象
        mResult.sendToTarget();//发送Message消息
    }
}
```

上面的代码逻辑，我们重点关注以下两点：

- 调用 CommandException.fromRilErrno 方法获取 CommandException 对象，逻辑非常简单，即根据异常编号构造不同 Error 类型的 CommandException 对象。
- 根据异常编号及对应的 CommandException 对象创建 AsyncResult 对象，更新 mResult 消息对象的 obj 属性，最后发出 Message 消息通知。

10.1.3　IRadio 关联的服务

android.hardware.radio.V1_0 包下的众多类，其中最关键的是 IRadio 类，其代码文件在 Android

源码编译输出的 out 目录下，详细的路径是：out/target/common/gen/JAVA_LIBRARIES/android.hardware.radio-V1.0-java-static_intermediates，可反推出 hardware/interfaces/radio/1.0/Android.mk 编译脚本中定义的 MODULE 的详情如下：

```
LOCAL_MODULE := android.hardware.radio-V1.0-java
LOCAL_MODULE := android.hardware.radio-V1.0-java-static
```

hardware/interfaces/radio/1.0 代码库下的文件列表的详情如下：

```
Android.bp
Android.mk
IRadio.hal
IRadioIndication.hal
IRadioResponse.hal
ISap.hal
ISapCallback.hal
types.hal
```

上面的代码列表中主要是.hal 类型文件，它是什么类型呢？

HIDL（HAL interface definition language，硬件抽象层接口定义语言）与 AIDL（Android interface definition language）类似。AIDL 的实现基于 Android Binder 跨进程服务访问，用来定义 Android 基于 Binder 通信的 Client 与 Service 之间的接口；HIDL 同样定义基于 Binder 通信的 Client 与 Service 之间的接口，用来定义与 HAL 层的交互接口，以及支持 Java、C/C++接口程序的生成。

注 意
Android 8.1 开放了 aidl 和 hidl 两个工具的源码，都在 system/tools 目录下。

接着查看 IRadio.hal 文件中的接口定义，都可以在 CommandsInterface 中找到对应的接口，如 dial、setupDataCall、getCurrentCalls 和 getVoiceRegistrationState 等接口。重点关注 IRadio.hal 中定义的第一个接口 setResponseFunctions，详情如下：

```
setResponseFunctions(IRadioResponse radioResponse,
        IRadioIndication radioIndication);
```

涉及另外两个接口 RadioIndication 和 IRadioResponse，对应的 HIDL 接口描述文件分别是 IRadioIndication.hal 和 IRadioResponse.hal。

继续分析 Android.mk 中的关键信息，主要有两个方面的定义：
● 编译实体类
● 编译接口文件

根据 types.hal 中的结构定义编译出对应的实体类 Java 程序。以 RadioResponseInfo 为例，定义详情如下：

```
struct RadioResponseInfo {
    RadioResponseType type;    //返回类型
    int32_t serial;            //请求序列号
    RadioError error;          //返回的异常信息
};
enum RadioResponseType : int32_t {
    SOLICITED,
    SOLICITED_ACK,
    SOLICITED_ACK_EXP,
};
```

```
enum RadioError : int32_t {
    NONE = 0,                        // 成功无异常
    RADIO_NOT_AVAILABLE = 1,  //Radio 未启动或重置中
    ......
}
```

上面的定义信息可分为结构体 struct 和枚举 enum 两种类型，并且可以嵌套使用，RadioResponseInfo 在 Android.mk 中的编译脚本的详情如下。

```
#指定生成的 Java 代码文件
GEN := $(intermediates)/android/hardware/radio/V1_0/RadioResponseInfo.java
$(GEN): $(HIDL)//使用 HIDL 工具
$(GEN): PRIVATE_HIDL := $(HIDL)
$(GEN): PRIVATE_DEPS := $(LOCAL_PATH)/types.hal//依赖 types.hal 定义
$(GEN): PRIVATE_OUTPUT_DIR := $(intermediates)
$(GEN): PRIVATE_CUSTOM_TOOL = \
        $(PRIVATE_HIDL) -o $(PRIVATE_OUTPUT_DIR) \
        -Ljava \    //Java 类型
        -randroid.hardware:hardware/interfaces \
        -randroid.hidl:system/libhidl/transport \
        android.hardware.radio@1.0::types.RadioResponseInfo//指定包路径和类名
$(GEN): $(LOCAL_PATH)/types.hal//来源
        $(transform-generated-source)
LOCAL_GENERATED_SOURCES += $(GEN)
```

编译出的 RadioResponseInfo.java 文件有三个 public int 属性：type、serial 和 error，与 types.hal 中的定义 struct RadioResponseInfo 相对应；并且生成了 readFromParcel 和 writeToParcel 等方法，可使用 HwParcel 进行序列化和反序列化操作。

编译接口文件。以 IRadio 为例，Android.mk 中的定义详情如下。

```
#指定生成的 Java 代码文件
GEN := $(intermediates)/android/hardware/radio/V1_0/IRadio.java
$(GEN): $(HIDL)
$(GEN): PRIVATE_HIDL := $(HIDL)
#依赖关系 IRadioIndication.hal、IRadioResponse.hal 和 types.hal
$(GEN): PRIVATE_DEPS += $(LOCAL_PATH)/IRadioIndication.hal
$(GEN): $(LOCAL_PATH)/IRadioIndication.hal
......
$(GEN): PRIVATE_OUTPUT_DIR := $(intermediates)
$(GEN): PRIVATE_CUSTOM_TOOL = \
        $(PRIVATE_HIDL) -o $(PRIVATE_OUTPUT_DIR) \
        -Ljava \    //Java 类型
        -randroid.hardware:hardware/interfaces \
        -randroid.hidl:system/libhidl/transport \
        android.hardware.radio@1.0::IRadio//指定包路径和类名
$(GEN): $(LOCAL_PATH)/IRadio.hal
        $(transform-generated-source)
LOCAL_GENERATED_SOURCES += $(GEN)
```

编译出的 IRadio.java 文件的代码结构详情如下：

```
package android.hardware.radio.V1_0;
public interface IRadio extends android.hidl.base.V1_0.IBase {
    public static final String kInterfaceName = "android.hardware.radio@1.0::IRadio";
    public static IRadio getService(String serviceName) throws
        android.os.RemoteException {
            return IRadio.asInterface(android.os.HwBinder.getService(
                "android.hardware.radio@1.0::IRadio",serviceName));
    }
```

```
        public static final class Proxy implements IRadio {......}
        public static abstract class Stub extends android.os.HwBinder implements IRadio
{......}
}
```

上面的代码框架,我们重点关注以下几点:
- 继承 android.hidl.base.V1_0.IBase
- 静态方法 getService 可获取对应服务调用的 Client 对象
- Proxy 作为代理,实现了 IRadio 接口
- 抽象类 Stub 同样实现了 IRadio 接口,并继承自 android.os.HwBinder 类

android.hardware.radio-V1.0-java 生成的 Java 接口文件主要是 IRadio、IRadioResponse 和 IRadioIndication。请读者对比 hidl 与 aidl 生成 Java 接口文件的异同。

Telephony 业务模型的编译脚本中定义了与 android.hardware.radio-V1.0-java 的依赖关系,详情如下:

```
LOCAL_STATIC_JAVA_LIBRARIES := android.hardware.radio-V1.1-java-static
```

创建 RILJ 对象时,将同步初始化 IRadio、IRadioResponse 和 IRadioIndication 三个关键的接口服务。

在 RIL 类的构造方法中,同步创建 mRadioRespons 和 mRadioIndication 两个对象,详情如下:

```
mRadioResponse = new RadioResponse(this);
mRadioIndication = new RadioIndication(this);

public class RadioResponse extends IRadioResponse.Stub
public class RadioIndication extends IRadioIndication.Stub
```

RadioResponse 和 RadioIndication 两个类分别继承自 IRadioResponse 和 IRadioIndication 两个接口类的内部 Stub 接口类;IRadioResponse 和 IRadioIndication 两个接口服务,将运行在 com.android.phone 进程空间,与 RILJ 对象相互引用。

在 RIL 类的构造方法中,调用 getRadioProxy 方法完成三个接口的初始化操作,关键逻辑详情如下:

```
private IRadio getRadioProxy(Message result) {
    if (mRadioProxy != null) {//已经初始化,直接返回 mRadioProxy
        return mRadioProxy;
    }
    try {
        //根据 mPhoneId 获取对应的 IRadio 系统服务
        mRadioProxy = IRadio.getService(HIDL_SERVICE_NAME[
            mPhoneId == null ? 0 : mPhoneId]);
        if (mRadioProxy != null) {
            mRadioProxy.linkToDeath(mRadioProxyDeathRecipient,
                mRadioProxyCookie.incrementAndGet());//Binder 异常的处理
            //设置回调的两个服务
            mRadioProxy.setResponseFunctions(mRadioResponse, mRadioIndication);
        } else {......}
    } catch (RemoteException | RuntimeException e) {......}
    if (mRadioProxy == null) {......}
    return mRadioProxy;
}
```

上面的代码逻辑中有两个非常关键的处理:

- IRadio.getService 获取 IRadio 服务对象
- 设置回调的两个服务对象 IRadioResponse 和 IRadioIndication

因此，在创建 RILJ 对象时，将同步完成 IRadio 相关服务的初始化操作。

10.1.4　RIL 消息分类

RIL 中的数据交互方式，根据消息来源并按照其处理方式可分为两大类：
- Solicited
- UnSolicited

常用的如 dial 拨号、answer 接听电话、hangup 挂断电话等这些 AP 侧主动请求的操作，称为 Solicited 消息。Solicited 请求类的 RIL 消息，根据其消息传递方向，可分为 Solicited Request 和 Solicited Response 两种；正常情况下，Solicited Request 与 Solicited Response 消息成对出现，请求和应答是一一对应的。

BP 侧主动上报的消息，如通话状态变化、新短信、新彩信、基站信息等，称为 UnSolicited 消息。非请求类的 RIL 消息只有上报流程。

Android 8.1 源代码支持 143 个标准 Solicited 类消息，消息编号从 1~143；支持 49 个标准 UnSolicited 消息，消息编号从 1000 到 1049；可在 frameworks/base/telephony/java/com/android/internal/telephony/RILConstants.java 程序文件中找到它们的定义详情，都是 RIL_REQUEST_XXX 和 RIL_UNSOL_XXX 类型的常量定义。

10.1.5　Solicited Request

表 10-2 中，RILJ 对象提供两类与 RIL 请求相关的接口：控制和查询接口，这些接口的处理逻辑非常相似。以 getDataRegistrationState 查询当前移动数据网络注册信息为例，代码逻辑的详情如下。

```
public void getDataRegistrationState(Message result) {
    IRadio radioProxy = getRadioProxy(result); //获取 IRadio 服务调用对象
    if (radioProxy != null) {
        //创建并记录 RILRequest 对象
        RILRequest rr = obtainRequest(RIL_REQUEST_DATA_REGISTRATION_STATE,
            result, mRILDefaultWorkSource);
        //记录 RILJ 发出的 RIL 请求日志
        if (RILJ_LOGD) riljLog(rr.serialString() + "> " + requestToString(rr.mRequest));
        try {
            radioProxy.getDataRegistrationState(rr.mSerial); //调用 IRadio 服务接口
        } catch (RemoteException | RuntimeException e) {
            handleRadioProxyExceptionForRR(rr, "getDataRegistrationState", e);
        }
    }
}
```

上面的代码逻辑可分为四个处理逻辑：
- 获取 IRadio 服务的调用对象
- 创建 RILRequest 对象

RILJ 接口与 RIL 请求类型在 RIL.java 代码中有静态的对应关系，如在 getDataRegistrationState 方法中创建 RIL_REQUEST_DATA_REGISTRATION_STATE 类型的 RILRequest 对象。在 obtainRequest 方法

中创建并保存 RIL_REQUEST_XXX 类型的 RILRequest 对象，代码逻辑的详情如下。

```
private RILRequest obtainRequest(int request, Message result, WorkSource workSource) {
    //调用 RILRequest 的静态方法创建 RILRequest 对象
    RILRequest rr = RILRequest.obtain(request, result, workSource);
    addRequest(rr);
    return rr;
}
private void addRequest(RILRequest rr) {
    acquireWakeLock(rr, FOR_WAKELOCK);
    synchronized (mRequestList) {//同步锁
        rr.mStartTimeMs = SystemClock.elapsedRealtime();//更新 start 时间
        mRequestList.append(rr.mSerial, rr);//保存 RILRequest 对象
    }
}
```

- 记录 RILJ 请求日志

RILJ 请求日志中重点关注 RILRequest 对象 mSerial（唯一编号）、消息请求方向>，以及通过 requestToString(rr.mRequest)调用将 RIL 请求编号转换成对应的字符串信息。

摘录一条 RILJ 发起 RIL 请求的日志，详情如下：

```
D/RILJ    ( 3669): [4461]> DATA_REGISTRATION_STATE [SUB0]
```

4461 是 RILRequest 对象 mSerial（唯一编号），>说明是 RILJ 对象发出的请求，DATA_REGISTRATION_STATE 是请求的类型。

4461 是 RILRequest 对象的 serialString 方法，根据 mSerial 取值进行了转换。

- 调用 IRadio 服务对应的请求接口

RILJ 响应的 RIL 请求最终会转换为 IRadio 服务对应接口的调用,参数是 RILRequest 对象 mSerial（唯一编号属性）和 RIL 请求相关数据对象（在 types.hal 中定义）。

IRadio 服务接口并未提供返回结果，也不传入 RILRequest 对象。那么，调用 IRadio 服务接口处理后的返回结果，RILRequest 对象的 Message 消息将怎么产生消息回调呢？且看下一节的内容。

10.1.6　Solicited Response

选择 DATA_REGISTRATION_STATE 类型的 RIL 请求，跟进其返回的日志信息，对应 RILJ 的完整日志详情如下：

```
D/RILJ ( 3669): [4461]> DATA_REGISTRATION_STATE [SUB0]
D/RILJ ( 3669): [4461]< DATA_REGISTRATION_STATE {.regState =
NOT_REG_MT_SEARCHING_OP, .rat = 0, .reasonDataDenied = 0,
.maxDataCalls = 20, .cellIdentity = {......}} [SUB0]
```

RILJ 的完整日志的正常处理逻辑是成对出现，匹配关系如下：
- 根据 RILRequest 对象的 mSerial 来匹配，上面的日志中是 4461。
- >是发送 RIL 请求，<是接收到 RIL 请求的返回信息。

requestToString 返回时，通过 DATA_REGISTRATION_STATE 信息可匹配到 RIL 请求类型为 RIL_REQUEST_DATA_REGISTRATION_STATE，RILJ 提供的接口是 getDataRegistrationState，在

RadioResponse 中可找到对应的返回接口 getDataRegistrationStateResponse，处理逻辑的详情如下。

```
public void getDataRegistrationStateResponse(RadioResponseInfo responseInfo,
                DataRegStateResult dataReg Response) {
    RILRequest rr = mRil.processResponse(responseInfo);//RILJ 对象响应
    if (rr != null) {
            if (responseInfo.error == RadioError.NONE) {//无异常
                sendMessageResponse(rr.mResult, dataRegResponse);//发送 Message 消息
            }
            mRil.processResponseDone(rr, responseInfo, dataRegResponse);//完成处理
    }
}
```

原来，RILJ 接收到的 RIL 请求响应是在 IRadio 服务中通过跨进程调用 RadioResponse 对应的服务接口来返回 IRadio 接口请求的处理结果。

注意

CommandsInterface 与 IRadioResponse 定义的接口能——对应，比如：getDataRegistrationState 和 getDataRegistrationStateResponse、dial 和 dialResponse 等的对应关系，RIL 请求接口与返回接口相对应。

继续跟进 mRil.processResponse 的处理逻辑，即 RIL 类的 processResponse 方法，详情如下。

```
RILRequest processResponse(RadioResponseInfo responseInfo) {
    int serial = responseInfo.serial;//RILRequest 对象的 mSerial
    int error = responseInfo.error;
    int type = responseInfo.type;

    RILRequest rr = null;
    rr = findAndRemoveRequestFromList(serial);
    ......//异常处理或特殊 Request 请求
    return rr;
}
private RILRequest findAndRemoveRequestFromList(int serial) {
    RILRequest rr = null;
    synchronized (mRequestList) {//同步
        rr = mRequestList.get(serial);//根据 mSerial 获取 RILRequest 对象
        if (rr != null) {
            mRequestList.remove(serial);//移除
        }
    }
    return rr;
}
```

RIL 类 processResponse 方法的主要逻辑可分为以下三个部分：
- 在 RadioResponseInfo 对象中获取 serial、error 和 type 信息。
- 使用唯一编号 serial 在 mRequestList 列表中获取 RILRequest 对象。
- 若 mRequestList 列表中移除对应的 RILRequest 对象，说明 RIL 请求已经完成。

RIL 类的 processResponse 方法，最终返回 RIL 请求时保存的 RILRequest 对象。再次回到 RadioResponse 的处理逻辑，调用 sendMessageResponse 方法发出 Message 消息通知。

```
static void sendMessageResponse(Message msg, Object ret) {
    if (msg != null) {
        AsyncResult.forMessage(msg, ret, null);//更新 Message 对象的 object 为 RIL 返回消息
        msg.sendToTarget();//发出 Message 消息通知
    }
}
```

Message 消息是 RILRequest 的 mResult 对象，在 ServiceStateTracker 中创建，详情如下：

```
mCi.getDataRegistrationState(obtainMessage(EVENT_POLL_STATE_GPRS,mPollingContext))
```

因此，ServiceStateTracker 在 handleMessage 中响应 EVENT_POLL_STATE_GPRS 类型的 Message 消息。

最后，调用 RILJ 对象的 processResponseDone 方法进行收尾工作，详情如下。

```
void processResponseDone(RILRequest rr, RadioResponseInfo responseInfo, Object ret)
{
    if (responseInfo.error == 0) {//无异常
        if (RILJ_LOGD) {//记录 RIL 响应消息
            riljLog(rr.serialString() + "< " + requestToString(rr.mRequest)
                + " " + retToString(rr.mRequest, ret));
        }
    } else {
        ......//异常处理，记录异常日志并调用 rr.onError
    }
    mMetrics.writeOnRilSolicitedResponse(mPhoneId, rr.mSerial, responseInfo.error,
        rr.mRequest, ret);
    if (rr != null) {//释放资源
        if (responseInfo.type == RadioResponseType.SOLICITED) {
            decrementWakeLock(rr);
        }
        rr.release();
    }
}
```

Solicited Request 和 Solicited Response 消息一一对应，总结如图 10-3 所示。

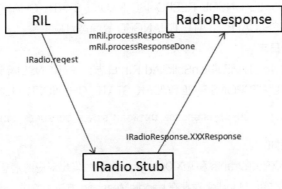

图 10-3　Solicited 消息处理

图 10-3 所示的 Solicited 消息处理，可重点掌握以下几点：

● IRadio 服务

IRadio.Stub 是 HAL 层实现了 IRadio.hal 接口定义的服务。

● Solicited Request

是 RILJ 对象向 HAL 层 IRadio 服务发起的接口调用，最主要的参数是 RILRequest 对象 mSerial（唯一编号）。

● Solicited Response

IRadio 服务处理完 RIL 请求后，调用 com.android.phone 进程中的 RadioResponse 服务对应的 XXXResponse 接口，将传递 HAL 层的处理结果 RadioResponseInfo 和数据对象。

- RadioResponse 服务处理逻辑

调用 RILJ 对象的 processResponse 和 processResponseDone 方法完成请求结果的回调，最关键的是通过 mSerial（唯一编号）在 mRequestList 列表中匹配 RILRequest 对象，发出 Message 消息通知。

IRadio 服务接口并未提供返回结果，也不传入 RILRequest 对象。那么，调用 IRadio 服务接口处理后的返回结果，RILRequest 对象的 Message 消息将怎么产生消息回调呢？现在可以回答这个问题了。

10.1.7 UnSolicited

com.android.phone 进程中的 RadioIndication 服务将接收 HAL 层发出的 UnSolicited 消息接口调用。以 radioStateChanged 为例，处理逻辑的详情如下：

```
public void radioStateChanged(int indicationType, int radioState) {
    mRil.processIndication(indicationType);//处理ACK消息
    //获取最新 Radio 状态
    CommandsInterface.RadioState newState = getRadioStateFromInt(radioState);
    if (RIL.RILJ_LOGD) {//记录 RIL_UNSOL_XXX 日志
        mRil.unsljLogMore(RIL_UNSOL_RESPONSE_RADIO_STATE_CHANGED,
                "radioStateChanged: " +  newState);
    }
    mRil.setRadioState(newState);//更新 Radio 状态
}
```

上面的代码逻辑，我们重点关注以下几点：

- RadioIndication 服务接口与 RIL_UNSOL_RESPONSE_XXX 消息在代码逻辑中的静态对应关系

radioStateChanged 与 RIL_UNSOL_RESPONSE_RADIO_STATE_CHANGED、callStateChanged 与 RIL_UNSOL_RESPONSE_CALL_STATE_CHANGED 等静态对应关系。

- UnSolicited RIL 日志

调用 mRil.unsljLogMore 方法记录 UnSolicited RIL 日志，注意[UNSL]标识和<消息传递方向，没有消息编号；摘录 UNSOL_RESPONSE_NETWORK_STATE_CHANGED 日志如下。

```
D/RILJ ( 3669): [UNSL]< UNSOL_RESPONSE_NETWORK_STATE_CHANGED [SUB0]
```

- RILJ 发出消息通知

使用 RILJ 对象的 mXXXRegistrants 和 mXXXRegistrant 属性发出对应的消息通知，主要是以 Tracker 为主的 Telephony 业务模型核心 Handler 对象在 handleMessage 方法中响应 Message 消息通知。

10.2 详解 rild

rild 的代码路径是 hardware/ril/rild，该代码库主要有三个代码文件：Android.mk、rild.c 和 rild.rc。首先查看 Android.mk 编译脚本，关键信息如下：

```
LOCAL_PATH:= $(call my-dir)
include $(CLEAR_VARS)
LOCAL_SRC_FILES:= \
    rild.c//仅有一个 C 代码文件
LOCAL_MODULE:= rild
```

```
LOCAL_INIT_RC := rild.rc//安装配置信息

include $(BUILD_EXECUTABLE)//编译生成 rild 可执行文件
```

rild.c 代码文件将编译出 rild 可执行文件，接着查看 rild.rc 的配置信息，详情如下：

```
service ril-daemon /vendor/bin/hw/rild
    class main
    user radio
    group radio cache inet misc audio log readproc wakelock
    capabilities BLOCK_SUSPEND NET_ADMIN NET_RAW
```

根据 rild.rc 配置信息，重点关注以下几点：
- init 进程将运行 rild 可执行文件，加载 ril-daemon service。
- ril-daemon service 进程使用 radio 用户。

查看 Nexus 6P 手机，关于 rild 进程的信息如下：

```
$ adb shell ps|grep rild
radio           602     1   66360   12728 binder_thread_read  0 S rild
```

从上面的信息中可获取 rild 进程的信息，使用 radio 用户，PID 为 602，PPID 为 1。

因此加载 rild 进程，将执行 rild.c 文件中的 main 方法，主要处理逻辑可分为三个部分：
- RIL_startEventLoop
- 获取 RIL_RadioFunctions
- 注册 RIL_RadioFunctions

10.2.1 RIL_startEventLoop

rild.c 文件中的 main 方法将调用 RIL_startEventLoop 函数，此函数在当前 C 文件中的声明如下：

```
extern void RIL_startEventLoop();
```

关键字 extern 修饰了该函数，说明 RIL_startEventLoop 函数在其他地方已经定义。查找 Android 源码，发现 hardware/ril/libril/ril.cpp C++代码中实现了此函数，其定义是：extern "C" void RIL_startEventLoop(void)，此处关注 extern "C"修饰符。

在 rild 的 Android.mk 文件中可找到 rild 与 libril 的关系，详情如下：

```
LOCAL_SHARED_LIBRARIES := \
    libcutils \
    libdl \
    liblog \
    libril
```

因此，rild 进程中调用的 RIL_startEventLoop 函数是 libril 中的实现，其处理逻辑将在下一节进行解析。

10.2.2 获取 RIL_RadioFunctions

接下来，解析 rild 加载过程中获取 RIL_RadioFunctions 的处理逻辑，详情如下：

```
//定义 rilInit 函数指针
const RIL_RadioFunctions *(*rilInit)(const struct RIL_Env *, int, char **);
```

```
const RIL_RadioFunctions *funcs;

dlHandle = dlopen(rilLibPath, RTLD_NOW);//打开动态链接库
if (dlHandle == NULL) {//失败处理
    RLOGE("dlopen failed: %s", dlerror());
    exit(EXIT_FAILURE);//异常退出
}

rilInit =
    (const RIL_RadioFunctions *(*)(const struct RIL_Env *, int, char **))
    dlsym(dlHandle, "RIL_Init");//获取动态链接库 RIL_Init 函数地址
if (rilInit == NULL) {//异常处理
    RLOGE("RIL_Init not defined or exported in %s\n", rilLibPath);
    exit(EXIT_FAILURE); //异常退出
}
//调用动态链接库 RIL_Init 函数，返回 funcs
funcs = rilInit(&s_rilEnv, argc, rilArgv);
```

上面的代码逻辑，我们重点关注以下几点：

- rilLibPath

main 函数传入的参数信息中，rilLibPath 是获取 Android 系统的 Properties：[rild.libpath]：[/vendor/lib64/libril-qc-qmi-1.so]。

- dlopen

以 RTLD_NOW 方式打开动态链接库文件 libril-qc-qmi-1.so。

- dlsym

在 libril-qc-qmi-1.so 库文件中获取 RIL_Init 函数地址并保存 rilInit 函数指针变量。

- rilInit 函数调用

调用 libril-qc-qmi-1.so 动态链接库中的 RIL_Init 函数，传入指向 RIL_Env 结构体的指针，返回指向 RIL_RadioFunctions 类型的指针。

RIL_Env 结构体在 ril.h 头文件中定义，详情如下：

```
struct RIL_Env {
    void (*OnRequestComplete)(RIL_Token t, RIL_Errno e, void *response,
        size_t responselen);
    void (*OnUnsolicitedResponse)(int unsolResponse,
            const void *data, size_t datalen, RIL_SOCKET_ID socket_id)
    void (*RequestTimedCallback) (RIL_TimedCallback callback,
            void *param, const struct timeval *relativeTime);
    void (*OnRequestAck) (RIL_Token t);
};
```

RIL_Env 结构体中定义了四个函数指针：OnRequestComplete、OnUnsolicitedResponse、RequestTimedCallback 和 OnRequestAck。在 rild.c 代码中创建 s_rilEnv 结构体，详情如下：

```
static struct RIL_Env s_rilEnv = {
    RIL_onRequestComplete,
    RIL_onUnsolicitedResponse,
    RIL_requestTimedCallback,
    RIL_onRequestAck
};
```

s_rilEnv 结构体中定义的四个函数在 rild.c 代码中均使用 extern 修饰，与 RIL_startEventLoop 函数相同，在 libril.cpp 文件中可找到这四个函数的实现，均使用了 extern "C"修饰符。

接下来，查看 rilInit 调用的返回值类型，在 hardware/ril/include/telephony/ril.h 头文件中可以找到

RIL_RadioFunctions 结构体的定义，详情如下：

```
typedef struct {
    int version;           /* set to RIL_VERSION */
    RIL_RequestFunc onRequest;
    RIL_RadioStateRequest onStateRequest;
    RIL_Supports supports;
    RIL_Cancel onCancel;
    RIL_GetVersion getVersion;
} RIL_RadioFunctions;

typedef void (*RIL_RequestFunc) (int request, void *data,
        size_t datalen, RIL_Token t, RIL_SOCKET_ID socket_id);
typedef RIL_RadioState (*RIL_RadioStateRequest)(RIL_SOCKET_ID socket_id);
......
```

在 RIL_RadioFunctions 结构体中首先是 RIL_VERSION 的定义，其他都是函数指针。

libril-qc-qmi-1.so 是高通平台实现的 RIL 动态链接库文件，在其中无法查看 RIL_Init 函数的处理逻辑。但是，我们可以参考 Android 源码中 reference-ril.c 中的实现，RIL_Init 函数处理逻辑的详情如下：

```
const RIL_RadioFunctions *RIL_Init(const struct RIL_Env *env, int argc, char **argv)
{
    s_rilenv = env;//保存指向 libril 的 RIL_Env 结构体指针
    pthread_attr_init (&attr);
    pthread_attr_setdetachstate(&attr, PTHREAD_CREATE_DETACHED);
    ret = pthread_create(&s_tid_mainloop, &attr, mainLoop, NULL);
    return &s_callbacks;//返回的是指针，注意&修饰符
}
static const RIL_RadioFunctions s_callbacks = {
    RIL_VERSION,
    onRequest,
    currentState,
    onSupports,
    onCancel,
    getVersion
};
```

上面的代码逻辑中，s_rilenv 保存指向 libril 中 RIL_Env 结构体的指针，将返回指向 RIL_RadioFunctions 结构体的指针；s_callbacks 中的函数指针都在 reference-ril.c 程序中实现了对应的函数，如 onRequest、currentState 等。

因此，rild 调用 RIL_Init 函数有两个重要目的：

- 动态链接库获取了 libril 的 RIL_onRequestComplete、RIL_onUnsolicitedResponse 等四个函数指针，通过名字可推测 RIL 请求之后回调的函数。
- libril 获取对应动态链接库文件中的 onRequest、currentState 等函数指针，通过名字可推测出是继续发起 RIL 请求的响应函数。

10.2.3 注册 RIL_RadioFunctions

rild 加载过程中最后的关键逻辑是注册 RIL_RadioFunctions，只有一行代码：RIL_register(funcs)，其参数是获取动态链接库文件中 RIL_RadioFunctions 结构体的地址，即指向 RIL_RadioFunctions 的指针。

RIL_register 函数在当前代码中的声明如下：

```
extern void RIL_register (const RIL_RadioFunctions *callbacks);
```

与 RIL_startEventLoop 函数的声明相同，同样是 hardware/ril/libril/ril.cpp C++代码中实现了此函数，其函数声明是：extern "C" void RIL_register (const RIL_RadioFunctions *callbacks)。

总结 rild 的加载过程，如图 10-4 所示。

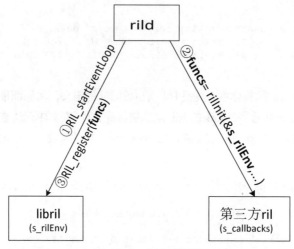

图 10-4　rild 加载过程

图 10-4 所示的 rild 加载过程，主要有三个步骤：

步骤 1：调用 libril 中的 RIL_startEventLoop 函数；

步骤 2：调用第三方库中的 RIL_Init 函数，传入参数是指向 RIL_Env 结构体的 s_rilEnv 指针，返回指向第三方库中 RIL_RadioFunctions 结构体的 s_callbacks 指针；

步骤 3：调用 libril 中的 RIL_RadioFunctions 函数，传入调用 RIL_Init 函数返回的指向第三方库中 RIL_RadioFunctions 结构体的 s_callbacks 指针。

10.3　libril 初始化流程

rild 进程在加载过程中，会调用 rild.c 代码中的 main 函数，而 main 函数中将调用三个 libril 中的函数：

- RIL_startEventLoop
- RIL_register
- rilc_thread_pool

10.3.1　RIL_startEventLoop

RIL_startEventLoop 函数的处理逻辑详情如下：

```
extern "C" void RIL_startEventLoop(void) {//extern "C"标识此方法可供 rild 调用
    /*挂起 eventLoop 线程等待获取启动信号 */
```

```
        s_started = 0; //启动标志
        pthread_mutex_lock(&s_startupMutex); //增加 pthread 的同步锁
        pthread_attr_t attr;
        pthread_attr_init(&attr); //初始化 pthread 参数
        pthread_attr_setdetachstate(&attr, PTHREAD_CREATE_DETACHED);
        //创建基于 eventLoop 函数调用的子线程
        int result = pthread_create(&s_tid_dispatch, &attr, eventLoop, NULL);
        if (result != 0) {//异常处理
            RLOGE("Failed to create dispatch thread: %s", strerror(result));
            goto done;
        }
        while (s_started == 0) {//pthread 启动标志, 会在 eventLoop 方法中设置为 1
            pthread_cond_wait(&s_startupCond, &s_startupMutex); //等待 s_startupCond 通知
        }
done:
    pthread_mutex_unlock(&s_startupMutex);
}
```

RIL_startEventLoop 函数的主要处理逻辑是：创建基于 eventLoop 函数调用的子线程。这里会使用到 pthread。Linux 系统下的多线程遵循 POSIX 线程接口，称为 pthread。编写 Linux 平台下的多线程程序时，需要引入头文件 pthread.h。

注 意

基于 Linux 的多线程编程基本都会使用非常成熟的 pthread 技术，pthread 编程技术和方法的难点在于多线程加锁、解锁的应用。感兴趣的读者请自行研究和学习。

接下来，进入 ril.cpp 代码中的 eventLoop 函数，其处理逻辑的详情如下。

```
static void *eventLoop(void *param) {
    int ret;
    int filedes[2];
    ril_event_init();//初始化 ril_event 双向链表
    pthread_mutex_lock(&s_startupMutex); //增加 pthread 同步锁
    s_started = 1; //修改启动状态为 1
    pthread_cond_broadcast(&s_startupCond); //发出 s_startupCond 通知
    pthread_mutex_unlock(&s_startupMutex); //释放 pthread 同步锁
    ret = pipe(filedes); //创建管道
    if (ret < 0) {//管道异常处理
        RLOGE("Error in pipe() errno:%d", errno);
        return NULL;
    }
    s_fdWakeupRead = filedes[0]; //输入的 fd
    s_fdWakeupWrite = filedes[1]; //输出的 fd
    fcntl(s_fdWakeupRead, F_SETFL, O_NONBLOCK);
    //创建 RIL 事件, 关注 s_fdWakeupRead 和回调函数 processWakeupCallback
    ril_event_set (&s_wakeupfd_event, s_fdWakeupRead, true,
                processWakeupCallback, NULL);
    rilEventAddWakeup (&s_wakeupfd_event); //增加 ril_event 节点并唤醒
    //有异常才会返回
    ril_event_loop();//开始循环监听和处理 ril_event 事件, 只有在异常情况下退出循环
    RLOGE ("error in event_loop_base errno:%d", errno);
    // 异常情况下 "杀死" 当前进程, init 进程将重新拉起 rild 进程
    kill(0, SIGKILL); //异常, "杀死" 当前进程, init 进程将重新拉起 rild 进程
    return NULL;
}
```

eventLoop 函数的处理逻辑主要分为以下几个部分。
- 修改 s_started 启动状态的取值为 1，并发出状态修改通知，由 RIL_startEventLoop 函数接收。
- 创建并激活 s_wakeupfd_event 的事件处理，此事件的发送和接收实现的方式基于 pipe 管道

通信：filedes[0]和filedes[1]。
- 调用 ril_event_loop 函数，循环接收和处理 ril_event 事件。

接下来，重点分析 s_wakeupfd_event 的事件处理和 ril_event_loop 函数。

s_wakeupfd_event 事件处理可总结为三个主要逻辑。
- 创建管道获取其输入和输出文件描述符 s_fdWakeupRead、s_fdWakeupWrite。
- 使用 s_fdWakeupRead 和 processWakeupCallback 创建 s_wakeupfd_event 事件。
- 增加并激活 s_wakeupfd_event 事件。

ril_event 双向链表中此时仅有一个节点，那就是 s_wakeupfd_event，此节点的 fd 文件描述符为 s_fdWakeupRead，RIL 事件的回调函数为：processWakeupCallback。

ril_event_loop 函数的处理逻辑的核心是 for (;;) 循环，只要循环中的处理逻辑不发生变化，ril_event_loop 函数调用是不会返回的。

RIL_startEventLoop 函数主要完成两个工作。
- 启动子线程，调用 ril_event_loop 函数，进入 for 循环监听 ril_event 事件。
- 完成 s_wakeupfd_event 事件的节点初始化和激活。

10.3.2 RIL_register

RIL_register 函数的处理逻辑的详情如下：

```
extern "C" void RIL_register (const RIL_RadioFunctions *callbacks) {
    ......//参数判断
    //拷贝 callbacks 到 s_callbacks 本地变量
    memcpy(&s_callbacks, callbacks, sizeof (RIL_RadioFunctions));

    s_registerCalled = 1;//更改调用状态标志
    //两个 for 循环检查 ril_commands.h 和 ril_unsol_commands.h 头文件中的 RIL 消息处理定义的正确
    //性，一旦有异常，则产生 assert 断言
    for (int i = 0; i < (int)NUM_ELEMS(s_commands); i++) {
        assert(i == s_commands[i].requestNumber);
    }
    for (int i = 0; i < (int)NUM_ELEMS(s_unsolResponses); i++) {
        assert(i + RIL_UNSOL_RESPONSE_BASE
                  == s_unsolResponses[i].requestNumber);
    }
    //注册服务
    radio::registerService(&s_callbacks, s_commands);
}
```

上面的代码逻辑有两个关键点：
- callbacks 参数

首先验证 callbacks 参数，是否为空和版本号等，然后是拷贝参数，即保存指向 RIL_RadioFunctions 结构体的指针。因此，libril 中就可以调用第三方动态链接库文件提供的 RIL 请求相关函数，保障第三方厂家代码的安全和保密。
- radio::registerService 调用

关注传入的参数 &s_callbacks 和 s_commands。

在 hardware/ril/libril/ril_service.cpp 文件中，可找到 radio::registerService 函数的实现逻辑，详情如下：

```
void radio::registerService(RIL_RadioFunctions *callbacks, CommandInfo *commands) {
    using namespace android::hardware;
    int simCount = 1;
    configureRpcThreadpool(1, true /* callerWillJoin */);//初始化远程调用进程
    for (int i = 0; i < simCount; i++) {
        pthread_rwlock_t *radioServiceRwlockPtr = getRadioServiceRwlock(i);
        int ret = pthread_rwlock_wrlock(radioServiceRwlockPtr);//同步锁
        assert(ret == 0);
        radioService[i] = new RadioImpl;//创建 RadioImpl 对象
        radioService[i]->mSlotId = i;
        oemHookService[i] = new OemHookImpl;//创建 OemHookImpl 对象
        oemHookService[i]->mSlotId = i;
        //注册系统服务
        android::status_t status = radioService[i]->registerAsService(serviceNames[i]);
        status = oemHookService[i]->registerAsService(serviceNames[i]);
        ret = pthread_rwlock_unlock(radioServiceRwlockPtr);  //释放同步锁
        assert(ret == 0);
    }
    s_vendorFunctions = callbacks;//第三方函数
    s_commands = commands;//本地 commands
}
```

上面的代码逻辑，我们重点关注以下几点：

- 支持多卡

radioService 和 oemHookService 数组的长度与支持的 SIM 卡数量一致，最多支持四张卡。

- RadioImpl 对象

RadioImpl 类在 ril_service.cpp 中定义：struct RadioImpl : public V1_1::IRadio，它继承自 IRadio 类。通过包含的头文件#include <android/hardware/radio/1.1/IRadio.h>，可推测 RadioImpl 类实现了 IRadio.hal 定义的接口。RILJ 对象发出的 IRadio 接口调用将在 RadioImpl 服务对象中响应，即 rild 进程响应 RILJ 发出的 IRadio 接口调用。

- registerAsService

为 HIDL 服务注册接口，关注 serviceNames 服务名。

- s_vendorFunctions 和 s_commands

rild 最后调用 rilc_thread_pool 完成 rild 进程加载，ril_service.cpp 中提供了对应的函数，详情如下：

```
void rilc_thread_pool() {
    joinRpcThreadpool();
}
```

最后，在 rild 进程中完成 IRadio 和 IOemHook 两个 HIDL 接口服务的启动。

总结 libril 的初始化流程可知，rild 在加载过程中调用 libril 中的 RIL_startEventLoop 和 RIL_register 两个函数。RIL_startEventLoop 函数加载 event 运行框架。RIL_register 函数将保存第三方动态链接库中 RIL_Init 函数返回的 RIL_RadioFunctions 结构体指针到 s_vendorFunctions，并启动 IRadio 和 IOemHook 两个 HAL 层服务。

10.4 扩展 hal 接口

为更好地理解 hal 接口的定义、语法、生成的模板代码以及客户端服务端的代码实现和交互机制，本节将扩展以 IRadio.hal 接口定义为中心的相关 hal 文件，并在 Nexus 6P 手机上验证。

10.4.1 增加接口定义

修改 IRadio.hal 和 IRadioResponse.hal 两个 HIDL 定义文件,分别增加一个方法,详情如下:

```
hardware/interfaces/radio/1.0/IRadio.hal 增加 testRilReq 接口
oneway testRilReq(int32_t serial, int32_t reqNum);

hardware/interfaces/radio/1.0/IRadioResponse.hal 增加 testRilResp 接口
oneway testRilResp(RadioResponseInfo info);
```

10.4.2 验证生成的代码

使用 mmm hardware/interfaces/radio/1.0/编译 radio1.0,根据 Android.mk 和 Android.bp 两个编译脚本,将输出 Java 和 C++对应的接口文件。

编译不会通过,将产生"This interface has been frozen. Do not change it!"异常提示。因为 HIDL 接口的哈希值保护机制———一种旨在防止意外更改接口并确保接口更改经过全面审查的机制。

首先,通过 hidl 工具生成修改后的 hal 文件的哈希值,操作详情和结果如下:

```
$ out/host/linux-x86/bin/hidl-gen -L hash -r .:hardware/interfaces android.hardware.radio@1.0
0a159f81359cd4f71bbe00972ee8403ea79351fb7c0cd48be72ebb3e424dbaef android.hardware.radio@1.0::types
c49f2d71ab2a5fb196ad16303e07f82a051e21c18fee52a71866c87e68895d39 android.hardware.radio@1.0::IRadio
5c8efbb9c451a59737ed2c6c20230aae4745839ca01d8088d6dcc9020e52d2c5 android.hardware.radio@1.0::IRadioIndication
......
```

有关 HIDL 工具的信息,重点关注以下几点:
- hidl-gen 是 system/tools/hidl 源码编译出的 HIDL 工具,在 Android 8.1 源代码根路径下运行。
- .:hardware/interfaces 中使用了相对路径,关注冒号前的.路径。
- 在 android.hardware.radio@1.0 后可指定具体的 hal 文件名,如::IRadio。

最后,将上面的哈希值保存到 hardware/interfaces/current.txt 文件的最后,也可使用以下命令进行更新。

```
$ out/host/linux-x86/bin/hidl-gen -L hash -r .:hardware/interfaces
android.hardware.radio@1.0::IRadio >> hardware/interfaces/current.txt
$ out/host/linux-x86/bin/hidl-gen -L hash -r .:hardware/interfaces
android.hardware.radio@1.0::IRadioResponse >> hardware/interfaces/current.txt
```

out/target/common/gen/JAVA_LIBRARIES/android.hardware.radio-V1.0-java-static_intermediates 目录作为 radio1.0 编译 Java 静态库文件的输出目录,主要有 IRadio.java、IRadioIndication.java、IRadioResponse.java 等接口文件,以及 CallState.java、GsmSignalStrength.java 和 GsmSmsMessage.java 等支持 HwParcel 序列化和反序列化的实体 Java 类文件。

查看 IRadio.java 文件和新增的 testRilReq 接口,详情如下:

```
public interface IRadio extends android.hidl.base.V1_0.IBase {
```

```java
public static final class Proxy implements IRadio {//Proxy 代理类实现了 IRadio 接口
    @Override
    public void testRilReq(int serial, int reqNum)
            throws android.os.RemoteException {
        android.os.HwParcel _hidl_request = new android.os.HwParcel();
        _hidl_request.writeInterfaceToken(IRadio.kInterfaceName);
        _hidl_request.writeInt32(serial);
        _hidl_request.writeInt32(reqNum);// HwParcel 序列化请求参数
        android.os.HwParcel _hidl_reply = new android.os.HwParcel();
        try {//发起远程 transact，注意 131 对应 testRilReq 请求
            mRemote.transact(131 /* testRilReq */, _hidl_request, _hidl_reply, android.os.IHwBinder.FLAG_ONEWAY);
            _hidl_request.releaseTemporaryStorage();
        } finally {
            _hidl_reply.release();
        }
    }
}
//内部抽象 Stub 类，继承 HwBinder 实现 IRadio 接口
public static abstract class Stub extends android.os.HwBinder implements IRadio {
    @Override//Proxy 类发起的远程调用
    public void onTransact(int _hidl_code, android.os.HwParcel _hidl_request,
final android.os.HwParcel _hidl_reply, int _hidl_flags) throws android.os.RemoteException {
        switch (_hidl_code) {
            case 131 /* testRilReq */://131 对应 testRilReq 接口调用
            {   //HwParcel 反序列化，获取请求参数
                _hidl_request.enforceInterface(IRadio.kInterfaceName);
                int serial = _hidl_request.readInt32();
                int reqNum = _hidl_request.readInt32();
                testRilReq(serial, reqNum);//调用 IRadio 远程实现的接口
                break;
            }
        }
}}
```

上面的代码框架均是 hidl-gen 通过 IRadio.hal 接口文件主动生成的，对比 IRadio.hal 接口文件中的内容，重点关注以下几点：

● package

IRadio.hal 定义的 "package android.hardware.radio@1.0;" 生成的 IRadio Java 类使用了相同的 package。

● 版本号

HIDL 要求每个使用 HIDL 编写的接口均必须带有版本编号；HAL 接口一经发布便会被冻结，如果要做任何进一步的更改，都只能在接口的新版本中进行。虽然无法对指定的已发布接口进行修改，但可以通过其他接口对其进行扩展。

Android 8.1 源码中的 radio 有两个版本：1.0 和 1.1。1.1 版本的 IRadio.hal 的定义是：interface IRadio extends @1.0::IRadio，扩展了 1.0 版本的 IRadio。

● Proxy 与 Stub 运行框架

使用了 HwBinder 机制完成跨进程的服务调用。

● IRadioResponse.java 文件中增加了 testRilResp 接口

out/soong/.intermediates/hardware/interfaces/radio/1.0/目录作为 radio1.0 编译 C++代码的输出目录，主要有两个目录，分别保存.h 头文件和.cpp 服务接口框架 C++程序。

主要的.h 头文件有：IRadio.h、IRadioIndication.h、IRadioResponse.h 等接口定义文件，以及与

IRadio 接口对应的 BpHwRadio.h、BnHwRadio.h 和 types.h 等类型定义文件。

关键的 C++ 代码有：RadioAll.cpp、RadioIndicationAll.cpp、RadioResponseAll.cpp 和 types.cpp，与 .h 头文件相对应。

接下来查看与 IRadio 接口相关的头文件关键信息，详情如下。

```
namespace android {
namespace hardware {
namespace radio {
namespace V1_0 {
struct IRadio : public ::android::hidl::base::V1_0::IBase {
    ......
    virtual ::android::hardware::Return<void> testRilReq(int32_t serial, int32_t reqNum) = 0;
}}}}}
```

对比 IRadio.hal 接口文件中的内容，重点关注以下几点：

- package。IRadio.hal 定义的 "package android.hardware.radio@1.0;" 生成的 IRadio.h，其命名空间是相同的。
- 版本号。1.1 版本的 IRadio.hal 生成的 IRadio.h 头文件的定义是：struct IRadio : public :: android:: hardware::radio::V1_0::IRadio。
- Proxy 与 Stub 运行框架。BpHwRadio.h 和 BnHwRadio.h 分别定义了 Proxy 与 Stub 的实现类，Bp 是 Binder Proxy 的缩写，而 Bn 是 Binder Native 的缩写。
- IRadio.h 文件中增加了 virtual testRilResp 接口定义。
- IRadio.h 定义了 linkToDeath、getService、registerAsService 等服务管理接口。

RadioAll.cpp 文件中实现了 IRadio 接口的相关逻辑，主要是 IRadio.h、IRadioIndication.h、IRadioResponse.h 三个类的实现。

关注 BpHwRadio::_hidl_testRilReq 函数的关键逻辑，详情如下：

```
_hidl_err = _hidl_data.writeInterfaceToken(BpHwRadio::descriptor);
//参数序列化操作
_hidl_err = _hidl_data.writeInt32(serial);
_hidl_err = _hidl_data.writeInt32(reqNum);
//远程接口 transact, 131
_hidl_err = ::android::hardware::IInterface::asBinder(_hidl_this)->transact(131 /* testRilReq */, _hidl_data, &_hidl_reply, ::android::hardware::IBinder::FLAG_ONEWAY);
```

与 IRadio.java 内部类 Proxy 的 testRilReq 方法中的处理逻辑相同：先序列化请求参数，发起 131 编号远程 transact 调用；由 BnHwRadio 本地响应 onTransact 远程调用。处理逻辑的详情如下。

```
::android::status_t BnHwRadio::onTransact(
        uint32_t _hidl_code,
        const ::android::hardware::Parcel &_hidl_data,
        ::android::hardware::Parcel *_hidl_reply,
        uint32_t _hidl_flags,
        TransactCallback _hidl_cb) {
    ::android::status_t _hidl_err = ::android::OK;
    switch (_hidl_code) {
        case 131 /* testRilReq */:
        {
            _hidl_err = ::android::hardware::radio::V1_0::BnHwRadio::_hidl_testRilReq(
                this, _hidl_data, _hidl_reply, _hidl_cb);
```

```
            break;
        }
    }
}
```

131 请求类型为 testRilReq，最后调用 BnHwRadio::_hidl_testRilReq 函数，关键逻辑详情如下：

```
::android::status_t BnHwRadio::_hidl_testRilReq(
        ::android::hidl::base::V1_0::BnHwBase* _hidl_this,
        const ::android::hardware::Parcel &_hidl_data,
        ::android::hardware::Parcel *_hidl_reply,
        TransactCallback _hidl_cb) {
    int32_t serial;
    int32_t reqNum;
    _hidl_err = _hidl_data.readInt32(&serial);//反序列化读取请求参数
    _hidl_err = _hidl_data.readInt32(&reqNum);
    //通过this调用testRilReq
    static_cast<BnHwRadio*>(_hidl_this)->_hidl_mImpl->testRilReq(serial, reqNum);
    ::android::hardware::writeToParcel(::android::hardware::Status::ok(), _hidl_reply);
    return _hidl_err;
}
```

与 IRadio.java 内部类 Stub 的 testRilReq 方法中的处理逻辑相同：先反序列化获取请求参数，再使用_hidl_impl 发起本地实现接口 testRilReq 的调用。

hardware/interfaces/radio/1.0/Android.bp 编译脚本是根据生成的.h 和.cpp 代码文件，最后编译出 android.hardware.radio@1.0.so 动态链接库。

hardware/interfaces/radio/1.1 中的接口继承自 1.0 接口，因此需要删除 1.1 的临时编译文件，再次重新编译。

vts 测试工具的依赖关系需要修改 hardware/interfaces/radio/1.0/vts 和 hardware/interfaces/radio/1.1/vts 中的代码来实现父类 RadioResponseAll.cpp 中的接口 testRilResp。

10.4.3 实现新增接口

libril 对 android.hardware.radio@1.0.so 动态链接库有依赖关系，RadioImpl 类继承自 IRadio，因此需要在 RadioImpl 类中完成 testRilReq 接口的实现，详情如下：

```
struct RadioImpl : public V1_1::IRadio {
    ......//类中声明 testRilReq 函数
    Return<void> testRilReq(int32_t serial, int32_t reqNum);
};
// RadioImpl 中实现 testRilReq 函数
Return<void> RadioImpl::testRilReq(int32_t serial, int32_t reqNum) {
    RLOGD("XXX testRilReq: serial:%d, reqNum:%d", serial, reqNum);
    if (radioService[0]->mRadioResponse != NULL) {
        RadioResponseInfo responseInfo = {};//构造默认返回对象
        responseInfo.serial = serial;
        responseInfo.type = RadioResponseType::SOLICITED;
        responseInfo.error = RadioError::NONE;
        //调用 phone 进程 RadioResponse 服务，返回处理结果 responseInfo
        Return<void> retStatus
                = radioService[0]->mRadioResponse->testRilResp(responseInfo);
        radioService[0]->checkReturnStatus(retStatus);
```

```
        }
        return Void();
}
```

Telephony 业务模型编译出的 telephony-common.jar 包对 android.hardware.radio.deprecated-V1.0-java-static 有依赖关系，com.android.phone 进程中的 RadioResponse 对象实现了 Iradio Response 接口，IRadioResponse 又新增了接口，因此需要在 RadioResponse.java 代码中实现新增接口，详情如下：

```
@Override//实现父类接口
public void testRilResp(RadioResponseInfo info) throws RemoteException {
    android.util.Log.d("XXX", "testRilResp:" + info.toString());
}
```

修改 RILJ 的构造方法，在构造方法的最后增加 testRilReq 接口调用，代码详情如下：

```
new Thread(new Runnable() {
    @Override
    public void run() {
        while (true) {
            try {//每10秒执行一次testRilReq调用
                Thread.sleep(10 * 1000);
                if (mRadioProxy != null) {
                    mRadioProxy.testRilReq(100, 1000);
                }
            } catch (Exception e) { e.printStackTrace();}
        }
    }
}).start();//启动子线程
```

10.4.4　运行结果验证

获取 Nexus 6P 手机上的 radio 日志，新增的日志输出详情如下：

```
D/RILC (603): XXX testRilReq: serial:100, reqNum:1000

1123   1266 D XXX: testRilResp:{.type = SOLICITED, .serial = 100, .error = NONE}

$ adb shell ps|grep radio
radio       603      1    65252  13100 binder_thread_read 73646ad4e8 S rild
radio      1123    585  4353544  68696 SyS_epoll_wait 76d4fbd3f8 S com.android.phone
```

以上面的运行日志中可得出以下几个结论：
- IRadio 服务

IRadio.hal 定义的接口在 RadioImpl 类中实现，运行在 603 进程，即 rild 进程中，在 init 进程的初始化过程中完成此服务的加载。
- IRadioResponse 服务

IRadioResponse.hal 定义的接口在 RadioResponse Java 类中实现，运行在 com.android.phone 进行中，在加载 Telephony 业务模型的过程中创建 RILJ 对象时，同步加载此服务，并调用 getRadioProxy 方法创建与 IRadio 服务的关系。
- Solicited 消息

再次验证 Solicited Request 和 Solicited Response 消息的处理机制，IRadio 服务接收 Solicited

Request 消息，IRadioResponse 服务接收 Solicited Response 消息。

注意 有关 HIDL 的更多信息参考 Android 官网。

10.5 RILC 运行机制

前面已经详细讲解了 libril 的初始化机制，本节从两个方面重点解析 RIL 运行机制。
- Solicited 消息处理机制
- UnSolicited 消息处理机制

10.5.1 Solicited 消息

通过前面的学习，已经知道 IRadio 服务接收 Solicited Request 消息，而 IRadio 服务的实现是在 RadioImpl 类中，RadioImpl 类提供的接口处理机制基本一致。以发起语音通话拨号业务为例，其代码逻辑的详情如下：

```
Return<void> RadioImpl::dial(int32_t serial, const Dial& dialInfo) {
    //保存 Request 请求信息，返回*RequestInfo
    RequestInfo *pRI = android::addRequestToList(serial, mSlotId, RIL_REQUEST_DIAL);
    //根据传入的拨号请求信息，初始化 RIL_Dial
    RIL_Dial dial = {};
    ......
    //继续发出 Request 请求
    CALL_ONREQUEST(RIL_REQUEST_DIAL, &dial, sizeOfDial, pRI, mSlotId);
    //收尾工作
    memsetAndFreeStrings(2, dial.address, uusInfo.uusData);
    return Void();
}
```

上面的代码逻辑，我们重点关注以下几点：
- IRadio 接口与 RIL 请求类型的静态对应关系

RILJ 中的对应关系延伸到了 libril 中，而 RIL_REQUEST_DIAL 等 RIL 请求类型的定义在 hardware/ril/include/telephony/ril.h 头文件中，与 RILConstants.java 文件中的定义是一致的。
- addRequestToList

传入 RIL 请求唯一编号（RILJ 中生成）和 RIL 请求类型，创建并保存 RequestInfo。
- CALL_ONREQUEST

宏定义详情如下：

```
#define CALL_ONREQUEST(a, b, c, d, e) s_vendorFunctions->onRequest((a), (b), (c), (d))
```

s_vendorFunctions 是第三方动态链接库中 RIL_Init 函数返回的 RIL_RadioFunctions 结构体指针，而 onRequest 是指向第三方动态链接库中 onRequest 函数的指针，因此，CALL_ONREQUEST 将调用第三方动态链接库中的 onRequest 函数。

addRequestToList 的实现逻辑如下：

```
RequestInfo *addRequestToList(int serial, int slotId, int request) {
    RequestInfo *pRI;
    int ret;
    RIL_SOCKET_ID socket_id = (RIL_SOCKET_ID) slotId;//关注 socketId 与 slotId 关系
    pthread_mutex_t* pendingRequestsMutexHook = &s_pendingRequestsMutex;
    RequestInfo**      pendingRequestsHook = &s_pendingRequests;//链表

    pRI = (RequestInfo *)calloc(1, sizeof(RequestInfo));//分配内存
    if (pRI == NULL) {
        RLOGE("Memory allocation failed for request %s", requestToString(request));
        return NULL;
    }
    pRI->token = serial;//token 与 serial 匹配关系
    pRI->pCI = &(s_commands[request]);//回调函数
    pRI->socket_id = socket_id;//也是 slotId

    ret = pthread_mutex_lock(pendingRequestsMutexHook);//加锁
    assert (ret == 0);
    //加入链表
    pRI->p_next = *pendingRequestsHook;
    *pendingRequestsHook = pRI;
    ret = pthread_mutex_unlock(pendingRequestsMutexHook);//解锁
    assert (ret == 0);
    return pRI;//返回指向 RequestInfo 的指针
}
```

上面的代码，我们重点关注以下几点：

- RIL_SOCKET_ID 可用 slotId 替换
- s_pendingRequests 链表保存 RequestInfo
- token 与 serial 的匹配关系
- s_commands[request]回调函数

那么 pRI→pCI = &(s_commands[request])是如何匹配回调函数的呢？首先确定 pCI 的定义，在 hardware/ril/libril/ril_internal.h 头文件中找到 RequestInfo 的定义，详情如下：

```
typedef struct RequestInfo {
    int32_t token;          //并非 RIL_Token 类型，而是 int32_t 类型
    CommandInfo *pCI;
    struct RequestInfo *p_next;
    char cancelled;
    char local;             //本地命令标识，不会进入命令进程后台执行
    RIL_SOCKET_ID socket_id;
    int wasAckSent;         // 发送 ack 标识
} RequestInfo;
```

RequestInfo 与 RILJ 中的 RILRequest 非常相似，关注 token、*pCI 和*p_next，出现新的类型 CommandInfo，详情如下：

```
typedef struct CommandInfo {
    int requestNumber;//请求编号
    int(*responseFunction) (int slotId, int responseType, int token,
            RIL_Errno e, void *response, size_t responselen);//函数指针
} CommandInfo;
```

pRI→pCI 就是 CommandInfo 结构体，s_commands[]数组中将保存 CommandInfo 结构体列表；ril_service.cpp 中的 s_commands 作为指针指向 ril.cpp 中的 s_commands 数组，其赋值逻辑的详情如下：

```
static CommandInfo s_commands[] = {
#include "ril_commands.h"
};
```

在 ril_commands.h 头文件中定义了 145 个 RIL 请求类型和回调函数,摘录部分定义如下。

```
{0, NULL},                                //none
{RIL_REQUEST_GET_CURRENT_CALLS, radio::getCurrentCallsResponse},
{RIL_REQUEST_DIAL, radio::dialResponse},
{RIL_REQUEST_RADIO_POWER, radio::setRadioPowerResponse},
{RIL_REQUEST_SEND_SMS, radio::sendSmsResponse},
{RIL_REQUEST_SETUP_DATA_CALL, radio::setupDataCallResponse},
```

ril_commands.h 头文件中的每一行记录将初始化为 s_commands 数组中的 CommandInfo 结构体,共有 145 + 1 个结构体,CommandInfo→requestNumber 取值从 0~145,而函数指针分别是 radio 命名空间的 XXXResponse 函数。

在 ril_commands.h 头文件中,固化了 Solicited Request 与 Solicited Response 回调函数的对应关系:Solicited Request 消息 RIL_REQUEST_XXX 类型和 Solicited Response 消息 XXXResponse 接口调用。

因此,对于 pRI→pCI = &(s_commands[request])匹配的回调函数,以拨号请求为例,request 取值为:RIL_REQUEST_DIAL,pRI→pCI→requestNumber 为 RIL_REQUEST_DIAL,RI→pCI→responseFunction 是 radio::dialResponse 函数。

注意 pRI 和 pCI 的简写,p 是 Pointer,RI 是 RequestInfo 的简写,CI 是 CommandInfo 的简写。

接下来,查看 reference-ril.c 文件中 onRequest 的参考实现,详情如下:

```
static void onRequest (int request, void *data, size_t datalen, RIL_Token t) {
    switch (request) {
        case RIL_REQUEST_DIAL:
            requestDial(data, datalen, t);
            break;
        ......
    }
}
static void requestDial(void *data, size_t datalen __unused, RIL_Token t) {
    RIL_Dial *p_dial;
    char *cmd;
    const char *clir;
    int ret;

    p_dial = (RIL_Dial *)data;
    ret = at_send_command(cmd, NULL);//发送 AT 指令
    free(cmd);
    RIL_onRequestComplete(t, RIL_E_SUCCESS, NULL, 0);//发起回调
}
#define RIL_onRequestComplete(t, e, response, responselen) s_rilenv->OnRequestComplete
(t,e, response, responselen)
```

reference-ril.c 文件中 onRequest 参考实现的处理逻辑主要是根据传入的参数 request,即 RIL 请求类型和请求数据创建 command AT 指令,并发送 AT 指令给 BP 处理;然后通过 RIL_onRequestComplete,发起 libril 中的 OnRequestComplete 函数调用,返回 RIL 请求的结果。

另外,需要注意一下 onRequest 请求参数 RIL_Token,OnRequestComplete 发起回调的参数同样使用了 RIL_Token,说明它们的调用是成对出现的,使用 RIL_Token 可匹配到对应的调用。

最后是 libril 响应 OnRequestComplete 函数调用，其处理逻辑的详情如下：

```
extern "C" void RIL_onRequestComplete(RIL_Token t, RIL_Errno e,
        void *response, size_t responselen) {
    RequestInfo *pRI;
    int ret;
    RIL_SOCKET_ID socket_id = RIL_SOCKET_1;
    pRI = (RequestInfo *)t;//强制类型转换

    if (!checkAndDequeueRequestInfoIfAck(pRI, false)) {//获取 RequestInfo
        RLOGE ("RIL_onRequestComplete: invalid RIL_Token");
        return;
    }

    socket_id = pRI->socket_id;
    if (pRI->cancelled == 0) {
        int responseType;
        ......//version ACK 判断处理逻辑
        int rwlockRet = pthread_rwlock_rdlock(radioServiceRwlockPtr);// rdlock
        assert(rwlockRet == 0);
        //调用 XXXResponse 函数
        ret = pRI->pCI->responseFunction((int) socket_id,
                responseType, pRI->token, e, response, responselen);
        rwlockRet = pthread_rwlock_unlock(radioServiceRwlockPtr);// unlock
        assert(rwlockRet == 0);
    }
    free(pRI);//释放 RequestInfo 内存
}
```

上面的代码逻辑中，调用 checkAndDequeueRequestInfoIfAck 函数找到 RIL 请求对应的 RequestInfo 结构体，然后通过 pRI→pCI→responseFunction 发起 XXXResponse 函数调用。

那么，checkAndDequeueRequestInfoIfAck 非常关键，其处理逻辑的详情如下：

```
static int checkAndDequeueRequestInfoIfAck(struct RequestInfo *pRI, bool isAck) {
    int ret = 0;
    pthread_mutex_t* pendingRequestsMutexHook = &s_pendingRequestsMutex;
    //s_pendingRequests 链表中匹配 RequestInfo
    RequestInfo ** pendingRequestsHook = &s_pendingRequests;

    pthread_mutex_lock(pendingRequestsMutexHook);//同步锁
    for(RequestInfo **ppCur = pendingRequestsHook
        ; *ppCur != NULL
        ; ppCur = &((*ppCur)->p_next)//循环 pendingRequests 链表
    ) {
        if (pRI == *ppCur) {//匹配的是 token
            ret = 1;
            *ppCur = (*ppCur)->p_next;//断开匹配到的元素
            break;
        }
    }
    pthread_mutex_unlock(pendingRequestsMutexHook);//释放锁
    return ret;
}
```

上面的代码逻辑中，关注到 if (pRI == *ppCur)匹配的内容是 token，因此 libril 在接收到 OnRequestComplete 函数调用将 RIL_Token 强制转换为 RequestInfo 时，RequestInfo 结构体的第一个元素保存 token，即 serial RIL 请求唯一编号。

接着，跟进 radio::dialResponse 函数的处理逻辑，详情如下。

```
int radio::dialResponse(int slotId,
                    int responseType, int serial, RIL_Errno e, void *response,
                    size_t responseLen) {
    if (radioService[slotId]->mRadioResponse != NULL) {
        RadioResponseInfo responseInfo = {};
        populateResponseInfo(responseInfo, serial, responseType, e);
        Return<void> retStatus = radioService[slotId]->mRadioResponse->
                dialResponse(responseInfo);
        radioService[slotId]->checkReturnStatus(retStatus);
    } else {
        RLOGE("dialResponse: radioService[%d]->mRadioResponse == NULL", slotId);
    }

    return 0;
}
int radio::dialResponse(int slotId, int responseType, int serial,
        RIL_Errno e, void *response, size_t responseLen) {
    if (radioService[slotId]->mRadioResponse != NULL) {
        RadioResponseInfo responseInfo = {};
        populateResponseInfo(responseInfo, serial, responseType, e);//构建返回结果
        Return<void> retStatus =//调用 Phone 进程 IRadioResponse 服务
radioService[slotId]->mRadioResponse->dialResponse(responseInfo);
        radioService[slotId]->checkReturnStatus(retStatus);
    } else {
        RLOGE("dialResponse: radioService[%d]->mRadioResponse == NULL", slotId);
    }
    return 0;
}
```

libril 中的 IRadio 服务响应 dial 接口调用,即接收到 Solicited Request 消息,接着调用第三方库中的 onRequest 函数,第三方库根据请求信息创建并发出 AT 指令给 BP 处理,并调用 libril 中的 OnRequestComplete 函数返回 RIL 请求结果。libril 根据传递和返回的 RIL_Token 找出对应的 RequestInfo,由 pRI→pCI→responseFunction 发起 radio::dialResponse 函数调用,最后通过 radioService[slotId] → mRadioResponse → dialResponse 调用 com.android.phone 进程中的 IRadioResponse 服务的 dialResponse 接口。

10.5.2 UnSolicited 消息

com.android.phone 进程中 UnSolicited 消息的处理机制:由于 UnSolicited 消息是 HAL 层主动上报的,BP 产生的通信状态变化的消息将发送给 AP,第三方 ril 库首先接收到,再发送给 libril。根据前面学习的内容,libril 中的 RIL_onUnsolicitedResponse 函数将响应 UnSolicited 消息请求的调用,其处理逻辑的详情如下:

```
extern "C" void RIL_onUnsolicitedResponse(int unsolResponse, const void *data,
        size_t datalen, RIL_SOCKET_ID socket_id) {
    ......//声明及异常处理
    //计算 s_unsolResponses 数组下标
    unsolResponseIndex = unsolResponse - RIL_UNSOL_RESPONSE_BASE;

    pthread_rwlock_t *radioServiceRwlockPtr = radio::getRadioServiceRwlock((int) soc_id);
    int rwlockRet = pthread_rwlock_rdlock(radioServiceRwlockPtr);//同步锁
    assert(rwlockRet == 0);
    //发起 responseFunction 调用
    ret = s_unsolResponses[unsolResponseIndex].responseFunction(
            (int) soc_id, responseType, 0, RIL_E_SUCCESS, const_cast<void*>(data),
            datalen);
```

```
        rwlockRet = pthread_rwlock_unlock(radioServiceRwlockPtr);//释放锁
        assert(rwlockRet == 0);
        return;
}
```

上面的代码逻辑有两个关键点：

- 计算 unsolResponseIndex

还记得 RILConstants 中有关 RIL_UNSOL_RESPONSE_XXX 的定义吗？从 1000 开始，RIL_UNSOL_RESPONSE_BASE 和 RIL_UNSOL_RESPONSE_RADIO_STATE_CHANGE 的取值都是 1000，在 ril.h 头文件中的定义是一致的。因此，相当于 unsolResponse – RIL_UNSOL_RESPONSE_BASE 减去了 1000 这个基数。

- 调用 s_unsolResponses[unsolResponseIndex].responseFunction

首先确定 s_unsolResponses 的定义，在 ril.cpp 文件中找到了 UnsolResponseInfo 结构体的定义，详情如下：

```
typedef struct {
    int requestNumber;//请求编号
    int (*responseFunction) (int slotId, int responseType, int token,
            RIL_Errno e, void *response, size_t responselen);
                // responseFunction 函数指针
    WakeType wakeType;//唤醒类型
} UnsolResponseInfo;
```

接着查看 s_unsolResponses 数组的初始化，详情如下：

```
static UnsolResponseInfo s_unsolResponses[] = {
#include "ril_unsol_commands.h"
};
```

上面的代码结构与 s_commands 数组的定义是相同的，在 ril_unsol_commands.h 头文件中定义了 50 个 UnSolicited 类型的 RIL 请求和调用函数，摘录部分定义如下。

```
{RIL_UNSOL_RESPONSE_RADIO_STATE_CHANGED,
        radio::radioStateChangedInd, WAKE_PARTIAL},
{RIL_UNSOL_RESPONSE_CALL_STATE_CHANGED,
        radio::callStateChangedInd, WAKE_PARTIAL},
{RIL_UNSOL_RESPONSE_VOICE_NETWORK_STATE_CHANGED,
        radio::networkStateChangedInd, WAKE_PARTIAL},
{RIL_UNSOL_RESPONSE_NEW_SMS, radio::newSmsInd, WAKE_PARTIAL},
```

ril_unsol_commands.h 头文件中的每一行记录将初始化为 s_unsolResponses 数组中的 UnsolResponseInfo 结构体，共有 50 个结构体，UnsolResponseInfo→requestNumber 取值从 0~49，而函数指针分别是 radio 命名空间的 XXXInd 函数。

在 ril_unsol_commands.h 头文件中，固化了 UnSolicited 消息类型与调用函数的对应关系：UnSolicited 消息 RIL_UNSOL_RESPONSE_XXX 类型和调用函数 XXXInd 的关系。

因此，s_unsolResponses[unsolResponseIndex].responseFunction 匹配函数调用。以 Call 状态变化为例，unsolResponse 为 RIL_UNSOL_RESPONSE_CALL_STATE_CHANGED，取值 1000，计算出 unsolResponseIndex 取值为 1，即 s_unsolResponses[1].responseFunction 对应的是 radio::callStateChangedInd 函数调用。

进入 ril_service.cpp 文件中的 radio::callStateChangedInd 函数，其处理逻辑的详情如下：

第 10 章 Radio Interface Layer

```
int radio::callStateChangedInd(int slotId, int indicationType,
        int token, RIL_Errno e, void *response, size_t responseLen) {
    if (radioService[slotId] != NULL && radioService[slotId]->mRadioIndication != NULL) {
        Return<void> retStatus = radioService[slotId]->mRadioIndication->callStateChanged(
                    convertIntToRadioIndicationType(indicationType));
        radioService[slotId]->checkReturnStatus(retStatus);
    } else {
        RLOGE("callStateChangedInd: radioService[%d]->mRadioIndication == NULL", slotId);
    }
    return 0;
}
```

上面的代码逻辑主要是发起 radioService[slotId] →mRadioIndication→callStateChanged 调用，将调用 com.android.phone 进程中 IRadioIndication 服务的 callStateChanged 接口。

到此，已完成 RIL 中两种消息运行处理机制的解析，Solicited 和 UnSolicited 消息处理和响应机制总结如图 10-5 所示。

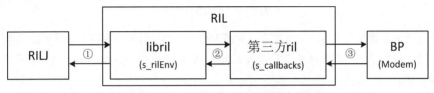

图 10-5　Solicited&UnSolicited 消息处理机制

图 10-5 所示的 Solicited&UnSolicited 消息处理机制，我们关注图中用数字标识的三种交互方式。

① phone 进程与 rild 进程交互，采用 HIDL 接口服务调用。com.android.phone 进程中提供 IRadioResponse 服务和 IRadioIndication 服务，rild 进程提供 IRadio 服务。

② RIL_register 调用建立了 libril 与等三方 ril 互相持有对方的函数指针；在 rild 进程中，使用函数指针发起进程内的函数调用。

③ 在第三方 ril 动态链接库中，实现了将 RIL 请求转换为与 BP 的交互机制，不同厂家实现的机制不同；Nexus 6P 手机基于高通平台，使用了 QMI（Qualcomm Message Interface）来完成 AP 与 BP 的通信。

本 章 小 结

总结 RIL 的系统架构，如图 10-6 所示。
图 10-6 所示的 RIL 系统架构，我们需要掌握以下几点：

- RIL 主要分 RILJ 和 RILC 两部分。RILJ 运行在 com.android.phone 进程空间的 Telephony Framework 框架层，RILC 运行在 User Libraries 系统运行库层中的 HAL 子层。
- rild、libril 和第三方 RIL 实现都运行在 rild 进程中，通过 rild.rc 配置文件由 Linux init 进程进行加载和管理。
- RILJ 与 RILC 通过 IRadio、IRadioIndication 和 IRadioResponse 服务接口调用，完成 Solicited 和 UnSolicited 消息交互。

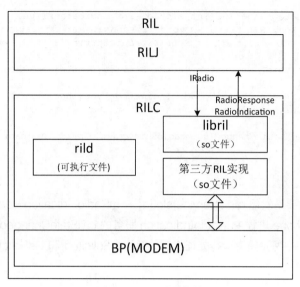

图 10-6　RIL 系统架构

- 第三方 RIL 库文件与 Modem 交互，完成 Modem 的操作控制和查询请求，以及接收 Modem 主动上报的消息。可参考完全符合 Android RIL 运行框架 libreference-ril、基于 AT 命令的实现方式。
- 掌握 Solicited 和 UnSolicited 消息在 RILJ 和 RILC 中的处理机制。

在 RILConstants.java 和 ril.h 中定义的 RILSolicited 和 UnSolicited 消息是一致的，分别是 RIL_REQUEST_XXX 和 RIL_UNSOL_XXX。

RILJ 对象 mRequestList 列表中的 RILRequest 对象与 libril 中 s_pendingRequests 链表中的 RequestInfo 具有相同的处理逻辑，都是完成 Solicited Request 和 Solicited Response 消息处理。

UnSolicited 消息的处理机制相对简单，主要是通过服务调用完成消息主动上报。